THE CONCISE
OXFORD DICTIONARY OF
PROVERBS

The Concise
Oxford Dictionary of
Proverbs

EDITED BY

J. A. SIMPSON

Oxford   New York   Toronto   Melbourne

OXFORD UNIVERSITY PRESS

1982

Oxford University Press, Walton Street, Oxford OX2 6DP

London Glasgow New York Toronto
Delhi Bombay Calcutta Madras Karachi
Kuala Lumpur Singapore Hong Kong Tokyo
Nairobi Dar es Salaam Cape Town
Melbourne Auckland
and associates in
Beirut Berlin Ibadan Mexico City Nicosia

© Oxford University Press 1982

British Library Cataloguing in Publication Data

The Concise Oxford dictionary of proverbs.
1. Proverbs, English—Dictionaries
I. Simpson, J. A.
398'.9'21   PN6421

ISBN 0–19–866131–2

Printed in the United States of America

PREFACE

The Concise Oxford Dictionary of Proverbs (*CODP*) is nominally an abridgement of the standard work, *The Oxford Dictionary of English Proverbs* (third
edition, 1970) (*ODEP*). Nevertheless, as the present volume was being
prepared it became clear that the two works served different purposes,
and that their range and content would differ accordingly. The parent
dictionary is a comprehensive survey of the proverb in Britain, and
concentrates particularly on the period up to the seventeenth century,
the heyday of the proverb as a vehicle for expressing unquestioned
moral truth. It contains proverbs and proverbial phrases both current
and obsolete, and illustrates entries only up to the late-nineteenth century. This concise version deals principally with proverbs known in the
twentieth century, especially in Britain and America, and it also gives full
documentary information from the point at which each proverb enters
the English language. In addition, copious modern examples have been
sought and included in order to establish evidence for each proverb's
present currency. As a result, this book is a completely revised version of
those parts of *ODEP* with which it is concerned. It is the only attempt that
has been made to record and describe the modern proverb principally in
Britain, but also in America and other important areas of the English-
speaking world.

I would like to thank the many people who have been kind enough to
offer assistance and comments during the preparation of this dictionary:
Mrs. Joanna M. Wilson, whose late husband Professor F. P. Wilson
edited the third edition of the *Oxford Dictionary of English Proverbs*,
generously made available to me her husband's valuable library of
proverb dictionaries and related works; the numerous readers who
provided quotations of proverbs from modern literature, principally
Miss Marghanita Laski; Iona and the late Peter Opie, for comments on an
early draft of the dictionary; and my colleagues on the staff of the Oxford
dictionaries, whose comments, support, and interest throughout the
work of preparation are greatly appreciated: especially the Chief Editor,
Dr. Robert Burchfield, Dr. R. E. Allen, Mrs. L. S. Burnett, and Mr.
E. S. C. Weiner; also Miss E. M. Knowles, Mrs. A. J. Burrell, and others,
whose meticulous checking of quotations in the major reference libraries
enabled me to scotch many myths which have grown up about the early
uses of proverbs, and to quote consistently from first editions. Lastly,
but by no means least, I am indebted to the editors of previous modern
dictionaries of proverbs who first made specialized information in their

own chosen areas available to the public: a list of these may be found in the Bibliography at the end of the book.

J. A. S.

Oxford
March, 1982

CONTENTS

INTRODUCTION

The Concise Oxford Dictionary of Proverbs provides a general history of all proverbs in common use in Britain in the twentieth century. Some of the proverbs have been in use throughout the English-speaking world for many years; others (especially Scottish proverbs) have spread from regional use to attain general currency in the nineteenth and twentieth centuries. Proverbs which originated in the United States and in other countries outside the British Isles, such as *If you don't like the heat, get out of the kitchen* or *The apple never falls far from the tree*, are included if they are now current in Britain, or if they are particularly prevalent in their region of origin.

A proverb is a traditional saying which offers advice or presents a moral in a short and pithy manner. Paradoxically, many phrases which are called 'proverbial' are not proverbs as we now understand the term. We might for instance refer to 'the proverbial fly on the wall' or say that something is 'as dead as the proverbial dodo', although neither of these phrases alludes to a proverb. The confusion dates from before the eighteenth century, when the term 'proverb' also covered metaphorical phrases, similes, and descriptive epithets, and was used far more loosely than it is today. Nowadays we would normally expect a proverb to be cast in the form of a sentence.

Proverbs fall readily into three main categories. Those of the first type take the form of abstract statements expressing general truths, such as *Absence makes the heart grow fonder* and *Nature abhors a vacuum*. Proverbs of the second type, which include many of the more colourful examples, use specific observations from everyday experience to make a point which is general; for instance, *You can take a horse to the water, but you can't make him drink* and *Don't put all your eggs in one basket*. The third type of proverb comprises sayings from particular areas of traditional wisdom and folklore. In this category are found, for example, the health proverbs *After dinner rest a while, after supper walk a mile* and *Feed a cold and starve a fever*. These are frequently classical maxims rendered into the vernacular. In addition, there are traditional country proverbs which relate to husbandry, the seasons, and the weather, such as *Red sky at night, shepherd's delight; red sky in the morning, shepherd's warning* and *When the wind is in the east, 'tis neither good for man nor beast*.

Several of the more common metaphorical phrases are included in the dictionary if they are also encountered in the form of a proverb. The phrases *to cut off your nose to spite your face* and *to throw the baby out with the bathwater*, for example, would not ordinarily qualify for inclusion, but have been admitted because they are often found in proverb form— *Don't cut off your nose to spite your face* and *Don't throw the baby out with the bathwater*. Other metaphorical phrases (*to win one's spurs, to throw in the towel*, etc.), similes (*as red as a rose, as dull as ditchwater*), and aphoristic quotations (*Power grows out of the barrel of a gun*) are not included. Nevertheless, proverbs which originated in English as quotations, such as *Hope springs eternal* or *Fools rush in where angels fear to tread*, are included when the origins of the quotations are no longer popularly remembered.

It is sometimes said that the proverb is going out of fashion, or that it has degenerated into the cliché. Such views overlook the fact that while the role of the proverb in English literature has changed, its popular currency has remained constant. In medieval times, and even as late as the seventeenth century, proverbs often had the status of universal truths and were used to confirm or refute an argument. Lengthy lists of proverbs were compiled to assist the scholar in debate; and many sayings from Latin, Greek, and the continental languages were drafted into English for this purpose. By the eighteenth century, however, the popularity of the proverb had declined in the work of educated writers, who began to ridicule it as a vehicle for trite, conventional wisdom. In Richardson's *Clarissa Harlowe* (1748), the hero, Robert Lovelace, is congratulated on his approaching marriage and advised to mend his foolish ways. His uncle writes: 'It is a long lane that has no turning.—Do not despise me for my proverbs.' Swift, in the introduction to his *Polite Conversation* (1738), remarks: 'The Reader must learn by all means to distinguish between Proverbs, and those polite Speeches which beautify Conversation: . . As to the former, I utterly reject them out of all ingenious Discourse.' It is easy to see how proverbs came into disrepute. Seemingly contradictory proverbs can be paired—*Too many cooks spoil the broth* with *Many hands make light work*: *Absence makes the heart grow fonder* with its opposite *Out of sight, out of mind*. Proverbs could thus become an easy butt for satire in learned circles, and are still sometimes frowned upon by the polished stylist. The proverb has none the less retained its popularity as a homely commentary on life and as a reminder that the wisdom of our ancestors may still be useful to us today. This shift is reflected in the quotations which accompany the entries in the dictionary: recent quotations are often taken from the works of minor writers, or from newspapers and magazines, while earlier quotations are more frequently from the works of major writers. It is a reflection of the proverb's vitality that new ones are continually being created as older ones fall into disuse. Surprisingly, neither *A change is as good as a rest* nor *A trouble shared is a trouble halved* are recorded before the twentieth century; the popular saying *A watched pot never boils* first occurs as late as

1848. The computer world has recently given us a potential classic, *Garbage in, garbage out,* and economics has supplied us with *There's no such thing as a free lunch.* Proverbs continue—as the early collectors never tired of stating—to provide the sauce to relish the meat of ordinary speech.

*

Proverb dictionaries differ in their manner of ordering material. There are a number of choices open to the compiler. One method favoured in early dictionaries was a straight alphabetical sequence, starting with all proverbs beginning with the word *a,* such as *A bird in the hand is worth two in the bush* and *A stern chase is a long chase,* and continuing in this rigid style until *z.* The problems caused by this system are manifold, the most apparent being the grouping of large numbers of unrelated proverbs under a few words such as *a, every, one,* and *the,* forcing the user to engage on a long search for the proverb of his choice. Another option is thematic presentation, whereby proverbs relating to cats, dogs, the Devil, Pride, etc., are each placed together. Despite the many advantages of this method, confusion can occur when there is no clear subject, as when a proverb falls under two or more thematic headings.

The manner of arrangement chosen here is that favoured by most major proverb collections of recent years, such as M. P. Tilley's *Dictionary of the Proverbs in England in the Sixteenth and Seventeenth Centuries* (1950) and B. J. Whiting's *Early American Proverbs and Proverbial Phrases* (1977). This method combines the advantages of alphabetical and thematic presentation by listing proverbs by the *first significant word;* thus *All cats are grey in the dark* may be found at *cats, You cannot put an old head on young shoulders* at *old,* while *Every picture tells a story* occurs at *picture.* Furthermore, a generous selection of cross-references is given in the text to assist the reader in cases of difficulty. The first of the three examples above, for example, is cross-referenced at *grey* and *dark,* the second at *head, young,* and *shoulder,* and the third at *every, tell,* and *story.* Variant forms are always noted at the main form when they are important enough to merit inclusion.

Illustrative quotations of proverbs are a major feature of the dictionary, as in *ODEP.* Accordingly, the earliest known example of each proverb's occurrence in literature is always given as the first quotation. Many of the proverbs were probably in common oral use before being recorded in print, but this dictionary clearly must rely upon the evidence of the printed word. When a proverb is known to have existed in another language before its emergence in English, this is indicated in a square-bracketed note preceding the quotations. For instance, although *There's many a slip 'twixt cup and lip* is first recorded in English in 1539, its parent form is found in both Greek and Latin, and this information is provided before the sixteenth-century English citation. Similarly, *Nothing succeeds like success*, first noted in English in 1867, was current in French some decades earlier. It is interesting to note that a high proportion of traditional English proverbs are of foreign origin. Like many of the words in our language, proverbs frequently passed into English from Latin or Greek, through the learned disciplines of medicine or the law, or from a knowledge of the classical authors; or they came into English from French in the years following the Conquest. A number of modern proverbs, such as *The opera isn't over till the fat lady sings* or *The family that prays together stays together*, originated in the United States. Predictably, one classic proverb of English origin is the old saying *It never rains but it pours*.

Each entry is provided with several illustrative quotations which show the contexts in which the proverb has been used, up to the present day. The standard form of a proverb often changes during its development: the first recorded use of the current form is always cited. Short notes are added when there is some obscurity in the meaning or use of a proverb which is not resolved in the quotations, or when there is some point of grammatical or syntactical interest which deserves mention. Thus, the legal implications of *Possession is nine points of the law* and *Every dog is allowed one bite* are explained, as are the historical origins of *Caesar's wife must be above suspicion* and *One might as well be hanged for a sheep as a lamb*. The original meanings of words such as *handsome* in *Handsome is as handsome does* are also discussed when necessary.

Much of the work involved in the compilation of the dictionary has concerned the verification of quotations. In the past, quotations have often been carried forward from one proverb dictionary to another without being checked; this is especially true of the older quotations. All quotations have been rechecked for this dictionary, and are quoted from the first edition of the relevant work, unless otherwise stated in the citation or in the Bibliography. Many quotations in other collections were found to have been wrongly dated, principally because they were taken from later (often bowdlerized Victorian) editions of the work in question, and frequently the true first edition contains a less-familiar version of the proverb, or no proverb at all.

Self-evident short titles are occasionally used in citations, but whenever possible the title and author of each work are given in full. Titles have been modernized, quotations (with the exception of Shakespeare) have not. Quotations are cited by reference to chapter; other styles are consistently employed when a work is not subdivided thus. Full references are given for the Bible, Shakespeare, and several other major writers; plays are cited by act and scene (failing scene, then page).

Biblical quotations are cited from the Authorized Version of 1611 unless otherwise stated: similar quotations may often be found in earlier translations, sermons, and homilies, but the modern form of a proverb usually reflects this translation. Contractions, which occur frequently in medieval sources, have been silently expanded.

ABBREVIATIONS
USED IN THE DICTIONARY

a	*ante* (before)	Jan.	January
Apr.	April	L.	Latin
Aug.	August	Mar.	March
c	*circa* (about)	med.	medieval
cent.	century	mod.	modern
cf.	*confer* (compare)	MS(S)	manuscript(s)
COD	Concise Oxford Dictionary	Nov.	November
Dec.	December	NY	New York
Dict.	dictionary (of)	Oct.	October
ed.	edition	OED	Oxford English Dictionary
EETS	Early English Text Society	Pt.	part
esp.	especially	quot.	quotation
et al.	*et alii* (and others)	rev.	revised
Feb.	February	Sept.	September
Fr.	French	Ser.	series
Ger.	German	St.	Saint
Gr.	Greek	STS	Scottish Text Society
Hist.	history (of), historical	tr.	translation (of)
Ibid.	*ibidem* (in the same place)	US	United States (of America)
Ital.	Italian	vol.	volume

A

abhors: *see* NATURE abhors a vacuum.

a-borrowing: *see* he that GOES a-borrowing, goes a-sorrowing.

abroad: *see* GO abroad and you'll hear news of home.

ABSENCE makes the heart grow fonder
    c **1850** in T. H. Bayly *Isle of Beauty* (rev. ed.) iii. Absence makes the heart grow fonder. **1923** *Observer* 11 Feb. 9 These saws are constantly cutting one another's throats. How can you reconcile the statement that 'Absence makes the heart grow fonder' with 'Out of sight, out of mind'? **1979** C. BRAND *Rose in Darkness* xi. 'Oh, I couldn't go *now*!' 'Absence makes the heart grow fonder.'

absolute: *see* POWER corrupts.

ACCIDENTS will happen (in the best-regulated families)
    1763 G. COLMAN *Deuce is in Him* I. 22 Accidents, accidents will happen—No less than seven brought into our infirmary yesterday. **1819** 'P. ATALL' *Hermit in America* i. Accidents will happen in the best regulated families. **1850** DICKENS *David Copperfield* xxviii. 'Copperfield,' said Mr. Micawber, 'accidents will occur in the best-regulated families; and in families not regulated by . . the influence of Woman, in the lofty character of Wife, they must be expected with confidence, and must be borne with philosophy.' **1939** W. S. MAUGHAM *Christmas Holiday* x. Accidents will happen in the best regulated families, and . . if you find you've got anything the matter with you, . . go and see a doctor right away. **1979** J. SCOTT *Angels in your Beer* xii. It would be so convenient if something happened to them . . . Accidents do happen, as they say.

There is no ACCOUNTING for tastes
    [A version of the Latin tag *de gustibus non est disputandum*, there is no disputing about tastes. **1599** J. MINSHEU *Dialogues in Spanish* 6 Against ones liking there is no disputing.] **1794** A. RADCLIFFE *Mysteries of Udolpho* I. xi. I have often thought the people he disapproved were much more agreeable than those he admired;—but there is no accounting for tastes. **1889** GISSING *Nether World* II. viii. There is no accounting for tastes. Sidney . . not once . . congratulated himself on his good fortune. **1974** T. SHARPE *Porterhouse Blue* x. 'He was in the grip of Mrs Biggs.' . . 'No accounting for tastes,' said the Dean.

accuser: *see* a GUILTY conscience needs no accuser.

acorn: *see* GREAT oaks from little acorns grow.

ACTIONS speak louder than words
    First recorded in its current form in the United States.
    1628 J. PYM *Speech* 4 Apr. in Hansard *Parliamentary Hist. England* (1807) II. 274 'A word spoken in season is like an Apple of Gold set in Pictures of Silver,' and actions are more precious than words. **1736** *Melancholy State of Province* in A. M. Davis *Colonial Currency* (1911) III. 137 Actions speak louder than Words, and are more to be regarded. **1856** A. LINCOLN *Works* (1953) II. 352 'Actions speak louder than words' is the maxim; and, if true, the South now distinctly says to the North, 'Give us the *measures*, and you take the *men*.' **1979** 'C. AIRD' *Some die Eloquent* xvii. 'He's very sorry about it all.' . . 'Actions speak louder than words.'

When ADAM delved and Eve span, who was then the gentleman?
    c **1340** HAMPOLE in G. G. Perry *Reli-*

gious Pieces (EETS) 88 When Adam dalfe[1] and Eue spane . . Whare was than the pride of man? **1381** in Brown & Robbins *Index Middle English Verse* (1943) 628 Whan adam delffid and eve span, Who was than a gentilman? **1562** J. PILKINGTON *Aggeus & Abdias* I. ii. When Adam dalve, and Eve span, Who was than a gentle man? Up start the carle, and gathered good, And thereof came the gentle blood. **1874** J. R. GREEN *Short Hist. English People* v. A spirit fatal to the whole system of the Middle Ages breathed in the popular rime which condensed the levelling doctrine of John Ball: 'When Adam delved and Eve span, who was then the gentleman?'[2] **1918** A. G. GARDINER *Leaves in Wind* 81 It is not only the humanising influence of the garden, it is the democratising influence too. When Adam delved and Eve span, Where was then the gentleman? **1979** C. E. SCHORSKE *Fin-de-Siècle Vienna* vi. When Adam delved and Eve span Who was then the gentleman? The question had ironic relevance for the *arrivé*.

[1] dug. [2] Reportedly used by Ball (d. 1381) to incite the multitude to rebel against their lords in the 'Peasants' Revolt' of 1381; cf. JACK *is as good as his master.*

ADVENTURES are to the adventurous
 1844 DISRAELI *Coningsby* III. I. 244 'I fear that the age of adventures is past.' . . 'Adventures are to the adventurous,' said the stranger. **1914** 'SAKI' *Beasts & Super-Beasts* 264 Adventures, according to the proverb, are to the adventurous. **1952** 'T. HINDE' *Mr Nicholas* iv. He told himself that adventure was to the adventurous . . . If he could not make the effort for the small he would miss the big adventure.

ADVERSITY makes strange bedfellows
 There is some variation in the first word of the proverb: cf. also POLITICS *makes strange bedfellows.*
 1611 SHAKESPEARE *Tempest* II. ii. 37 My best way is to creep under his gaberdine; there is no other shelter hereabout. Misery acquaints a man with strange bedfellows. **1837** DICKENS *Pickwick Papers* xli. (*heading*) Illustrative . . of the old proverb, that adversity brings a man acquainted with strange bedfellows. **1886** H. JAMES *Princess Casamassima* I. I. x. She loathed them [the people] with the outspoken violence of one who had known poverty, and the strange bedfellows it makes. **1927** *Times* 27 Aug. 12 The . . alliance of 1923–5 was an illustration of the adage that adversity makes strange bedfellows. **1982** *Times* 15 Mar. 9 (*heading*) Poverty makes strange bedfellows.

afraid: *see* he who RIDES a tiger is afraid to dismount.

AFTER a storm comes a calm
 Cf. the phrase *the calm after the storm.*
 a **1250** *Ancrene Riwle* (1962) 191 Iblescet ibeo thu laverd the makest stille efter storm [blessed are you, Lord, who makes a calm after the storm]. **1377** LANGLAND *Piers Plowman* B. XVIII. 409 After sharpe shoures . . moste shene[1] is the sonne. **1576** C. HOLYBAND *French Littleton* E1ᵛ After a storme commeth a calme. **1655** T. FULLER *Church Hist. Britain* IX. viii. After a storm comes a calm. Wearied with a former blustering they began now to repose themselves in a sad silence. **1851** MELVILLE *Moby Dick* III. xxviii. The mingled, mingling threads of life are woven by warp and woof—calms crossed by storms, a storm for every calm. **1979** 'J. LE CARRÉ' *Smiley's People* i. For the next two weeks nothing happened . . . After the storm had come the calm.

[1] bright.

AFTER dinner rest a while, after supper walk a mile
 The sense turns on the fact that dinner is a heavy meal, while supper is a light one.
 [Med. L. health proverb *post prandium stabis, post cœnam ambulabis*, after luncheon you will stand still, after supper you will walk about.] **1582** G. WHETSTONE *Heptameron of Civil Discourses* E3

After dynner, talke a while, After supper, walke a mile. **1584** T. COGAN *Haven of Health* ccxi. That olde English saying: After dinner sit a whyle, and after supper walke a myle. **1876** BLACKMORE *Cripps* III. xvi. He neighed . . for he felt quite inclined for a little exercise . . . 'After supper, trot a mile.' **1979** *Daily Telegraph* 24 Dec. 3 'The physiological reaction to a heavy indigestible meal . . seems to be to sleep it off.' What it all seems to boil down to is the old adage: After dinner rest a while, after supper walk a mile.

after: *see also* it is easy to be WISE after the event.

Agamemnon: *see* BRAVE men lived before Agamemnon.

age: *see* the age of MIRACLES is past.

agree: *see* BIRDS in their little nests agree; TWO of a trade never agree.

ALL good things must come to an end
The addition of 'good' is a recent development. The earlier forms may be compared with EVERYTHING *has an end.*
c **1440** *Partonope of Blois* (EETS) l. 11144 Ye wote[1] wele of all thing moste be an ende. **1562** G. LEGH *Accidence of Armoury* 182 All worldly thinges haue an ende (excepte the housholde wordes, betwene man & wife). **1738** SWIFT *Polite Conversation* I. 85 All Things have an End, and a Pudden[2] has two. **1857** H. H. RILEY *Puddleford Papers* xxiii. All things must have an end, and the grand caravan, in time, came to its end. **1904** E. M. FORSTER in *Independent Review* June 128 'En route!' said the shrill voice of Mrs. Forman. 'Ethel! Mr. Graham! The best of things must end.' **1967** RIDOUT & WITTING *English Proverbs Explained* 16 All good things come to an end. Pleasures cannot go on for ever. **1980** B. PAUL. *First Gravedigger* (1982) vi. Life with you . . is nirvana itself. But all good things must come to an end.

[1] know. [2] i.e. a kind of sausage, as a *black pudding.*

It takes ALL sorts to make a world
1620 T. SHELTON tr. *Cervantes' Don Quixote* II. vi. In the world there must bee of all sorts. **1767** S. JOHNSON *Letter* 17 Nov. (1952) I. 194 Some Lady surely might be found . . in whose fidelity you might repose. The World, says Locke, has people of all sorts. **1844** D. W. JERROLD *Story of Feather* xxviii. Click can't get off this time? . . Well, it takes all sorts to make a world. **1975** J. I. M. STEWART *Young Pattullo* iii. 'My father's a banker during the week and a country gent at week-ends. Takes all sorts, you know.' 'Takes all sorts?' 'To make a world.'

ALL things are possible with God
[MATTHEW xix. 26 Iesus . . said vnto them, With men this is vnpossible, but with God al things are possible; cf. HOMER *Odyssey* x. 306 θεοὶ δέ τε πάντα δύνανται, with the gods all things can be done.] **1694** P. A. MOTTEUX tr. *Rabelais' Pantagruel* V. xliii. Drink . . and you shall find its taste and flavor to be exactly that on which you shall have pitched. Then never presume to say that anything is impossible to God. **1712** C. MATHER *Letter* 22 Nov. (1971) 117 However, take it again; all things are possible with God. **1826** L. BEECHER *Letter* 11 June in *Autobiography* (1865) II. viii. Sometimes it seems as if persons had too much . . intellect to be converted easily. But all things are possible with God. **1965** M. SPARK *Mandelbaum Gate* vi. It would be interesting, for a change, to prepare and be ready for possibilities of, I don't know what, since all things are possible with God and nothing is inevitable.

ALL things come to those who wait
[Cf. Fr. *tout vient à celui qui sait attendre,* all comes to him who knows how to wait.] **1530** A. BARCLAY *Eclogues* (EETS) II. 843 Somewhat shall come who can his time abide. **1642** G. TORRIANO *Select Italian Proverbs* 26 He who can wait, hath what he desireth. **1847** DISRAELI *Tancred* II. IV. viii. I have got it at last, everything comes if a man will only

wait. **1863** LONGFELLOW *Poems* (1960) 402 All things come to him who will but wait. **1872** V. FANE *Tout vient à qui sait Attendre* in *From Dawn to Noon* II. 85 Ah! 'All things come to those who wait.' . . They *come*, but often come *too late*. **1931** E. F. BENSON *Mapp & Lucia* vi. There . . was a gay striped figure . . skipping away like mad . . . Miss Mapp gave a shrill crow of triumph. All came to him who waited. **1980** M. SELLERS *Leonardo & Others* viii. Everything comes to those who wait. The theory fitted well into my lazy way of thinking.

all: *see also* all's for the BEST in the best of all possible worlds; all CATS are grey in the dark; DEATH pays all debts; why should the DEVIL have all the best tunes?; don't put all your EGGS in one basket; all's FAIR in love and war; all is FISH that comes to the net; all that GLITTERS is not gold; all is GRIST that comes to the mill; HEAR all, see all, say nowt; there is MEASURE in all things; MODERATION in all things; to the PURE all things are pure; all ROADS lead to Rome; the THIRD time pays for all; all's WELL that ends well; you can't WIN them all; all WORK and no play makes Jack a dull boy.

alone: *see* he TRAVELS fastest who travels alone.

alter: *see* CIRCUMSTANCES alter cases.

always: *see* ONCE a —, always a —; there is always ROOM at the top; the UNEXPECTED always happens.

and: *see* if IFS and ands were pots and pans, there'd be no work for tinkers' hands.

angel: *see* FOOLS rush in where angels fear to tread.

angry: *see* a HUNGRY man is an angry man.

answer: *see* a SOFT answer turneth away wrath.

anvil: *see* the CHURCH is an anvil which has worn out many hammers.

ANY port in a storm

1749 J. CLELAND *Memoirs of Woman of Pleasure* II. 133 It was going by the right door, and knocking desperately at the wrong one . . . I told him of it: 'Pooh,' says he, 'my dear, any port in a storm.' **1821** SCOTT *Pirate* I. iv. As the Scotsman's howf[1] lies right under your lee, why, take any port in a storm. **1965** J. PORTER *Dover Three* ii. It was not quite the sort of company with which Dover would mix from choice but, as the jolly sailors say, any port in a storm.

[1] refuge, haunt.

If ANYTHING can go wrong, it will

Commonly known as *Murphy's Law*, the saying has numerous variations. It is said to have been invented by George Nichols in 1949. Nichols was then a project manager working in California for the American firm of Northrop, and developed the maxim from a remark made by a colleague, Captain E. Murphy, of the Wright Field-Aircraft Laboratory. The contexts of the early quotations appear to support this explanation.

1955 *Aviation Mechanics Bulletin* May–June 11 Murphy's Law: If an aircraft part can be installed incorrectly, someone will install it that way. **1956** *Scientific American* Apr. 166 Dr. Schaefer's observation confirms this department's sad experience that editors as well as laboratory workers are subject to Murphy's Laws, to wit: 1. If something can go wrong it will, [etc.]. **1958** *Product Engineering* 21 Apr. 32 If anything can go wrong with an experiment—it will. **1961** LEEDS & WEINBERG *Computer Programming Fundamentals* viii. What we desire is the presentation of the information in . . an accurate and complete form . . . Recalling 'Murphy's law'—'If something can go wrong or be misinterpreted, it will'—should be enough stimulus for the goals we desire. **1974** *New York Times Magazine* 8 Sept. 33 'If anything can go wrong, it will,' says Murphy's law. In this computer age, the 'law' has been helped along by clever con men. **1980** A. E. FISHER

Midnight Men vii. Of course, the up train was delayed. There was some vast universal principle. If anything can go wrong it will.

An APE's an ape, a varlet's a varlet, though they be clad in silk or scarlet
[LUCIAN *Adversus Indoctum* 4 πίθηκος ὁ πίθηκον . . κἂν χρυσέα ἔχῃ σύμβολα, an ape is an ape . . even if it has gold markings; ERASMUS *Adages* I. vii. *simia simia est, etiamsi aurea gestet insignia.*] **1539** R. TAVERNER tr. *Erasmus' Adages* 21 An ape is an ape although she weare badges of golde. **1586** J. CASE *Praise of Music* ii. You may cloath an Ape in golde, and an Infant in Hercules armour: doth an infant therfore chaunge his age, or an Ape forgoe his nature? **1659** J. HOWELL *Proverbs* (English) 1 An Ape's an Ape, A Varlett's a Varlett,[1] Though they be cladd in silk, or scarlett.[2] **1732** T. FULLER *Gnomologia* no. 6391 An Ape's an Ape: a Varlet's a Varlet, Tho' they be clad in Silk or Scarlet. **1967** D. MORRIS *Naked Ape* i. The naked ape is in danger of . . forgetting that beneath the surface gloss he is still very much a primate. ('An ape's an ape, a varlet's a varlet, though they be clad in silk or scarlet.') Even a space ape must urinate.
[1] servant; rogue.   [2] the colour of the official or ceremonial costume of a judge, cardinal, etc.

ape: *see also* the HIGHER the monkey climbs the more he shows his tail.

appear: *see* TALK of the Devil, and he is bound to appear.

APPEARANCES are deceptive
A common US form is *appearances are deceiving.*
1666 G. TORRIANO *Italian Proverbs* 12 Appearance oft deceives. **1748** SMOLLETT *Gil Blas* (1749) III. VII. i. Egad, appearances are very deceitful. **1784** in *Collections of Massachusets Hist. Society* (1877) III. 186 The appearances in those mountainous regions are extremely deceptive. **1846** MELVILLE *Typee* xxiv. Appearances . . are deceptive. Little men are sometimes very potent, and rags sometimes cover very extensive pretensions. **1927** E. F. BENSON *Lucia in London* v. Mr. Merriall . . watched the three figures at Georgie's door. 'Appearances are deceptive,' he said. 'But isn't that Olga Shuttleworth and Princess Isabel?' **1976** L. ALTHER *Kinflicks* (1977) ii. Apparently she looked lost and in need . . . A ready convert. And in this case, appearances weren't deceiving.

APPETITE comes with eating
Desire or facility increases as an activity proceeds.
[**1534** RABELAIS *Gargantua* I.5 *l'appétit vient en mangeant,* appetite comes with eating.] **1653** URQUHART & MOTTEUX tr. *Rabelais' Gargantua* I. v. Appetite comes with eating. *a* **1721** M. PRIOR *Dialogues of Dead* (1907) 227 But as we say in France, the Appetite comes in Eating; so in Writing You stil found more to write. **1906** W. MAXWELL *From Yalu to Port Arthur* i. Appetite comes with eating. Having absorbed Port Arthur and begun on Manchuria, Russia saw no reason why she should not have Korea also. **1943** S. CLOETE *Congo Song* xxiv. The appetite came with eating. The more he had of her, the more he wanted.

appetite: *see also* HUNGER is the best sauce.

An APPLE a day keeps the doctor away
1866 *Notes & Queries* 3rd Ser. IX. 153 A Pembrokeshire Proverb.—'Eat an apple on going to bed, And you'll keep the doctor from earning his bread.' **1913** E. M. WRIGHT *Rustic Speech* xiv. Ait a happle avore gwain to bed, An' you'll make the doctor beg his bread (Dev.); or as the more popular version runs: An apple a day keeps the doctor away. **1972** I. CHACE *Tartan Touch* iv. He gave me a truly wicked look . . . 'An apple a day keeps the doctor away!' he taunted me.

The APPLE never falls far from the tree
Frequently used to assert the continuity

of family characteristics; cf. *Like*
FATHER, *like son.* Quot. 1839 implies
return to one's original home.
[Apparently of Eastern origin; cf.
16th-cent. Ger. *der Apfel fellt nicht gerne
weit vom Baume*, the apple does not
usually fall far from the tree.] **1839**
EMERSON *Letter* 22 Dec. (1939) II. 243 As
men say the apple never falls far from
the stem, I shall hope that another year
will draw your eyes & steps to this old
dear odious haunt of the race. **1939**
H. W. THOMPSON *Body, Boots & Britches*
xix. As a . . farmer remarked, 'If you
breed a pa'tridge, you'll git a pa'tridge.'
Another way of setting that truth forth
is, . . 'An apple never falls far from the
tree.' **1981** *Women's Journal* Apr. 179
He's a fool, Muffie, as his father was.
The apple never falls far from the tree.

apple: *see also* the ROTTEN apple injures its
neighbours; STOLEN fruit is sweet.

APRIL showers bring forth May flowers
   c **1560** in T. Wright *Songs & Ballads*
(1860) 213 Aprell sylver showers so
sweet Can make May flowers to
sprynge. **1570** T. TUSSER *Husbandry*
(rev. ed.) 22 Swete Aprill showers, Do
spring the May flowers. **1670** J. RAY
English Proverbs 41 April showers bring
forth May flowers. **1821** SCOTT *Kenilworth* III. vii. I believe . . if showers fall
in April, that we shall have flowers in
May. **1846** M. A. DENHAM *Proverbs relating to Seasons, &c.* 36 March winds
and April showers bring forth May
flowers. **1921** *Sphere* 14 May 152 If
there was anybody left to believe in the
saying that 'April showers bring forth
May flowers' their simple faith must
have been rudely shattered by May's
behaviour this year. **1931** K. BOYLE *Pla-
gued by Nightingale* xxxii. 'Dear, dear,'
he said, '''April showers bring May
flowers.'''

architect: *see* EVERY man is the architect of
his own fortune.

arm: *see* KINGS have long arms; STRETCH
your arm no further than your sleeve

will reach; YORKSHIRE born and Yorkshire bred, strong in the arm and weak
in the head.

An ARMY marches on its stomach
The proverb has been attributed to both
Napoleon and Frederick the Great; this
figurative use of (*on one's*) *stomach* is unusual in English.
   1904 *Windsor Magazine* Jan. 268 'An
army marches on its stomach.' *'C'est la
soupe qui fait le soldat.'* These Napoleonic
aphorisms . . have been increasingly
appreciated by our War Office. **1911**
F. W. HACKWOOD *Good Cheer* xxvi. 'An
army marches on its stomach,' says the
old proverb—and, of course, fights on
it too. **1977** J. B. HILTON *Dead-Nettle* x.
'They say an army marches on its stomach,' Gilbert Slack began to say. 'You
mean that Frank was a cook?'

arrive: *see* it is BETTER to travel hopefully
than to arrive.

ART is long and life is short
Hippocrates compares the difficulties
encountered in learning the art of medicine or healing with the shortness of
human life. *Art* is now commonly
understood in the proverb in a less specific sense. In quot. 1958, it refers to (the
durability of) a work of art.
   [HIPPOCRATES *Aphorisms* I. 1. ὁ βίος
βραχύς, ἡ δὲ τέχνη μακρή, life is short,
but art is long; L. *ars longa, vita brevis*, art
is long, life is short.] *c* **1380** CHAUCER
Parliament of Fowls l.1 The lyf so short,
the craft so long to lerne. **1558** W. BUL-
LEIN *Government of Health* 5ᵛ And
although oure life be shorte, yet the arte
of phisicke is long. **1581** G. PETTIE tr. *S.
Guazzo's Civil Conversation* I. 16 An art is
long and life is short. **1710** S. PALMER
Proverbs 380 Art is Long, Life Short. Our
Philosophical Meditations on Time are
very Obscure and Confus'd. **1869**
M. ARNOLD *Culture & Anarchy* vi. If . .
we take some other criterion of man's
well-being than the cities he has built . .
our Liberal friends . . take us up very
sharply. 'Art is long,' says the *Times*,

'and life is short.' **1958** L. DURRELL
Balthazar IV. xiii. The shapely hand on
his shoulder still wore the great ring
taken from the tomb of a Byzantine
youth. Life is short, art long.

ash: *see* when the OAK is before the ash,
then you will only get a splash; beware
of an OAK it draws the stroke.

ASK no questions and hear no lies

1773 GOLDSMITH *She stoops to Conquer*
III. 51 Ask me no questions and I'll tell
you no fibs. **1818** SCOTT *Heart of Midlothian* I. ix. If ye'll ask nae questions, I'll
tell ye nae lees. **1900** H. LAWSON *Over
Sliprails* 135 'Where did you buy the
steer, father?' she asked. 'Ask no questions and hear no lies.' **1906** KIPLING
Puck of Pook's Hill 252 Them that asks no
questions isn't told a lie—Watch the
wall, my darling, while the Gentlemen
go by! **1970** V. CANNING *Great Affair* xii.
'What has happened to Sarah?' . . 'Ask
no questions hear no lies.'

a-sorrowing: *see* he that GOES a-borrowing,
goes a-sorrowing.

ATTACK is the best form of defence

[**1775** W. H. DRAYTON in R. W.
Gibbes *Documentary Hist. American
Revolution* (1855) I. 174 It is a maxim,
that it is better to attack than to receive
one.] **1799** G. WASHINGTON *Writings*
(1940) XXXVII. 250 Make them believe,
that offensive operations, often times, is
the surest, if not the only . . means of
defence. **1941** *Times* 26 Nov. 8 The British People had always believed that the
best form of defence was attack on the
enemy. **1965** N. S. GRAY *Apple-Stone* xi.
'Attack', she said, 'is the best means of
defence.' She sounded so smug that I
told her the thought was not original.
1980 F. OLBRICH *Desouza in Stardust* iv.
Attack is the best form of defence, they
say, and when politicians lose their
principles they play a dirty game.

away: *see* when the CAT's away, the mice
will play.

B

babe: *see* out of the MOUTHS of babes —.

baby: *see* don't THROW the baby out with the bathwater.

back: *see* GOD makes the back to the burden; what is GOT over the Devil's back is spent under his belly; it is the LAST straw that breaks the camel's back.

A BAD excuse is better than none

1551 T. WILSON *Rule of Reason* S6 This is as thei saie in English, better a badde excuse, then none at all. **1579** GOSSON *School of Abuse* 24 A bad excuse is better, they say, then none at all. Hee, because the Frenchmen paid tribute every moneth, into xiii monethes devided the yeere. **1686** J. DUNTON in *Publications of Prince Society* (1867) IV. 30 Philaret . . being loth to dye so early in the morning, would not leave his Cabin . . till he had found his Ruffles: (a bad Excuse, you know, Brother, is better than none). **1821** W. WIRT *Letter* 29 Aug. in J. P. Kennedy *Memoirs* (1849) II. vii. The old fellow's look had a glimpse of passing cunning as much as to say, 'A bad excuse is better than none.' **1981** P. VAN GREENAWAY *'Cassandra' Bill* xiii. What excuse is better than none?

BAD money drives out good

Commonly known as Gresham's Law, after Sir Thomas Gresham (*c* 1519–79), founder of the Royal Exchange. Gresham saw the economic need to restore the purity of the coinage, though there is no evidence that he actually used this expression. Quot. 1902 states that the principle, not the proverb, is mentioned in Gresham's letter to the Queen.

[**1858** H. D. MACLEOD *Elements of Political Economy* 477 He [Gresham] was the first to perceive that a bad and debased currency is the cause of the disappearance of the good money.] **1902** *New English Dictionary* VI. 116 Gresham's *law*, the principle, involved in Sir Thomas Gresham's letter to Q. Elizabeth in 1558, that 'bad money drives out good'. **1933** A. HUXLEY *Letter* 18 Nov. (1969) 438 Gresham's Law holds good in every field . . and bad politics tends to drive out good politics just as bad money drives out good money. **1952** R. CROSSMAN *Journal* 16 June in *Backbench Diaries* (1981) 109 The one thing we all know is that, if you have Government radio and sponsored radio side by side, the bad currency drives out the good. **1979** *Times* 12 Dec. 15 Bad money drives out good, by which is meant that a man who has both good and bad money will keep the good and use the bad to settle transactions.

BAD news travels fast

[**1539** R. TAVERNER tr. *Erasmus' Adages* II. A4 Sad and heuy tydynges be easly blowen abroade be they neuer so vaine and false and they be also sone beleued.] **1592** KYD *Spanish Tragedy* I. B2[v] Euill newes[1] flie faster still than good. **1694** *Terence's Comedies made English* 46 Bad News always fly faster than good. **1792** T. HOLCROFT *Road to Ruin* II. i. All these bills . . brought . . this morning. Ill news travels fast. **1935** W. IRWIN *Julius Caesar Murder Case* xxv. 'Where'd you get it [a knife]?' 'On the Plains of Philippi.' 'Bad news travels fast,' said Hercules. **1976** 'J. HERRIOT' *Vets might Fly* xxiii. They say bad news travels fast and I had hardly started my return journey when . . the loudspeaker asked me to report to the manager's office.

[1] in early use, *news* was construed as a plural noun.

A BAD penny always turns up

The proverb, also used allusively in simile and metaphor (see quots. 1766 and 1979, second sentence), usually refers to the predictable, and often unwanted, return of a disreputable or pro-

digal person to his place of origin after
some absence.

[**1766** A. ADAMS in L. H. Butterfield et
al. *Adams Family Correspondence* (1963)
I.55 Like a bad penny it returnd, to me
again.] **1824** SCOTT *Redgauntlet* II. ii.
Bring back Darsie? little doubt of that—
the bad shilling is sure enough to come
back again. **1884** R. H. THORPE *Fenton
Family* iii. Just like as not he'll be coming
back one of these days, when he's least
wanted. A bad penny is sure to return.
1941 A. UPDEGRAFF *Hills look Down* vi. 'I
miss Bart.' 'Oh, a bad penny always
turns up again.' **1979** G. MITCHELL
Mudflats of Dead iii. 'Stop worrying. The
bad pennies always turn up.' 'Oh,
Adrian, I don't think she's a bad penny,
not really.'

A BAD workman blames his tools

[Late 13th-cent. Fr. *mauvés ovriers
ne trovera ja bon hostill*, a bad workman
will never find a good tool.] **1611**
R. COTGRAVE *Dict. French & English* s.v.
Outil, A bungler cannot find (or fit
himselfe with) good tooles. **1640**
G. HERBERT *Outlandish Proverbs* no. 67
Never had ill workeman good tooles.
1859 S. SMILES *Self-Help* iv. It is proverbial that the bad workman never yet had
a good tool. **1940** J. G. COZZENS *Ask Me
Tomorrow* vii. I've read somewhere that
a poor workman quarrels with his
tools. **1979** A. FOX *Threat Signal Red* xv.
Damn! Dropped the screwdriver. . .
Bad workmen blame their tools.

bad: *see also* give a DOG a bad name and
hang him; a GOOD horse cannot be of a
bad colour; HARD cases make bad law;
HOPE is a good breakfast but a bad supper; NOTHING so bad but it might have
been worse; THREE removals are as bad
as a fire.

bag: *see* EMPTY sacks will never stand
upright; there's many a GOOD cock come
out of a tattered bag.

bairn: *see* FOOLS and bairns should never
see half-done work; the SHOEMAKER's
son always goes barefoot.

As you BAKE, so shall you brew

As you begin, so shall you proceed (cf.
as you MAKE *your bed, so you must lie upon
it*). Complementary to *As you* BREW, *so
shall you bake.*

c **1577** *Misogonus* III. i. As thou bakst,
so shat brewe. **1775** D. GARRICK *May-
Day* ii. To keep . . My bones whole and
tight, To speak, nor look, would I dare;
As they bake they shall brew. **1909**
W. DE MORGAN *It never can happen Again*
I. v. Each one [i.e. young person] . .
was . . the centre of an incubation of
memories that were to last a lifetime.
'As they bake, so they will brew,' philo-
sophized Mr. Challis to himself.

bake: *see also* as you BREW, so shall you
bake.

bare: *see* there goes more to MARRIAGE
than four bare legs in a bed.

barefoot: *see* the SHOEMAKER's son always
goes barefoot.

bargain: *see* it takes TWO to make a bargain.

bark: *see* DOGS bark, but the caravan goes
on; why KEEP a dog and bark yourself?

A BARKING dog never bites

Bark is contrasted (both as a noun and
verb) with *bite* in other phrases, esp. *his
bark is worse than his bite.*

[13th-cent. Fr. *chascuns chiens qui abaie
ne mort pas*, the dog that barks does not
bite.] c **1550** *Thersytes* E1 Great barking
dogges, do not most byte And oft it is
sene that the best men in the hoost Be
not suche, that vse to bragge moste.
1595 *Locrine* (1908) IV. i. Soft words
good sir . . . A barking dog doth sil-
dome strangers bite. **1629** *Book of Merry
Riddles* 22 A barking dog seldome
bites. **1730** SWIFT *Traulus* I. 5 When
Towzer snaps At People's Heels . . The
Neighbours all cry, shoot him dead . . .
You mistake the Matter quite; Your
barking Curs will seldom bite. **1837**
F. CHAMIER *Arethusa* III. x. Our dogs
which bark, Abdallah, seldom bite.
1980 *Daily Telegraph* 1 May 18 A can-

vassing candidate came to a house where there was an Alsatian who barked ferociously. His agent said: 'Just go in. Don't you know the proverb "A barking dog never bites"?' 'Yes,' said the candidate, 'I know the proverb, you know the proverb, but does the dog know the proverb?'

BARNABY bright, Barnaby bright, the longest day and the shortest night
St. Barnabas' Day, 11 June, in Old Style reckoned the longest day of the year.

[**1595** SPENSER *Epithalamion* l. 266 This day the sunne is in his chiefest hight, With Barnaby the bright.] **1659** J. HOWELL *Proverbs* (English) 20 Barnaby bright, the longest day and shortest night. **1858** *Notes & Queries* 2nd Ser. VI. 522 In some parts of the country the children call the lady-bird Barnaby Bright, and address it thus: —'Barnaby Bright, Barnaby Bright, The longest day and the shortest night.' **1906** E. HOLDEN *Country Diary of Edwardian Lady* (1977) 72 Barnaby bright All day and no night. **1978** R. WHITLOCK *Calendar of Country Customs* vii. Barnaby bright, Barnaby bright, The longest day and the shortest night, is a reminder that, before the change in the calendar in 1752, 11 June *was* the longest day of the year.

basket: *see* don't put all your EGGS in one basket.

bathwater: *see* don't THROW the baby out with the bathwater.

battalion: *see* PROVIDENCE is always on the side of the big battalions.

battle: *see* the RACE is not to the swift, nor the battle to the strong.

BE what you would seem to be
c **1377** LANGLAND *Piers Plowman* B.X. 253 Suche as thow semest in syghte, be in assay[1] y-founde. **1547** W. BALDWIN *Treatise of Moral Philosophy* II. xi. Be the selfe same that thou pretendest. **1640** G. HERBERT *Outlandish Proverbs* no. 724

Be what thou wouldst seeme to be. **1721** J. KELLY *Scottish Proverbs* 68 Be what you seem, and seem what you are. The best way! for Hypocrisy is soon discovered. **1865** 'L. CARROLL' *Alice's Adventures in Wonderland* ix. It's a vegetable. It doesn't look like one, but it is . . . The moral of that is—'Be what you would seem to be.' **1980** G. SIMS in H. Watson *Winter Crimes 12* 158 The Benningworth family motto *Esse quam videri*, 'To be rather than to seem to be'.[2]

[1] trial, experiment. [2] There is a difference of emphasis in this quotation.

bean: *see* CANDLEMAS day, put beans in the clay, put candles and candlesticks away.

BEAR and forbear
[EPICTETUS *Fragments* x. ἀνέχου καὶ ἀπέχου, be patient and endure; ERASMUS *Adages* II. vii. 13 *sustine et abstine*.] **1573** T. TUSSER *Husbandry* (rev. ed.) II. 12[v] Both beare and forbeare, now and then as ye may, then wench God a mercy,[1] thy husband will say. **1688** BUNYAN *Discourse of Building, &c. House of God* 53 To bear and forbear here, will tend to rest. **1761** LADY M. W. MONTAGU *Letter* 20 Feb. (1967) 253, I know that in this world one must bear and forbear. **1871** S. SMILES *Character* xi. The golden rule of married life is, 'Bear and forbear'. **1940** H. W. THOMPSON *Body, Boots & Britches* xix. You must take two bears to live with you—Bear and Forbear.

[1] *God a mercy,* God have mercy (used in the sense 'God reward you').

beard: *see* it is MERRY in hall when beards wag all.

beast: *see* when the WIND is in the east, 'tis neither good for man nor beast.

If you can't BEAT them, join them
Beat is usually replaced by *lick* in the US.
1941 Q. REYNOLDS *Wounded don't Cry* i. There is an old political adage which says 'If you can't lick 'em, jine 'em'. **1953** P. GALLICO *Foolish Immortals* xvii. It was

vital to him to get the reins back into his own hands again. He remembered an old adage: 'If you can't lick 'em, join 'em.' **1979** D. LESSING *Shikasta* 266, I said, Running things, what's the point? He said, If you can't beat them, join them!

beat: *see also* one ENGLISHMAN can beat three Frenchmen; it is easy to find a STICK to beat a dog; a WOMAN, a dog, and a walnut tree, the more you beat them the better they be.

beautiful: *see* SMALL is beautiful.

BEAUTY is in the eye of the beholder
Beauty is not judged objectively, but according to the beholder's estimation.
[**1742** HUME *Essays Moral & Political* II. 151 Beauty, properly speaking, lyes . . in the Sentiment or Taste of the Reader.] **1769** F. BROOKE *Hist. Emily Montague* IV. 205 You should remember, my dear, that beauty is in the lover's eye. **1847** C. BRONTË *Jane Eyre* II. ii. Most true is it that 'beauty is in the eye of the gazer'. **1878** M. W. HUNGERFORD *Molly Bawn* I. xii. 'I have heard she is beautiful—is she?' 'Beauty is in the eye of the beholder,' quotes Marcia. **1940** G. SEAVER *Scott of Antarctic* II. 48 'Beauty is in the eye of the beholder.' The eye, which is the reflector of the external world, is also the mirror of the soul within. **1980** J. R. RICHARDS *Sceptical Feminist* vii. It does not matter if beauty is in the eye of the beholder: you can always find out what the beholder likes.

BEAUTY is only skin-deep
Physical beauty is no guarantee of good character, temperament, etc.
[*a* **1613** T. OVERBURY *Wife* (1614) B8[v] All the carnall beautie of my wife, Is but skinne-deep.] **1616** J. DAVIES *Select Second Husband* B3 Beauty's but skin-deepe. **1742** RICHARDSON *Pamela* IV. lx. Beauty is but . . a mere skin-deep perfection. **1829** COBBETT *Advice to Young Men* III. cxxix. The less favoured part of the sex say, that 'beauty is but skin deep' . . but it is very agree-able though, for all that. **1882** E. M. INGRAHAM *Bond & Free* xiii. Mother used to say that beauty was only skin deep, but I never before realized that bones could be so fearfully repulsive. **1921** W. H. HUDSON *Traveller in Little Things* iv. It is only the ugly (and bad) who fondly cherish the delusion that beauty . . is only skin-deep and the rest of it. **1978** A. PRICE *'44 Vintage* xix. Beauty is only skin-deep, but it's only the skin you see.

bed: *see* EARLY to bed and early to rise, makes a man healthy, wealthy, and wise; as you MAKE your bed, so you must lie upon it; there goes more to MARRIAGE than four bare legs in a bed.

bedfellow: *see* ADVERSITY makes strange bedfellows; POLITICS makes strange bedfellows.

beer: *see* LIFE isn't all beer and skittles; TURKEY, heresy, hops, and beer came into England all in one year.

beget: *see* LENGTH begets loathing; LOVE begets love.

Set a BEGGAR on horseback, and he'll ride to the Devil
A proverb (now frequently used elliptically) with many variations, meaning that one unaccustomed to power or luxury will abuse or be corrupted by it.
1576 G. PETTIE *Petit Palace* 76 Set a Beggar on horsebacke, and he wyl neuer alight. **1591** SHAKESPEARE *Henry VI, Pt.* 3 I. iv. 127 It needs not . . proud queen; Unless the adage must be verified, That beggars mounted run their horse to death. **1592** NASHE *Pierce Penniless* I. 174 These whelpes . . drawne vp to the heauen of honor from the dunghill of abiect fortune, haue long been on horsebacke to come riding to your Diuelship. **1616** T. ADAMS *Sacrifice of Thankfulness* 6 He that serues the Flesh serues his fellow: And a Beggar mounted on the backe of Honour, rides post to the Diuell. **1669** W. WINSTANLEY *New Help to Discourse* 151 Set a Beggar

on Horse-back, and he will ride to the Devil. **1855** GASKELL *North & South* I. x. You know the proverb. . . 'Set a beggar on horseback, and he'll ride to the devil,'—well, some of these early manufacturers did ride to the devil in magnificent style. **1923** C. WELLS *Affair at Flower Acres* ii. I should think your early days of forced economy would have taught you not to be quite so extravagant. But there's an old proverb—'Set a beggar on horseback—' and so forth, that jolly well fits you. **1945** R. HARGREAVES *Enemy at Gate* 21 The plebians . . eventually attained to that comfortless pre-eminence . . which can only be experienced by beggars on horseback devoid of all aptitude for . . the precarious art of equitation.

beggar: *see also* if WISHES were horses, beggars would ride.

BEGGARS can't be choosers
The replacement of *can't* for *must not* is a recent development.

1546 J. HEYWOOD *Dialogue of Proverbs* I. x. D1 Folke say alwaie, beggers shulde be no choosers. **1579** GOSSON *Apology of School of Abuse* in *Ephemerides of Phialo* 90ᵛ Beggars, you know, muste bee no chosers. **1728** VANBRUGH *Journey to London* III. i. My Lords, says I, Beggars must not be Chusers; but some Place about a thousand a Year . . might do pretty weel. **1863** C. READE *Hard Cash* xxiii. The dustman . . grumbled at the paper and the bones, he did. So I told him beggars mustn't be choosers. **1888** N. J. CLODFELTER *Snatched from Poor House* iv. Crawl out o' that bed! I 'spose you do feel a little bad, but 'beggars can't be choosers!' **1939** J. SHEARING *Blanche Fury* 72 'I suppose . . you would marry any man with a good character and a fine estate.' . . 'Beggars can't be choosers, you mean!' **1967** O. MILLS *Death enters Lists* xvii. I don't pretend I don't miss my garden. But there, beggars can't be choosers.

begin: *see* CHARITY begins at home; LIFE begins at forty; when THINGS are at the worst they begin to mend; *also* BEGUN.

beginning: *see* a GOOD beginning makes a good ending.

begun: *see* WELL begun is half done.

beholder: *see* BEAUTY is in the eye of the beholder.

BELIEVE nothing of what you hear, and only half of what you see
[*a* **1300** *Proverbs of Alfred* (1907) 35 Gin thu neuere leuen alle monnis spechen, Ne alle the thinge that thu herest singen. **1770** C. CARROLL *Letter* 4 Sept. in *Maryland Hist. Magazine* (1918) XIII. 58 You must not take Everything to be true that is told to you.] **1858** D. M. MULOCK *Woman's Thoughts about Women* viii. 'Believe only half of what you see, and nothing that you hear,' is a cynical saying, and yet less bitter than at first appears. **1979** D. KYLE *Green River High* ii. I listened with the old magician's warning lively in my mind; believe nothing of what you hear—and only half of what you see!

believing: *see* SEEING is believing.

A BELLOWING cow soon forgets her calf
An excessive show of grief (at a bereavement) quickly passes.

[**1553** T. WILSON *Art of Rhetoric* 42 The Cowe lackyng her Caulfe, leaueth Loweyng within three or foure daies at the farthest.] **1895** S. O. ADDY *Household Tales* 142 In the East Riding they say, 'A bletherin' coo soon forgets her calf,' meaning that excessive grief does not last long. **1928** *London Mercury* Feb. 439 Common proverb in the West Country is 'A belving cow soon forgets her calf'. **1945** F. THOMPSON *Lark Rise* xxxiv. When a woman, newly widowed, had tried to throw herself into her husband's grave at his funeral . . some one . . said drily . .' Ah, you wait. The bellowing cow's always the first to forget its calf.'

belly: *see* what is GOT over the Devil's back is spent under his belly.

bent: *see* as the TWIG is bent, so is the tree inclined.

All's for the BEST in the best of all possible worlds

[**1759** W. RIDER tr. *Voltaire's Candide* xxx. Pangloss would sometimes say to Candidus: 'All Events are linked together in this best of all possible Worlds.] **1911** G. B. SHAW *Shewing-up of Blanco Posnet* 299 The administrative departments were consuming miles of red tape in the correctest forms of activity, and everything was for the best in the best of all possible worlds. **1943** A. CHRISTIE *Moving Finger* xv. I agreed with happy Miss Emily that everything was for the best in the best of possible worlds. **1961** WODEHOUSE *Ice in Bedroom* ii. Fate had handed him the most stupendous bit of goose[1] and . . all was for the best in this best of all possible worlds.

[1] luck.

The BEST is the enemy of the good

Cf. *The* GOOD *is the enemy of the best.*

1861 R. C. TRENCH *Commentary on Epistles to Seven Churches in Asia* p. v, 'The best is oftentimes the enemy of the good'; and . . many a good book has remained unwritten . . because there floated before the mind's eye . . the ideal of a better or a best. **1925** *Times* 1 Dec. 16 This is not the first time in the history of the world when the best has been the enemy of the good; . . one single step on . . solid ground may be more profitable than a more ambitious flight. **1960** D. JONES *Letter* 1 June in R. Hague *Dai Greatcoat* (1980) III. 182 Tom told me a very good Spanish proverb: 'The best is the enemy of the good.' **1981** *Times* 2 Mar. 13 To maintain that all that a school provides must be provided free makes the best the enemy of the good.

The BEST-laid schemes of mice and men gang aft agley

Often used allusively in shortened form (see quot. 1911).

1786 BURNS *Poems* 140 The best laid schemes o' Mice an' Men, Gang aft agley.[1] **1911** D. H. LAWRENCE *Letter* 21 Sept. (1979) I. 305, I am sorry the bookbinding has gone pop. But there 'The best laid schemes' etc. etc. **1976** A. PRICE *War Game* I. vii. Such a curious thing, utterly unforeseeable, had made the best-laid scheme go agley.

[1] *gang aft agley*, often go awry.

The BEST of friends must part

[*c* **1385** CHAUCER *Troilus & Criseyde* v. 343 Alwey frendes may nat ben yfeere.[1]] **1611** G. CHAPMAN *May-Day* IV. 70 Friends must part, we came not all together, and we must not goe all together. **1685** J. DUNTON in *Publications of Prince Society* (1867) 10 But the dearest friends must part. **1784** J. F. D. SMYTH *Tour in USA* I. xxxvii. Sooner or later, all, even the dearest of friends, must part. **1821** SCOTT *Kenilworth* I. xi. 'You are going to leave me, then?'. . 'The best of friends must part, Flibbertigibbet.' **1910** G. W. E. RUSSELL *Sketches & Snapshots* 212 But the best of friends must part, and it is time to take our leave of this . . high-souled cavalier. **1979** W. GOLDING *Darkness Visible* ii. 'Aren't there going to be any more lessons?' . . 'The best of friends must part.'

[1] *may nat ben yfeere*, may not be together.

The BEST of men are but men at best

1680 J. AUBREY *Letter* 15 June in *Brief Lives* (1898) I. 12, I remember one sayeing of generall Lambert's, that 'the best of men are but men at best'. **1885** T. HARLEY *Moon Lore* 191 We can but repeat to ourselves the saying, 'The best of men are but men at best'.

The BEST things come in small packages

Parcels sometimes replaces *packages.*

[Cf. 13th-cent. Fr. *menue[s] parceles ensemble sunt beles,* small packages considered together are beautiful. **1659** J. HOWELL *Proverbs* (French) 10 The best ointments are put in little boxes.] **1877** B. FARJEON *Letter* 22 Jan. in E. Farjeon *Nursery in Nineties* (1935) v. As the best things are (said to be) wrapped in small parcels (proverb), I select the smallest sheet of paper I can find . . to make you acquainted with the . . state of affairs. **1979** R. THOMAS *Eighth Dwarf* xviii. 'The little gentleman.' . . 'The best things sometimes come in small packages,' Jackson said, wincing at his own banality.

The BEST things in life are free

1927 B. G. DE SILVA et al. *Best Things in Life are Free* (song) 3 The moon belongs to ev'ryone, The best things in life are free, The stars belong to ev'ryone, They gleam there for you and me. **1948** in B. Stevenson *Home-Book of Proverbs* 887 In gloomy tones we need not cry: 'How many things there are to buy!' Here is a thought for you and me: 'The best things in life are free.' **1955** W. GADDIS *Recognitions* II. ii. Someone once told them the best things in life are free, and so they've got in the habit of not paying. **1980** *Listener* 29 May 698 The best things in life may be free, but the better ones cost you.

It is BEST to be on the safe side

1668 DRYDEN & NEWCASTLE *Sir Martin Mar-all* v. i. I'm resolv'd to be on the sure side. **1811** J. AUSTEN *Sense & Sensibility* III. iv. Determining to be on the safe side, he made his apology in form as soon as he could say any thing. **1847** MARRYAT *Children of New Forest* I. xi. Be on the safe side, and do not trust him too far. **1935** L. I. WILDER *Little House on Prairie* iii. Best to be on the safe side, it saves trouble in the end.

best: *see also* ACCIDENTS will happen (in the best-regulated families); ATTACK is the best form of defence; why should the DEVIL have all the best tunes?; the best DOCTORS are Dr. Diet, Dr. Quiet, and Dr. Merryman; EAST, west, home's best; EX-PERIENCE is the best teacher; the GOOD is the enemy of the best; HONESTY is the best policy; HOPE for the best and prepare for the worst; HUNGER is the best sauce; he LAUGHS best who laughs last; it is best to be OFF with the old love before you are on with the new; an old POACHER makes the best gamekeeper;

SECOND thoughts are best; SILENCE is a woman's best garment.

BETTER a dinner of herbs than a stalled ox where hate is

1560 BIBLE (Geneva) *Proverbs* xv. 17 Better is a dinner of grene herbes[1] where loue is, then a stalled[2] oxe and hatred therewith. **1611** BIBLE *Proverbs* xv. 17 Better is a dinner of herbes where loue is, then a stalled oxe, and hatred therewith. **1817** S. SMITH *Letter* 13 Mar. in S. Holland *Memoir* (1855) II. 138 When you think of that amorous and herbivorous parish of Covent Garden, and compare it with my agricultural benefice, you will say, 'Better is the dinner of herbs where love is, than the stalled OX,' etc. etc. **1914** 'SAKI' *Beasts & Super-Beasts* 227 The ox had finished the vase-flowers . . and appeared to be thinking of leaving its rather restricted quarters. . . I forget how the proverb runs. . . Something about 'better a dinner of herbs than a stalled ox where hate is'. **1979** J. DRUMMOND *I saw Him Die* viii. Lunch was a silent affair. . . I said, '"Better a dinner of herbs than a stalled ox where hate is."'

[1] used in the archaic sense 'plants of which the leaves are used as food'. [2] fattened in a stall for killing.

BETTER a good cow than a cow of a good kind

A good character is better than a distinguished family.

1922 J. BUCHAN *Huntingtower* x. I'm no weel acquaint wi' his forbears, but I'm weel eneuch acquaint wi' Sir Erchie, and 'better a guid coo than a coo o' a guid kind', as my mither used to say.

BETTER be an old man's darling, than a young man's slave

1546 J. HEYWOOD *Dialogue of Proverbs* II. vii. 13ᵛ Many yeres sens, my mother seyd to me, Hyr elders wold saie, it ys better to be An olde mans derlyng, then a yong mans werlyng.[1] **1721** J. KELLY *Scottish Proverbs* 74 Better an old Man's Darling, than a young Man's Wonder-

ling, say the Scots, Warling, say the
English. **1859** J. R. PLANCHÉ *Love & Fortune* 8 Let defeated rivals snarling, Talk
of one foot in the grave. Better be an old
man's darling, Than become a young
man's slave. **1885** E. J. HARDY *How to be
Happy though Married* v. Perhaps the
majority of girls would rather be a
young man's slave than an old man's
darling. **1980** J. MARCUS *Marsh Blood* ix.
Find yourself an older man. Much better
to be an old man's darling, than a young
man's slave.

[1] one who is despised or disliked.

BETTER be envied than pitied

[PINDAR *Pythian Odes* I. 163 κρέσσων
γὰρ οἰκτιρμοῦ φθόνος, envy is stronger
than pity; HERODOTUS *Hist.* iii.
52 φθονέεσθαι κρέσσον ἐστὶ ἢ
οἰκτείρεσθαι, it is better to be envied
than to be pitied; ERASMUS *Adages* IV. iv.
87 *praestat invidiosum esse quam
miserabilem*.] **1546** J. HEYWOOD *Dialogue
of Proverbs* I. xi. D2[v] Sonne, better be
envied then pitied, folke sey. *a* **1631**
DONNE *Poems* (1633) 94 Men say, and
truly, that they better be Which be en-
vyed then pittied. **1902** G. W. E. RUS-
SELL *Onlooker's Note-Book* xxxiii. Her
friend responded sympathetically, 'My
dear, I'd much rather be envied than
pitied.'

BETTER be out of the world than out of the fashion

1639 J. CLARKE *Parœmiologia Anglo-
Latina* 171 As good out of th' world as
out o' th' fashion. **1738** SWIFT *Polite Conversation* II. 117 'Why, Tom, you are
high in the Mode.'. . 'It is better to be
out of the World, than out of the
Fashion.' **1903** E. F. MAITLAND *From
Window in Chelsea* iv. Women seem seldom hindered by lack of money when it
is a case of follow-my-leader. 'Better be
out of the world than out of the
fashion.' **1935** J. MAXTON *If I were Dictator* i. Dictatorships are fashionable just
now. There was an old-time song which
said 'If you are out of the fashion you
had better leave the world.'

BETTER be safe than sorry

1837 S. LOVER *Rory O'More* II. xxi. 'Jist
countin' them,—is there any harm in
that?' said the tinker: 'it's betther be sure
than sorry'. **1933** *Radio Times* 14 Apr.
125 Cheap distempers very soon crack
or fade. Better be safe than sorry. Ask
for Hall's. **1972** J. WILSON *Hide & Seek*
vii. It's not that I want to shut you in . .
but—well, it's better to be safe than
sorry.

BETTER late than never

[DIONYSIUS OF HALICARNASSUS *Roman
Antiquities* ix. 9 κρεῖττον ἐστὶν ἄρξασθαι
ὀψὲ τὰ δέοντα πράττειν ἢ μηδέποτε, it is
better to start doing what one has to late
than not at all; LIVY *Hist.* IV. xxiii. *potius
sero quam nunquam*.] *c* **1330** in C. Keller
Die Mittelenglische Gregoriuslegende
(1941) 146 A. Better is lat than neuer
blinne[1] Our soules to maken fre. *c* **1450**
LYDGATE *Assembly of Gods* (EETS) l. 1204
Vyce to forsake ys bettyr late then
neuer. **1546** J. HEYWOOD *Dialogue of
Proverbs* I. x. C4 Things done, can not be
vndoone, . . But better late then neuer
to repent this. **1708** S. OCKLEY *Conquest
of Syria* I. 276 Whilst he was murdering
the unhappy Aleppians, Caled (better
late than never) came to their Relief.
1852 C. M. YONGE *Two Guardians* xviii.
She obtained from Agnes some admiration for Caroline's conduct, though in
somewhat of the 'better late than never
style'. **1954** A. HUXLEY *Letter* 16 Sept.
(1969) 711, I am sorry your holiday will
have to be postponed so long; but better
late than never. **1980** *Listener* 7 Aug 186
Peter Finch['s] . . Oscar was awarded
posthumously. Oh well, better late than
never.

[1] cease, desist.

BETTER one house spoiled than two

Said of two foolish or wicked people
joined in marriage and troubling only
themselves. *Spoiled* or (*spilled*[1]) is sometimes contrasted with *filled* (see quots.
1670 and 1805).

1586 T.B. tr. *de la Primaudaye's French
Academy* xlvi. The wicked and repro-
bate, of whom that common proverbe is

spoken, that it is better one house be troubled with them than twaine. **1587** GREENE *Penelope's Web* V. 162 The old prouerb is fulfild, better one house troubled than two. **1670** J. RAY *English Proverbs* 51 Better one house fill'd then two spill'd. This we use when we hear of a bad Jack who hath married as bad a Jyll. **1805** W. BENTLEY *Diary* 28 May (1911) III. 161 One of the company discovering a disposition to speak much of his own wife . . the Gen. observed . . One house filled was better than two spoiled. **1924** *Folk-Lore* XXXV. 358 Better one house spoilt than two (said when a witless man marries a foolish woman).

¹ destroyed.

The BETTER the day, the better the deed
Frequently used to justify working on a Sunday or Holy Day.
[Cf. early 14th-cent. Fr. *a bon jour bone euvre*, for a good day, a good deed.] **1607** MIDDLETON *Michaelmas Term.* III. i. Why, do you work a' Sundays, tailor? The better day the better deed, we think. **1721** J. KELLY *Scottish Proverbs* 328 The better Day, the better Deed. I never heard this used but when People say that they did such an ill thing on Sunday. **1896** J. C. HUTCHESON *Crown & Anchor* xiii. The better the day, the better the deed . . It was only the Pharises who objected to any necessary work being done on the Sabbath. **1938** N. STREATFEILD *Circus is Coming* x. It was Good Friday. . . 'Us for the station to fetch that box. . . The better the day the better the deed.' **1976** L. DEIGHTON *Twinkle, twinkle, Little Spy* ix. 'Merry Christmas,' I said. 'The better the day, the better the deed.'

BETTER the devil you know than the devil you don't know
[**1539** R. TAVERNER tr. *Erasmus' Adages* 48 *Nota res mala, optima.* An euyl thynge knowen is best. It is good kepyng of a shrew¹ that a man knoweth. **1576** G. PETTIE *Petit Palace* 84 You had rather keepe those whom you know, though with some faultes, then take those

whom you knowe not, perchaunce with moe faultes. **1586** D. ROWLAND tr. *Lazarillo de Tormes* H6ᵛ The olde prouerbe: Better is the euill knowne, than the good which is yet to knowe.] **1857** TROLLOPE *Barchester Towers* II. vii. 'Better the d— you know than the d— you don't know,' is an old saying . . but the bishop had not yet realised the truth of it. **1937** W. H. SAUMAREZ SMITH *Letter* 16 May in *Young Man's Country* (1977) ii. Habit has practically made me resigned to Madaripur—'Better the devil you know than the devil you don't.' **1973** J. WAINWRIGHT *Devil You Don't* 138 'Better the devil you know than the devil you don't.'. . It makes good sense. Take your lot out, and we could have a right bastard move in and set up shop.

¹ a scolding or ill-tempered wife.

It is BETTER to be born lucky than rich
1639 J. CLARKE *Parœmiologia Anglo-Latina* 49 Better to have good fortune then be a rich mans child. **1784** *New Foundling Hospital for Wit* (new ed.) IV. 128 Estate and honours!—mere caprich! Better be fortunate than rich: Since oft me find . . Is verify'd what proverbs prate. **1846** M. A. DENHAM *Denham Tracts* (1892) I. 224 Better to be born lucky than rich. **1852** A. CARY *Clovernook* 248 What good luck some people always have. . . It's better to be born lucky than rich. **1980** T. MORGAN *Somerset Maugham* vx. This was Maugham at his most lighthearted, exposing the fallacy of the moralist position. 'I'm glad to be able to tell you that it has a moral,' he said, 'and that is: it's better to be born lucky than to be born rich.'

It is BETTER to give than to receive
The Biblical form is also used.
[ACTS xx. 35 It is more blessed to giue, then to receiue.] c **1390** GOWER *Confessio Amantis* v. 7725 Betre is to yive than to take. c **1527** T. BERTHELET tr. *Erasmus' Sayings of Wise Men* B2 It is better to gyue than to take, for he that takethe a gyfte of another is bonde to quyte¹ it, so that his lyberte is gone. **1710** S. PALMER

Proverbs 351 'Tis better to Give than to Receive, but yet 'tis Madness to give so much Charity to Others, as to become the Subject of it our Selves. **1968** 'E. LATHEN' *Come to Dust* xxiii. John Thatcher . . had been on the receiving end of too much discomfort lately. It is always more blessed to give than to receive. **1980** *Times* (Christmas Supplement) 15 Nov. p. i, There is no harm in reminding your relatives and friends that it is better to give than to receive.
[1] requite, repay.

'Tis BETTER to have loved and lost, than never to have loved at all

1700 CONGREVE *Way of World* II. i. Say what you will, 'tis better to be left, than never to have lov'd. **1812** G. CRABBE *Tales* xiv. Better to love amiss than nothing to have lov'd. **1850** TENNYSON *In Memoriam* XXVII. 44, I hold it true, whate'er befall . . 'Tis better to have loved and lost Than never to have loved at all. **1980** G. LYALL *Secret Servant* xxviii. In politics, it is better never to have loved at all than to have loved and lost.

It is BETTER to travel hopefully than to arrive

1881 R. L. STEVENSON *Virginibus Puerisque* IV. 190 To travel hopefully is a better thing than to arrive, and the true success is to labour. **1918** D. H. LAWRENCE in *English Review* Jan. 29 Love is strictly a travelling. 'It is better to travel than to arrive,' somebody has said. **1959** 'J. DUNCAN' *My Friend Muriel* II. 83 Remember, . . it is better to travel hopefully than to arrive. The satisfaction lies mainly in the travelling. **1981** *Times Literary Supplement* 7 Aug. 904 Faculty councils and the like—whose motto seems to be that it is better to travel hopefully than to arrive—. . don't want to hear about *any* change.

BETTER to wear out than to rust out

It is better to remain active than to succumb to idleness: used particularly with reference to elderly people. Frequently attributed in its current form to Bishop R. Cumberland (d. 1718).

[**1557** R. EDGEWORTH *Sermons* A1ᵛ Better it is to shine with laboure, then to rouste for idlenes.] **1598** SHAKESPEARE *Henry IV, Pt. 2* I. ii. 206, I were better to be eaten to death with a rust than to be scoured to nothing with perpetual motion.[1] **1820** in Southey *Life of Wesley* II. xxv. I had rather wear out than rust out. **1834** M. EDGEWORTH *Helen* II. xiii. Helen . . trembled for her health . . but she repeated her favourite maxim—'Better to wear out, than to rust out.' **1947** S. BELLOW *Victim* xvii. It was better to wear out than to rust out, as was often quoted. He was a hard worker himself. **1972** *Times* 24 May 16 'A man will rust out sooner'n he'll wear out' is one of his oft-repeated maxims.
[1] the sense is inverted in this quotation.

BETTER wed over the mixen than over the moor

It is better to marry a neighbour than a stranger.

a **1628** in M. L. Anderson *Proverbs in Scots* (1957) no. 320 Better to wow[1] over middin, nor[2] over mure. *a* **1661** T. FULLER *Worthies* (Cheshire) 174 Better Wed over the Mixon[3] then over the Moor . . that is, hard by or at home, Mixon being that heap of Compost which lyeth in the yards of good husbands. **1818** SCOTT *Heart of Midlothian* III. vi. He might hae dune waur[4] than married me. . . Better wed over the mixen as over the moor, as they say in Yorkshire. **1874** HARDY *Far from Madding Crowd* I. xxii. 'That means matrimony.'. . 'Well, better wed over the mixen than over the moor,' said Laban Tall.
[1] woo. [2] than. [3] midden, dunghill. [4] worse.

better: *see also* DISCRETION is the better part of valour; the GREY mare is the better horse; a LIVE dog is better than a dead lion.

BETWEEN two stools one falls to the ground

Inability to choose between, or accommodate oneself to, alternative viewpoints or courses of action may end in disaster. Now more common in the metaphorical phrase *to fall between two stools*.

[Med. L. *labitur enitens sellis herere duabus*, he falls trying to sit on two seats.] *c* **1390** GOWER *Confessio Amantis* IV. 626 Thou farst[1] as he betwen tuo stoles That wolde sitte and goth to grounde. *c* **1530** R. HILL *Commonplace Book* (EETS) 129 Betwen two stolis, the ars goth to grwnd. **1731** FIELDING *Tom Thumb* II. x. While the two Stools her Sitting Part confound, Between 'em both fall Squat upon the Ground. **1841** DICKENS *Old Curiosity Shop* I. xxxiii. She was . . still in daily occupation of her old stool opposite to that of her brother Sampson. And equally certain it is, by the way, that between these two stools a great many people had come to the ground. **1907** W. DE MORGAN *Alice-for-Short* xvi. Your mother wants to put it off on me. . . But I won't be let into saying anything. . . Charles saw that between the two stools the young couple wouldn't fall to the ground, but would go to the altar. **1979** A. CHISHOLM *Nancy Cunard* xxi. Politically, Nancy had fallen between stools.

[1] farest, (modern) fares.

beware: *see* let the BUYER beware; beware of an OAK it draws the stroke.

BIG fish eat little fish

a **1200** *Old English Homilies* (EETS) 2nd Ser. 179 The more[1] fishes in the se eten the lasse.[2] *c* **1300** in J. Small *English Metrical Homilies* (1862) 136 Al this werld es bot a se, . . And gret fisches etes the smale. For riche men of this werd[3] etes, That pouer wit thair travail getes.[4] **1608** SHAKESPEARE *Pericles* II. i. 27 Master, I marvel how the fishes live in the sea.—Why, as men do a–land—the great ones eat up the little ones. **1979** *New Society* 6 Dec. 557 The state today . . seems like nothing so much [as] a huge aquarium. . . Big fish eat little fish, and the great fish eat the big.

[1] bigger. [2] smaller. [3] world. [4] *that pouer wit thair travail getes*, what the poor get with their labour.

BIG fleas have little fleas upon their backs to bite them, and little fleas have lesser fleas, and so *ad infinitum*

1733 SWIFT *Poems* II. 651 The Vermin only teaze and pinch Their Foes superior by an Inch. So Nat'ralists observe, a Flea Hath smaller Fleas that on him prey, And these have smaller Fleas to bite 'em, And so proceed *ad infinitum*. **1872** A. DE MORGAN *Budget of Paradoxes* 377 Great fleas have little fleas upon their backs to bite 'em, And little fleas have lesser fleas, and so *ad infinitum*. And the great fleas themselves, in turn, have greater fleas to go on; While these again have greater still, and greater still, and so on. **1979** R. BARNARD *Posthumous Papers* ii. There will be a long article in the *Sunday Chronicle* . . and I'm afraid the *Sunday Grub* has got onto the story as well. Big fleas and little fleas, you know—.

big: *see also* PROVIDENCE is always on the side of the big battalions.

The BIGGER they are, the harder they fall

Commonly attributed in its current form to the boxer Robert Fitzsimmons, prior to a fight *c* 1900 (see quot. 1902). *Come* is sometimes used instead of *are*.

[A similar form is found in earlier related proverbs, such as: **1493** H. PARKER *Dives & Pauper* R7[v] It is more synne in the man For the higher degre[1] the harder is the fal; **1670** J. RAY *English Proverbs* 102 The higher standing the lower fall.] **1902** *National Police Gazette* 27 Sept. 6 'If I can get close enough,' he [Fitzsimmons] once said, 'I'll guarantee to stop almost anybody. The bigger the man, the heavier the fall.' **1929** *American Speech* IV. 464 The bigger they are the harder they fall. **1942** M. WILSON *Footsteps behind Her* IV. 157 Those guys don't bother me. The bigger they are, the harder they fall. **1971** J. CLIFF (*song-title*) The bigger they come the harder they fall.

[1] degree, standing, position.

bill: *see* DEATH pays all debts.

billet: *see* every BULLET has its billet.

bind: *see* SAFE bind, safe find.

binding: *see* you can't tell a BOOK by its cover.

A BIRD in the hand is worth two in the bush
It is better to accept what one has than to try to get more and risk losing everything.
[Med. L. *plus valet in manibus avis unica fronde duabus*, one bird in the hands is worth more than two amongst the foliage.] *c* **1450** J. CAPGRAVE *Life of St. Katharine* (EETS) II. iii. It is more sekyr[1] a byrd in your fest, Than to haue three in the sky a-boue. *c* **1470** *Harley MS 3362* f.4 Betyr ys a byrd in the hond than tweye in the wode. **1581** N. WOODES *Conflict of Conscience* IV. i. You haue spoken reasonably, but yet as they say, One Birde in the hande, is worth two in the bush. **1678** BUNYAN *Pilgrim's Progress* I. 42 That Proverb, A Bird in the hand is worth two in the Bush, is of more Authority with them, then are all . . testimonies of the good of the world to come. **1829** P. EGAN *Boxiana* 2nd Ser. II. 507 We have stated thus much to show how the London Fancy[2] were 'thrown out' of the above fight; likewise, to bear in mind, in future, 'that a bird in the hand is worth two in the bush'. **1973** G. GREENE *Honorary Consul* II. iii. We have an expression in English — A bird in the hand is worth two in the bush. I don't know anything about that 'afterwards'. I only know I would like to live another ten years.
[1] certain, sure. [2] boxing enthusiasts.

A BIRD never flew on one wing
Mainly Scottish and Irish. Now frequently used to justify a further gift, esp. another drink.
1721 J. KELLY *Scottish Proverbs* 308 The Bird must flighter[1] that flies with one Wing. Spoken by them who have Interest only in one side of the House. **1824** S. FERRIER *Inheritance* III. xxxii. 'The bird maun flichter that flees wi' ae wing'—but ye's haud up your head yet in spite o' them a'. **1914** K. F. PURDON *Folk of Furry Farm* ii. He held out a shilling to Hughie. 'A bird never yet flew upon the one wing, Mr. Heffernan!' said Hughie, that was looking to get another shilling. **1925** S. O'CASEY *Juno & Paycock* III. 89 Fourpence, given to make up the price of a pint, on th' principle that no bird ever flew on wan wing. **1980** J. O'FAOLAIN *No Country for Young Men* iii. I'll just have another quick one. A bird never flew on wan wing.
[1] flutter.

bird: *see also* you cannot CATCH old birds with chaff; the EARLY bird catches the worm; FINE feathers make fine birds; it's an ILL bird that fouls its own nest; in vain the NET is spread in the sight of the bird.

There are no BIRDS in last year's nest
Circumstances have altered.
1620 T. SHELTON tr. *Cervantes' Don Quixote* II. lxxiv. I pray you go not on so fast, since that in the nests of the last yeere, there are no birds of this yeere. Whilom[1] I was a foole, but now I am wise. **1732** T. FULLER *Gnomologia* no. 4863 There are no Birds this Year, in last year's Nest. **1845** LONGFELLOW *Poems* 62 All things are new. . even the nest beneath the eaves;—There are no birds in last year's nest. **1926** *Times* 19 Jan. 15 Things may not be as they were; 'there are no birds in last year's nest', and there may be no fish in the old rivers.
[1] formerly.

BIRDS in their little nests agree
A nursery proverb, also used as a direction, which states that young children should not argue among themselves.
1715 I. WATTS *Divine Songs* 25 Birds in their little Nests agree; And 'tis a shameful Sight, When Children of one Family Fall out, and chide, and fight. **1941** L. I. WILDER *Little Town on Prairie* xiv. 'Birds in their little nests agree,' she said, smiling. . . She knew nothing at all about birds. **1961** J. STEINBECK *Winter of our Discontent* I. i. 'Birds in their little nests

agree,' he said. 'So why can't we? . . You kids can't get along even on a pretty morning.' **1980** A. T. ELLIS *Birds of Air* 52 Her mother used to say to her and Mary: 'Birds in their little nests agree.'

BIRDS of a feather flock together
People of the same (usually, unscrupulous) character associate together.

[Cf. **1611** BIBLE *Ecclesiasticus* xxvii. 9 The birds will resort vnto their like, so will truth returne vnto them that practise in her.] **1545** W. TURNER *Rescuing of Romish Fox* B8 Byrdes of on kynde and color flok and flye allwayes together. **1599** J. MINSHEU *Spanish Grammar* 83 Birdes of a feather[1] will flocke together. **1660** W. SECKER *Nonsuch Professor* 81 Our English Proverb . . That birds of a feather will flock together. To be too intimate with sinners, is to intimate that you are sinners. **1828** BULWER-LYTTON *Pelham* III. xv. It is literally true in the systematised roguery of London, that 'birds of a feather flock together'. **1974** W. FOLEY *Child in Forest* 14 Birds of a feather flock together, and in our village the few feckless . . tended to live at one end; the prim . . and prosperous at the other.

[1] *of a feather*, of the same species.

Little BIRDS that can sing and won't sing must be made to sing
1678 J. RAY *English Proverbs* (ed. 2) 343 The bird that can sing and will not sing must be made to sing. **1846** DICKENS *Cricket on Hearth* ii. 'The bird that can sing and won't sing, must be made to sing, they say,' grumbled Tackleton. 'What about the owl that can't sing, and oughtn't to sing, and will sing?' **1888** A. QUILLER-COUCH *Troy Town* i. 'A little music might perhaps leave a pleasant taste.'. . 'Come, Sophy! Remember the proverb about little birds that can sing and won't sing?'

bite: *see* (noun) a BLEATING sheep loses a bite; every DOG is allowed one bite; (verb) a BARKING dog never bites; BIG fleas have little fleas upon their backs to bite them; DEAD men don't bite.

bitten: *see* ONCE bitten, twice shy.

black: *see* the DEVIL is not so black as he is painted; TWO blacks don't make a white.

blame: *see* a BAD workman blames his tools.

A BLEATING sheep loses a bite
Opportunities are missed through too much chatter.

1599 J. MINSHEU *Dialogues in Spanish* 20 That sheepe that bleateth looseth a bit.[1] **1659** G. TORRIANO *English & Italian Dict.* 37 A bleating sheep loseth her pasture. **1861** T. HUGHES *Tom Brown at Oxford* II. vii. He said something about a bleating sheep losing a bite; but I should think this young man is not much of a talker. **1978** R. V. JONES *Most Secret War* xlv. I thought of reminding him [Churchill] of an adage that I had learnt from my grandfather: 'Every time a sheep bleats it loses a nibble.'

[1] a portion of food bitten off at one time; a small piece (of food): now replaced by *bite*.

BLESSED is he who expects nothing, for he shall never be disappointed
1727 POPE *Letter* 6 Oct. (1956) II. 453, I have. . repeated to you, a ninth Beatitude. . 'Blessed is he who expects nothing, for he shall never be disappointed.' **1739** B. FRANKLIN *Poor Richard's Almanack* (May) Blessed is he that expects nothing, for he shall never be disappointed. **1911** *Times Literary Supplement* 5 Oct. 359 Evidently Sir Edwin's hope is not too roseate, and he is among those who are accounted blessed because they expect little. **1931** A. R. & R. K. WEEKES *Emerald Necklace* xix. 'When I get back. . I shall expect to find *all* our luggage in the hall.' 'Blessed is he that expecteth nothing,' said Louis, 'for he shall not be disappointed.'

blessed: *see also* it is BETTER to give than to receive; blessed are the DEAD that the rain rains on.

BLESSINGS brighten as they take their flight

[**1732** T. FULLER *Gnomologia* no. 989 Blessings are not valued, till they are gone.] **1742** YOUNG *Night Thoughts* II. 37 How blessings brighten as they take their flight. **1873** 'S. COOLIDGE' *What Katy Did* xi. Blessings brighten as they take their flight. Katy began to appreciate for the first time how much she had learned to rely on her aunt.

There's none so BLIND as those who will not see

Parallel to *There's none so* DEAF *as those who will not hear.*

1546 J. HEYWOOD *Dialogue of Proverbs* II. ix. K4 Who is so deafe, or so blynde, as is hee, That wilfully will nother here nor see. **1551** CRANMER *Answer to Gardiner* 58 There is no manne so. . blynd as he that will not[1] see, nor so dull as he that wyll not vnderstande. **1659** P. HEYLYN *Examen Historicum* 145 Which makes me wonder. . that having access to those Records. . he should declare himself unable to decide the doubt. . . But none so blind as he that will not see. **1738** SWIFT *Polite Conversation* III. 191 You know, there's none so blind as they that won't see. **1852** E. FITZGERALD *Polonius* 58 'None so blind as those that won't see.'. . A single effort of the will was sufficient to exclude from his view whatever he judged hostile to his immediate purpose. **1942** M. YATES *Murder by Yard* i. There's none so blind as he who will not see any one except his wife. **1980** J. PORTER *Dover beats Band* xv. 'If a highly trained copper. . can't tell who the murderer is—' 'There's none so blind.'

[1] *will not*, does not wish to, refuses to.

When the BLIND lead the blind, both shall fall into the ditch

Now more common in the metaphorical phrase, *the blind leading the blind.*

[MATTHEW xv. 14 Let them alone: they be blinde leaders of the blinde. And if the blinde lead the blinde, both shall fall into the ditch.] *c* **897** ALFRED *Gregory's Pastoral Care* (EETS) i. Gif se blinda thone blindan læt, he feallath begen[1] on ænne pytt. *c* **1300** *Body & Soul* (1889) 49 Ac hwanne the blinde lat the blinde, In dike he fallen bothe two. **1583** B. MELBANCKE *Philotimus* 165 In the ditch falls the blind that is led by the blind. **1678** BUNYAN *Pilgrim's Progress* I. 99 That Ditch is it into which the blind have led the blind in all Ages, and have both there miserably perished. **1836** CARLYLE *Sartor Resartus* II. iii. It is written, When the blind lead the blind, both shall fall into the ditch. . . May it not sometimes be safer, if both leader and led simply—sit still? **1947** L. P. HARTLEY *Eustace & Hilda* xvi. To get Hilda out of the house was a step forward, even in a Bath chair. . even if they could see nothing beyond their noses, the blind leading the blind.

[1] both.

A BLIND man's wife needs no paint

1659 J. HOWELL *Proverbs* (Spanish) 4 The blind mans wife needs no painting. **1736** B. FRANKLIN *Poor Richard's Almanack* (June) God helps them that help themselves. Why does the blind man's wife paint her self? **1892** C. M. YONGE *Old Woman's Outlook in Hampshire Village* 166 His [the schoolmaster's] copies too were remarkable. One was 'A blind man's wife needs no paint.' 'Proverbs, sir, Proverbs,' he answered, when asked where it came from.

blind: *see also* in the COUNTRY of the blind, the one-eyed man is king; a DEAF husband and a blind wife are always a happy couple; LOVE is blind; a NOD's as good as a wink to a blind horse; NOTHING so bold as a blind mare.

bliss: *see* where IGNORANCE is bliss, 'tis folly to be wise.

You cannot get BLOOD from a stone

Frequently used, as a resigned admission, to mean that it is hopeless to try extorting money, etc., from those who have none.

[c **1435** LYDGATE *Minor Poems* (EETS)

666 Harde to likke hony out of a marbil stoon, For ther is nouthir licour nor moisture.] **1666** G. TORRIANO *Italian Proverbs* 161 There's no getting of bloud out of that wall. **1850** DICKENS *David Copperfield* xi. Blood cannot be obtained from a stone, neither can anything on account be obtained . . from Mr. Micawber. **1945** F. THOMPSON *Lark Rise* xix. They can't take nothing away from us, for you can't get blood from a stone. **1979** *Daily Telegraph* 9 Mar. 3 In a paradoxical variation of the adage that you cannot get blood from a stone, the . . Revenue Service wants to extract money from a Miami woman's rare and lucrative blood.

BLOOD is thicker than water

Predominantly used to mean that a family connection will outweigh other relationships.

[12th-cent. Ger. *ouch hoer ich sagen, daz sippebluot von wassere niht verdirbet*, also I hear it said that kin-blood is not spoiled by water. **1412** LYDGATE *Troy Book* (EETS) III. 2071 For naturely blod will ay of kynde Draw vn-to blod, wher he may it fynde.] **1813** J. RAY *English Proverbs* (ed. 5) 281 Blood's thicker than water. **1815** SCOTT *Guy Mannering* II. xvii. Weel—blood's thicker than water—she's welcome to the cheeses. **1895** G. ALLEN *Woman who Did* xi. At moments of unexpected danger, angry feelings between father and son are often forgotten, and blood unexpectedly proves itself thicker than water. **1933** A. POWELL *From View to Death* iv. Really . . And then they say that blood is thicker than water. They know perfectly well that I have had hayfever. **1960** A. CHRISTIE *Adventures of Christmas Pudding* 240 It's exactly like a serial . . Reconciliation with the nephew, blood is thicker than water.

The BLOOD of the martyrs is the seed of the Church

The Church has thrived on persecution.

[TERTULLIAN *Apologeticus* l. *semen est sanguis Christianorum*, the blood of Christians is seed.] **1560** J. PILKINGTON *Aggeus the Prophet* U4ᵛ Cipriane wrytes that the bloud of Martirs is the seede of the Church. **1562** J. WIGAND *De Neutralibus & Mediis* M8ᵛ It is a very goodly & a most true saying: Christian mennes bloud is a sede, and in what felde so euer is sowed, ther spring vp Christian men most plenteously thick. **1655** T. FULLER *Church Hist. Britain* I. iv. Of all Shires in England, Stafford-shire was . . the largest sown with the Seed of the Church, I mean, the bloud of primitive Martyrs. **1889** J. LUBBOCK *Pleasures of Life* II. xi. The Inquisition has even from its own point of view proved generally a failure. The blood of the martyrs is the seed of the Church. **1979** *Church Times* 15 June 10 It is not merely that 'the blood of the martyrs is the seed of the Church'; it is that a little persecution is good for you.

BLOOD will have blood

Violence begets violence.

[GENESIS ix. 6 Who so sheddeth mans blood, by man shall his blood be shed.] *a* **1449** LYDGATE *Minor Poems* (EETS) 512 Blood will have wreche,[1] that wrongfully is spent. **1559** *Mirror for Magistrates* (1938) 99 Blood wyll haue blood, eyther fyrst or last. **1605–6** SHAKESPEARE *Macbeth* III. iv. 122 It will have blood; they say blood will have blood. **1805** SOUTHEY *Madoc* I. vii. Blood will have blood, revenge beget revenge. **1974** A. FOREST *Cricket Term* i. 'Blood will have blood,' quoted Lawrie smugly.

[1] vengeance.

BLOOD will tell

Family characteristics or heredity cannot be concealed.

1850 G. H. BOKER *World a Mask* IV. in S. Bradley *Glaucus* (1940) 38 He looked like the tiger in the Zoological, when I punch him with my stick. . .Game to the backbone—blood will tell. **1897** 's. GRAND' *Beth Book* v. Blood will tell, sir. Your gentleman's son is a match for any ragamuffin. **1979** D. ANTHONY *Long Hard Cure* xv. He likes to say that blood will tell, meaning that Mayhew blood will triumph.

bloom: *see* when the FURZE is in bloom, my love's in tune; when the GORSE is out of bloom, kissing's out of fashion.

blow: *see* it's an ILL wind that blows nobody any good; SEPTEMBER blow soft, till the fruit's in the loft; STRAWS tell which way the wind blows.

BLUE are the hills that are far away
A northern proverb comparable to DIS-TANCE *lends enchantment to the view.*

1887 T. H. HALL CAINE *Deemster* I. v. 'What's it sayin',' they would mutter, 'a green hill when far away from me; bare, bare, when it is near.' **1902** J. BUCHAN *Watcher by Threshold* IV. 236 'Blue are the hills that are far away' is an owercome[1] in the countryside. **1914** *Spectator* 6 June 955 It is the habit of the Celt to create fanciful golden ages in the past— 'Blue are the faraway hills,' runs the Gaelic proverb.

[1] a common or hackneyed expression (Scottish).

body: *see* CORPORATIONS have neither bodies to be punished nor souls to be damned.

boil: *see* a WATCHED pot never boils.

bold: *see* NOTHING so bold as a blind mare.

bolted: *see* it is too late to shut the STABLE-door after the horse has bolted.

bond: *see* an ENGLISHMAN's word is his bond.

bone: *see* what's BRED in the bone will come out in the flesh; you BUY land you buy stones, you buy meat you buy bones; a DOG that will fetch a bone will carry a bone; HARD words break no bones; the NEARER the bone, the sweeter the meat; STICKS and stones may break my bones, but words will never hurt me.

You can't tell a BOOK by its cover
1929 *American Speech* IV. 465 You can't judge a book by its binding. **1954** R. HAYDN *Journal of Edwin Carp* 131 This is a nice respectable street, wouldn't you say, sir? . . Unfortunately, sir, you can't tell a book by its cover. **1969** S. MAY et al. *You can't judge Book by its Cover* (song) 1, I can't let you know you're getting to me . . 'cause you can't judge a book by its cover. My pappa used to say, look, child, look beyond a tender smile.

book: *see also* a GREAT book is a great evil.

If you're BORN to be hanged then you'll never be drowned
Commonly used to qualify another's apparent good luck.

[Early 14th-cent. Fr. *ne peut noier qui doit pendre,* he who must hang cannot drown. *c* **1503** A. BARCLAY tr. *Gringore's Castle of Labour* (1506) A8 He that is drowned may no man hange.] **1593** J. ELIOT *Ortho-Epia Gallica* 127 He thats borne to be hangd shall neuer be drownde. **1723** DEFOE *Colonel Jack* (ed.2) 126 He had a Proverb in his Favour, and he got out of the Water . . not being born to be drown'd, as I shall observe after-wards in its place. **1884** BLACKMORE *Sir Thomas Upmore* I. viii. Don't tumble into it . . though you never were born to be drowned, that I'll swear. **1933** C. ROB-BINS *Mystery of Mr. Cross* 343 You have a proverb . . that those who are to hang will not drown. **1956** H. LEWIS *Witch & Priest* v. There is another picture, and underneath it says . . *If you're born to be hanged, then you'll never be drowned.*

born: *see also* it is BETTER to be born lucky than rich; the MAN who is born in a stable is a horse; YORKSHIRE born and Yorkshire bred, strong in the arm and weak in the head.

borrow: *see* the EARLY man never borrows from the late man.

borrowing: *see* he that GOES a-borrowing, goes a-sorrowing.

bottle: *see* you can't put NEW wine in old bottles.

bottom: *see* TRUTH lies at the bottom of a well; every TUB must stand on its own bottom.

bought: *see* GOLD may be bought too dear.

bowls: *see* those who PLAY at bowls must look out for rubbers.

Two BOYS are half a boy, and three boys are no boy at all
The more boys that help, the less work they do.

c 1930 F. THOMPSON *Country Calendar* 114 Their parents do not encourage the joining of forces. . .We have a proverb here: 'Two boys are half a boy, and three boys are no boy at all.' 1971 *New York Times* 31 Jan. IV. 12 Rural New England of the mid-nineteenth century, not commonly sophisticated in mathematics but witty enough about man's condition, used to [say] . . 'One boy helping, a pretty good boy; two boys, half a boy; three boys, no boy.'

BOYS will be boys
Occasionally *girls will be girls*.

1601 A. DENT *Plain Man's Pathway* 64 Youth will be youthfull, when you haue saide all that you can. 1826 T. H. LISTER *Granby* II. vii. Girls will be girls. They like admiration. 1848 THACKERAY *Vanity Fair* xiii. As for the pink bonnets . . why boys will be boys. 1964 WODE-HOUSE *Frozen Assets* iii. I tried to tell him that boys will be boys and you're only young once.

brae (slope, hill-side): *see* put a STOUT heart to a stey brae.

BRAG is a good dog, but Holdfast is better
Tenacity and quietness of manner are preferable to ostentation.

[1580 A. MUNDAY *Zelanto* 146 Brag is a good Dogge, whyle he will holde out: but at last he may chaunce to meete with his matche. 1599 SHAKESPEARE *Henry V*

II. iii. 52 And Holdfast is the only dog, my duck.] 1709 O. DYKES *English Proverbs* 123 Brag is a good Dog, but Hold-fast is a Better. . . Nothing edifies less in an ingenuous Conversation, than Boasting and Rattle. 1752 JOHNSON *Rambler* 4 Feb. VIII. 92 When I envied the finery of any of my neighbours, [my mother] told me, that 'brag was a good dog, but holdfast was a better'. 1889 *Pictorial Proverbs for Little People* 11 Brag's a good dog, but Holdfast is better. 1937 R. W. WINSTON *It's Far Cry* xxi. In golf, as in life . . the exceptional has no staying qualities. To quote a Southern [US] saying, 'Brag is a good dog, but Holdfast is better'.

brain: *see* an IDLE brain is the Devil's workshop.

brass: *see* where there's MUCK there's brass.

None but the BRAVE deserve the fair
1697 DRYDEN *Poems* (1958) III. 148 Happy, happy, happy Pair! . . None but the Brave deserves the Fair. 1829 P. EGAN *Boxiana* 2nd Ser. II. 354 The tender sex . . feeling the good old notion that 'none but the brave deserve the fair', were sadly out of temper. 1873 TROL-LOPE *Phineas Redux* II. xiii. All the proverbs were on his side. 'None but the brave deserve the fair,' said his cousin. 1978 F. WELDON *Praxis* xii. She frequented the cafe where the Rugger set hung out, and on a Saturday, after closing hours, could be seen making for the downs, laughing heartily and noisily in the company of one or other of the brave, who clearly deserved the fair.

BRAVE men lived before Agamemnon
[HORACE *Odes* IV. ix. *vixere fortes ante Agamemnona multi*, many brave men lived before Agamemnon. 1616 JON-SON *Forest* VIII. 114 There were braue men, before Aiax or Idomen, or all the store That Homer brought to Troy.] 1819 BYRON *Don Juan* I. v. I want a hero: an uncommon want, When every year and month sends forth a new

one. . .Brave men were living before Agamemnon And since, exceeding valorous and sage. **1980** *Times* 23 June 16 Brave men lived before Agamemnon, lots of them. But on all of them . . eternal night lies heavy, because they have left no records behind them.

brave: *see also* (adjective used as noun) FORTUNE favours the brave; (verb) ROBIN Hood could brave all weathers but a thaw wind.

The BREAD never falls but on its buttered side
Cf. *If* ANYTHING *can go wrong, it will* ('Murphy's Law').
1867 A. D. RICHARDSON *Beyond Mississippi* iii. *His* bread never fell on the buttered side. **1891** J. L. KIPLING *Beast & Man* x. We express the completeness of ill-luck by saying, 'The bread never falls but on its buttered side.' **1929** A. GRAY *Dead Nigger* xix. Didn't her bread and butter always fall butter downwards? **1980** *Guardian* 3 Dec. 12 Murphy's (or Sod's) Law. . . Murphy's many relatives always quote it as 'Buttered bread falls buttered side down—and if it's a sandwich it falls open.'

bread: *see also* HALF a loaf is better than no bread; MAN cannot live by bread alone.

break: *see* (noun) never give a SUCKER an even break; (verb) HARD words break no bones; if it were not for HOPE, the heart would break; it is the LAST straw that breaks the camel's back; STICKS and stones may break my bones, but words will never hurt me; *also* BREAKING, BROKEN.

breakfast: *see* HOPE is a good breakfast but a bad supper.

breaking: *see* IGNORANCE of the law is no excuse for breaking it; you cannot make an OMELETTE without breaking eggs.

What's BRED in the bone will come out in the flesh
Lifelong habits or inherited characteris-

tics cannot be concealed (cf. BLOOD *will tell*). The form and emphasis of the proverb have been altered in recent years by the omission of a negative.
[Med. L. *osse radicatum raro de carne recedit,* that which is rooted in the bone rarely comes out from the flesh.] *c* **1470** MALORY *Morte d'Arthur* (1947) I. 550 Sir Launcelot smyled and seyde, Harde hit ys to take oute off the fleysshe that ys bredde in the bone. **1546** J. HEYWOOD *Dialogue of Proverbs* II. viii. K2 This prouerbe prophecied many yeres agone, It will not out of the fleshe, thats bred in the bone. **1603** FLORIO tr. *Montaigne's Essays* III. xiii. They are effects of custome and vse: and what is bred in the bone will never out of the flesh. **1832** J. P. KENNEDY *Swallow Barn* III. v. What is bred in the bone—you know the proverb. **1909** A. MACLAREN *Epistle to Romans* 231 'You cannot expel nature with a fork,' said the Roman. 'What's bred in the bone won't come out of the flesh,' says the Englishman. **1923** *Chambers's Journal* June 432, I know the clan. . . What's bred in the bone will out, and they'll be high-steppers to the end. *a* **1957** L. I. WILDER *First Four Years* (1971) iv. We'll always be farmers, for what is bred in the bone *will* come out in the flesh. **1981** B. HEALEY *Last Ferry* iv. There's bad blood there. . . What's bred in the bone comes out in the flesh.

bred: *see also* YORKSHIRE born and Yorkshire bred, strong in the arm and weak in the head.

breed: *see* FAMILIARITY breeds contempt; LIKE breeds like.

BREVITY is the soul of wit
1600-1 SHAKESPEARE *Hamlet* II. ii. 90 Since brevity is the soul of wit[1] . . I will be brief. **1833** M. SCOTT *Tom Cringle's Log* II. v. Brevity is the soul of wit,— ahem. **1946** D. WELCH *Maiden Voyage* x. I will not repeat myself, since brevity is the soul of wit. **1981** *Times* 18 Sept. 27 Mr Carrott . . has yet to learn that brevity is often the soul of wit, but he has

little to learn about the vast vocabulary of body language.

[1] *soul of wit*, essence of wisdom; *wit* is now commonly understood in its modern sense, the '(power of giving sudden intellectual pleasure by) unexpected combining or contrasting of previously unconnected ideas or expressions' (COD).

As you BREW, so shall you bake

Cf. *As you* BAKE, *so shall you brew*.

[**1264** in C. Brown *English Lyrics of XIIIth Century* (1932) 131 Let him habbe ase he brew, bale to dryng.[1] *a* **1325** *Cursor Mundi* (EETS) l. 2848 Nathing of that land vn-sonken,[2] Suilk[3] als thai brued now ha thai dronken. *c* **1450** *Towneley Play of Second Shepherd* (EETS) l. 501 Bot we must drynk as we brew And that is bot reson.] *c* **1570** T. INGELEND *Disobedient Child* D8ᵛ As he had brewed, that so shulde bake. **1766** COLMAN & GARRICK *Clandestine Marriage* I. 3 As you sow, you must reap—as you brew, so you must bake. **1922** S. J. WEYMAN *Ovington's Bank* xxiii. No, you may go, my lad. As you ha' brewed you may bake.

[1] literally, 'let him have as he brews, bale [i.e. evil, injury, misery] to drink'. [2] not sunken or submerged in water. [3] such.

brew: *see also* as you BAKE, so shall you brew.

You cannot make BRICKS without straw

Nothing can be made or performed without the necessary materials. Frequently used as a metaphorical phrase, *to make bricks without straw*. A (misapplied) allusion to EXODUS v. 7: Yee shall no more giue the people straw to make bricke, as heretofore: let them goe and gather straw for themselues.

[Cf. **1624** BURTON *Anatomy of Melancholy* (ed. 2) I. ii. (Hard taske-masters as they [patrons] are) they take away their straw, & compell them to make their number of bricke.] **1658** T. HYDE *Letter* in *Verney Memoirs* (1904) II. xxxviii. I have made the enclosed. It is an hard task to make bricks without straw, but I have raked together some rubbish. **1737** in *Publications of Prince Society*

(1911) III. 170 Let Men be never so willing & industrious, they can't make Brick without Straw. **1883** M. BETHAM-EDWARDS *Disarmed* I. i. Your task from today will be to make bricks without straw. **1909** A. BENNETT *Literary Taste* iv. You can only acquire really useful general ideas by first acquiring particular ideas. . . You cannot make bricks without straw. **1934** F. W. CROFTS *12.30 from Croydon* xxiv. 'I never thought we had any chance,' Heppenstall declared. 'Can't make bricks without straw,' Quilter pointed out cheerily.

Happy is the BRIDE that the sun shines on

Cf. *Blessed are the* DEAD *that the rain rains on*.

1648 HERRICK *Hesperides* 129 Blest is the Bride, on whom the Sun doth shine. And thousands gladly wish You multiply, as doth a fish. **1787** F. GROSE *Provincial Glossary* (Superstitions) 61 It is reckoned a good omen, or a sign of future happiness, if the sun shines on a couple coming out of the church after having been married. . . Happy is the bride that the sun shines on. **1979** *Guardian* 16 June 8 Happy is the bride on whom the sun shines. We nod assent.

It is good to make a BRIDGE of gold to a flying enemy

An enemy closely pursued may become desperate. . . By all means, then, let the vanquished have a free course: T. Fielding *Proverbs of all Nations* (1824) 14.

[The idea is attributed to Aristides (480 BC), who warned Themistocles not to destroy the bridge of boats which Xerxes had built across the Hellespont in order to invade Greece (PLUTARCH *Themistocles* 16); ERASMUS *Apophthegms* viii. *Hostibus fugientibus pontem argentuem exstraendum esse*, for a fleeing enemy one should construct a bridge of silver.] **1576** W. LAMBARDE *Perambulation of Kent* 323 It was well sayde of one . . If thine enemie will flye, make him a bridge of Golde. **1642** T. FULLER *Holy State* IV. xvii. He [the good general] makes his flying enemy a bridge of gold, and disarms them of their best weapon, which is necessity to fight whether they will or no. **1889** R. L. STEVENSON *Master of Ballantrae* iv. You may have heard a military proverb: that it is a good thing to make a bridge of gold to a flying enemy. I trust you will take my meaning.

bridge: *see also* don't CROSS the bridge till you come to it; everyone SPEAKS well of the bridge which carries him over.

brighten: *see* BLESSINGS brighten as they take their flight.

bring: *see* the WORTH of a thing is what it will bring.

broken: *see* PROMISES, like pie-crust, are made to be broken.

broom: *see* NEW brooms sweep clean.

broth: *see* TOO many cooks spoil the broth.

build: *see* FOOLS build houses and wise men live in them; where GOD builds a church, the Devil will build a chapel; it is easier to PULL down than to build up.

built: *see* ROME was not built in a day.

Every BULLET has its billet
Fate determines who shall be killed; quot. 1922 implies more generally that fate plays a part in all human affairs.
1575 G. GASCOIGNE *Fruits of War* I. 155 Suffiseth this to proove my theame withall, That every bullet hath a lighting place. **1765** WESLEY *Journal* 6 June (1912) V. 130 He never received one wound. So true is the odd saying of King William, that 'every bullet has its billet'. **1846** J. GRANT *Romance of War* II. iii. 'Tis the fortune of war:—every bullet has its billet,—their fate to-day may be ours tomorrow. **1922** JOYCE *Ulysses* 366 The ball rolled down to her as if it understood. Every bullet has its billet.

A BULLY is always a coward
1817 M. EDGEWORTH *Ormond* in *Harrington & Ormond* III. xxiv. Mrs.

M'Crule, who like all other bullies was a coward, lowered her voice. **1826** LAMB *Elia* in *New Monthly Magazine* XVI. 25 A Bully is always a coward. . . Confront one of the silent heroes with the swaggerer of real life, and his confidence in the theory quickly vanishes. **1853** T. C. HALIBURTON *Wise Saws* iv. I never saw a man furnished with so much pleasure in my life. A brave man is sometimes a desperado. A bully is always a coward. **1981** *Times* 9 May 2 The old adage holds good: all bullies are cowards, and most cowards are bullies.

bung-hole: *see* SPARE at the spigot, and let out at the bung-hole.

burden: *see* GOD makes the back to the burden.

A BURNT child dreads the fire
Cf. ONCE *bitten twice shy*.
c **1250** *Proverbs of Hending* in *Anglia* (1881) IV. 199 Brend child fuir fordredeth.[1] *c* **1400** *Romaunt of Rose* l. 1820 'For evermore gladly,' as I rede, 'Brent child of fier hath mych drede.' **1580** LYLY *Euphues* II. 92 A burnt childe dreadeth the fire. . . Thou mayst happely forsweare thy selfe, but thou shalt neuer delude me. **1777** P. THICKNESSE *Journey* I. xviii. He then observed, that a burnt child dreads the fire; . . that a Jew had lately passed thro' France, who had put off false Bank notes, and that I might . . have taken some. **1889** *Pictorial Proverbs for Little People* 5 She will not touch a match or a lighted candle . . which proves that the proverb is true which says: a burnt child dreads the fire. **1948** WODEHOUSE *Uncle Dynamite* II. vii. The burnt child fears the fire, and bitter experience had taught Pongo Twistleton to view with concern the presence in his midst of Ickenham's fifth earl.
[1] is in dread of, fears.

burnt: *see also* if you PLAY with fire you get burnt.

bury: *see* let the DEAD bury the dead.

bush: *see* a BIRD in the hand is worth two in the bush; GOOD wine needs no bush.

The BUSIEST men have the most leisure

Complementary to IDLE *people have the least leisure.*

[**1866** S. SMILES *Self-Help* (new ed.) i. Those who have most to do . . will find the most time.] **1884** J. PAYN *Canon's Ward* II. xxxiv. It is my experience that the men who are really busiest have the most leisure for everything. **1911** *Times Literary Supplement* 6 Oct. 365 The busiest men have always the most leisure; and while discharging the multifarious duties of a parish priest and a guardian he found time for travelling.

BUSINESS before pleasure

The two nouns are frequently contrasted in other, non-proverbial, expressions.

[c **1640** *Grobiana's Nuptials* (*MS Bodley 30*) 15 Well to the buisnesse.—On; buisnesse is senior to complement. **1767** T. HUTCHINSON *Diary & Letters* (1883) I. v. Pleasure should always give way to business.] **1837** C. G. F. GORE *Stokeshill Place* III. vi. 'Business before pleasure' is a golden rule which most of us regard as iron. **1943** S. STERLING *Down among Dead Men* v. This is business, Sarge. You know what business comes before. **1976** HOOKER & BUTTERWORTH *M.A.S.H. goes to San Francisco* xiii. 'Oh, how nice! And I think about *you*, too. But business before pleasure, as I always say.'

business: *see also* EVERYBODY's business is nobody's business; PUNCTUALITY is the soul of business.

butter: *see* (noun) there are more WAYS of killing a dog than choking it with butter; (verb) FINE words butter no parsnips.

buttered: *see* the BREAD never falls but on its buttered side.

BUY in the cheapest market and sell in the dearest

1595 T. LODGE *Fig for Momus* H2 Buy cheape, sell deare. **1862** RUSKIN *Unto this Last* II. 60 Buy in the cheapest market?—yes; but what made your market cheap? . . Sell in the dearest? . . But what made your market dear? **1880** J. A. FROUDE *Bunyan* vii. 'To buy in the cheapest market and sell in the dearest' was Mr. Badman's common rule in business. . . In Bunyan's opinion it was knavery in disguise.

You BUY land, you buy stones; you buy meat, you buy bones

[**1595** *Pedler's Prophesy* B4^v You shall be sure to haue good Ale, for that haue no bones.] **1670** J. RAY *English Proverbs* 211 He that buys land buys many stones; He that buys flesh buys many bones; He that buys eggs buys many shells, But he that buys good ale buys nothing else. **1721** J. KELLY *Scottish Proverbs* 172 He that buys Land, buys Stones; He that buys Beef, buys Bones; He that buys Nuts, buys shells: He that buys good Ale, buys nought else. **1970** *Countryman* Autumn 172 Welsh butcher to customer complaining of bony meat: 'Well, missus, you buy land, you buy stones; buy meat, you buy bones.'

buy: *see also* why buy a COW when milk is so cheap?

Let the BUYER beware

A warning that the buyer must satisfy himself of the nature and value of a purchase before proceeding with the transaction. The Latin tag *caveat emptor* is also frequently found. See also the next proverb.

[L. *caveat emptor, quia ignorare non debuit quod jus alienum emit,* let the purchaser beware, for he ought to be ignorant of the nature of the property which he is buying from another party.] **1523** J. FITZHERBERT *Husbandry* 36 And[1] he [a horse] be tame and haue ben rydden vpon than caveat emptor be ware thou byer. **1592** NASHE *Pierce Penniless* I. 155 Sed caueat emptor, Let the interpreter beware. **1607** E. SHARPHAM *Fleire* II. C4 They are no prouerb breakers: beware the buyer say they. **1927** *Times* 29 Sept. 10 We dislike very much, whether it is put in Latin or in English, the phrase 'Let the buyer beware!' **1974** D. FRANCIS *Knock Down* xi. *'Caveat emptor,'* I said. 'What does that mean?' 'Buyer beware.' 'I know one buyer who'll beware for the rest of his life.'

¹ if.

The BUYER has need of a hundred eyes, the seller of but one

[It. *chi compra ha bisogna di cent' occhi; chi vende n' ha assai di uno,* he who buys needs a hundred eyes; he who sells not one.] **1640** G. HERBERT *Outlandish Proverbs* no. 390 The buyer needes a hundred eyes, the seller not one. **1745** B. FRANKLIN *Poor Richard's Almanack* (July) He who buys had need have 100 Eyes, but one's enough for him that sells the Stuff. **1800** M. EDGEWORTH *Parent's Assistant* (ed. 3) III. 86 He taught him . . to get . . from customers by taking advantage of their ignorance. . . He often repeated . . 'The buyer has need of a hundred eyes; the seller has need of but one.' **1843** R. S. SURTEES *Handley Cross* I. xiii. The buyer has need of a hundred eyes, the seller of but one, says another equestrian conjurer. **1928** *Illustrated Sporting & Dramatic News* 7 Jan. 27 *(caption)* The buyer has need of a hundred eyes. The seller of but one.

C

CAESAR's wife must be above suspicion
Julius Caesar replied thus (according to Plutarch) when asked why he had divorced his wife Pompeia. He considered his honour and position compromised, since she was indirectly associated with Publius Clodius' trial for sacrilege.

[1580 LYLY *Euphues* II. 101 Al women shal be as Caesar would haue his wife, not onelye free from sinne, but from suspition.] 1779 A. ADAMS *Letter* 4 Jan. in L. H. Butterfield et al. *Adams Family Correspondence* (1973) III. 148 It is a very great misfortune that persons imployed in the most important Departments should . . have seperate interests from the publick whom they profess to serve. Caesars wife ought not to be suspected. 1847 J. C. & A. W. HARE *Guesses at Truth* (ed. 3) 1st Ser. 263 Caesar's wife ought to be above suspicion. . . Caesar himself ought to be so too. 1930 D. L. SAYERS *Strong Poison* xxi. You've got a family and traditions, you know. Caesar's wife and that sort of thing. 1965 O. MILLS *Dusty Death* xxi. Policemen . . are like . . candidates for the Church of England ministry, . . and Caesar's wife. . . Not only they, but all their relations, must be above suspicion.

cake: *see* you cannot HAVE your cake and eat it.

calf: *see* a BELLOWING cow soon forgets her calf.

call: *see* call no man HAPPY till he dies; he who PAYS the piper calls the tune.

calm: *see* AFTER a storm comes a calm.

camel: *see* it is the LAST straw that breaks the camel's back.

He who CAN does; he who cannot, teaches
There are many (frequently humorous) variations on this proverb.

1903 G. B. SHAW *Maxims for Revolutionists* in *Man & Superman* 230 He who can, does. He who cannot, teaches. 1979 *Daily Telegraph* 6 Aug. 8 A version of an old adage came to me—those who can, do, those who can't, attend conferences. 1981 P. SOMERVILLE-LARGE *Living Dog* i. He who can, does sang the train wheels, he who cannot, teaches.

candle: *see* CANDLEMAS day, put beans in the clay, put candles and candlesticks away.

candlelight: *see* never CHOOSE your women or your linen by candlelight.

If CANDLEMAS day be sunny and bright, winter will have another flight; if Candlemas day be cloudy with rain, winter is gone, and won't come again
[1523 SKELTON *Works* I. 418 Men were wonte for to discerne By candlemas day[1] what wedder shulde holde. 1584 R. SCOT *Witchcraft* XI. xv. If Maries purifieng daie, Be cleare and bright with sunnie raie, The frost and cold shalbe much more, After the feast than was before.] 1678 J. RAY *English Proverbs* (ed. 2) 51 If Candlemas day be fair and bright Winter will have another flight: If on Candlemas day it be showre and rain, Winter is gone and will not come again. This is a translation . . of that old Latin Distich; *Si Sol splendescat Maria purificante, Major erit glacies post festum quam fuit ante.* 1906 E. HOLDEN *Country Diary of Edwardian Lady* (1977) 13 If Candlemas Day be fair & bright Winter will have another flight But if Candlemas Day be clouds & rain Winter is gone and will not come again. 1980 *Times* 2 Feb. 11 Today is Candlemas Day. So let us see if the old legend holds good again as it did last year. 'If Candlemas day be sunny and bright, winter will have another flight; if Candlemas day be cloudy with rain; winter is gone and won't come again.

[1] 2 February.

CANDLEMAS day, put beans in the clay;
put candles and candlesticks away
   1678 J. RAY *English Proverbs* (ed. 2) 344
On Candlemas day throw candle and
candle-stick away. **1876** T. F. THISELTON-
DYER *British Popular Customs* 55 From
Candlemas the use of tapers at vespers
and litanies, which had continued
through the whole year, ceased until the
ensuing All Hallow Mass . . On Candlemas Day, throw candle and candlestick
away. **1948** F. THOMPSON *Still glides
Stream* ii. Broad beans were planted . .
on Candlemas Day. *Candlemas Day, stick
beans in the clay, Throw candle and candle-
stick right away,* they would quote. **1974**
K. BRIGGS *Folklore of Cotswolds* ii. Candlemas Day was the time . . when lights
were extinguished. . . An old rhyme
said: Candlemas Day, put beans in
the clay: Put candles and candlesticks
away.

If the CAP fits, wear it
Used with reference to the suitability of
names or descriptions as demonstrated
by the behaviour of the person concerned.
   [**1600** N. BRETON *Pasquil's Fools-Cap* A3
Where you finde a head fit for this
Cappe,[1] either bestowe it vpon him in
charity, or send him where he may haue
them for his money.] **1732** T. FULLER
Gnomologia no. 2670 If any Fool finds the
Cap fit him, let him wear it. **1750**
RICHARDSON *Clarissa* (ed. 3) VII. ii. If
indeed thou findest . . that the cap fits
thy own head, why then . . clap it on.
1854 DICKENS *Hard Times* II. vii.
'Mercenary. . . Who is not mercenary?' . . 'You know whether the cap fits
you. . . If it does, you can wear it.' **1979**
Guardian 8 Sept. 3 Was he talking about
Paisley? 'I just set out my views,' he
said. 'If the cap fits, wear it.'
   [1] a dunce's cap.

capacity: *see* GENIUS is an infinite capacity
   for taking pains.

caravan: *see* DOGS bark, but the caravan
   goes on.

Where the CARCASE is, there shall the
eagles be gathered together
   [MATTHEW xxiv. 28 Wheresoeuer the
carkeise is, there will the Eagles[1] bee
gathered together.] *c* **1566** W. P. tr.
Curio's Pasquin in Trance 33 Where the
caraine[2] is, thither do the Eagles resort.
1734 B. FRANKLIN *Poor Richard's Almanack* (Jan.) Where carcasses are, eagles
will gather, And where good laws are,
much people flock thither. **1979** 'S.
WOODS' *Proceed to Judgement* 190 [He]
was surprised to find Sir Nicholas and
Vera, as well as Roger and Meg, having
tea with Jenny. 'Where the carcase is,
there shall the eagles be gathered
together,' he commented, not very
politely.
   [1] i.e. vultures.   [2] carrion.

card: *see* LUCKY at cards, unlucky in love.

CARE killed the cat
Cf. CURIOSITY *killed the cat.*
   1598-9 SHAKESPEARE *Much Ado about
Nothing* v. i. 133 Though care kill'd a cat,
thou hast mettle enough in thee to kill
care. **1726** SWIFT *Poems* II. 761 Then,
who says care will kill a cat? Rebecca
shews they're out in that. **1890** 'R. BOL-
DREWOOD' *Miner's Right* II. xxiii. He was
always ready to enjoy himself. . . 'Care
killed a cat.' **1949** S. SMITH *Holiday* xii.
We must be careful of that. Care killed
the cat, said Caz. **1962** A. CHRISTIE *Mirror Crack'd* xxii. Care[1] killed the cat, they
say. . . You don't want kindness rubbed into your skin, . . do you?
   [1] i.e. care lavished on the cat rather than (as
elsewhere) exercised by it.

care: *see also* CHILDREN are certain cares,
   but uncertain comforts; the DEVIL looks
   after his own; take care of the PENCE and
   the pounds will take care of themselves.

careful: *see* if you can't be GOOD, be careful.

carry: *see* a DOG that will fetch a bone will
   carry a bone; everyone SPEAKS well of
   the bridge which carries him over.

case: *see* CIRCUMSTANCES alter cases; HARD

cases make bad law; no one should be JUDGE in his own cause.

Ne'er CAST a clout till May be out

A warning not to leave off old or warm clothes until the end of May. The proverb does not refer to May blossom, as is sometimes assumed.

1706 J. STEVENS *Spanish & English Dict.* s.v. Mayo, *Hasta passado Mayo no te quites el sayo*, Do not leave off your Coat till May be past. **1732** T. FULLER *Gnomologia* no. 6193 Leave not off a Clout,[1] Till May be out. **1832** A. HENDERSON *Scottish Proverbs* 154 Cast ne'er a clout till May be out. **1948** R. GRAVES *White Goddess* x. In ancient Greece, as in Britain, this [May] was the month in which people went about in old clothes—a custom referred to in the proverb 'Ne'er cast a clout ere May be out,' meaning 'do not put on new clothes until the unlucky month is over.' **1970** N. STREATFEILD *Thursday's Child* xxv. I still wear four petticoats . . Ne'er cast a clout till May be out.

[1] article of clothing.

cast: *see also* COMING events cast their shadows before; OLD sins cast long shadows; do not throw PEARLS to swine.

castle: *see* an ENGLISHMAN's home is his castle.

A CAT in gloves catches no mice

Restraint and caution (or 'pussyfooting') achieve nothing.

[14th-cent. Fr. *chat engaunté ne surrizera ja bien*, a gloved cat will never mouse well.] **1573** J. SANFORDE *Garden of Pleasure* 105 A gloued catte can catche no myse. **1592** G. DELAMOTHE *French Alphabet* II. 1, A mufled Cat is no good mouse hunter. **1758** B. FRANKLIN *Poor Richard's Almanack* (Preface) Handle your Tools without Mittens; remember that the Cat in Gloves catches no Mice. **1857** DICKENS *Little Dorrit* II. xiv. Mrs. General, if I may reverse the common proverb . . is a cat in gloves who *will* catch mice. That woman . . will be our

mother-in-law. **1979** *Country Life* 21 June 2047 There is hardly one [cat] but flings back the lie in the face of the old saying that a cat in gloves catches no mice. Why dirty your paws when your servants will do it for you?

A CAT may look at a king

1546 J. HEYWOOD *Dialogue of Proverbs* II. v. H3 What, a cat maie looke on a king, ye know. **1590** GREENE *Never too Late* VIII. 181 A Cat may look at a King, and a swaynes eye hath as high a reach as a Lords looke. **1721** N. BAILEY *English Dictionary* s.v. Cat, A Cat may look upon a King. This is a saucy Proverb, generally made use of by pragmatical Persons. **1935** I. COMPTON-BURNETT *House & its Head* xi. There is no harm in that, dear. A cat may look at a king; and it is only in that spirit that my poor brother looks at Alison.

When the CAT's away, the mice will play

[c **1470** *Harley MS 3362* in *Retrospective Review* (1854) May 309 The mows lordchypythe[1] ther a cat ys nawt. **1599** SHAKESPEARE *Henry V* I. ii. 172 To her unguarded nest the weasel Scot Comes sneaking, and so sucks her princely eggs, Playing the mouse in absence of the cat.] **1607** T. HEYWOOD *Woman killed with Kindness* II. 135 Mum; there's an old prouerbe, when the cats away, the mouse may play. **1670** J. RAY *English Proverbs* 68 When the cat is away, the mice play. **1876** I. BANKS *Manchester Man* III. xiv. Mrs. Ashton, saying 'that when the cat's away the mice will play', had decided on remaining at home. **1925** S. O'CASEY *Juno & Paycock* I. 13 It's a good job she has to be so often away, for when the cat's away, the mice can play! **1979** G. PETRIE *Hand of Glory* iii. 'Cat's away, mouse doth play,' he said. 'The cat's been away too long.'

[1] rules.

The CAT, the rat, and Lovell the dog, rule all England under the hog

The allusion is explained in the second quotation.

1516 R. FABYAN *New Chronicles of Eng-*

land & France VIII. 219[v] The Catte the
Ratte And Louell our dogge Rulyth all
Englande under a hogge. The whiche
was ment that Catisby Ratclyffe And the
Lord Louell Ruled the lande under the
kynge. **1586** R. HOLINSHED *Chronicles*
III. 746 [Richard III executed] a poore
gentleman called Collingborne [in
1484], for making a small rime of three of
his . . councellors, . . lord Louell, sir
Richard Ratcliffe . . and sir William
Catesbie. . . The Cat, the Rat, and
Louell our dog, Rule all England vnder
an hog. Meaning by the hog, the . .
wild boare, which was the Kings
cognisance.[1] **1816** SCOTT *Antiquary* ii.
'His name . . was Lovel.' 'What! the
cat, the rat, and Lovel our dog? Was
he descended from King Richard's
favourite?' **1931** J. BUCHAN *Blanket of
Dark* viii. This Francis Lovell, With
Catesby and Ratcliffe he ruled the land
under King Richard. You have heard
the country rhyme: The Cat, the Rat,
and Lovell our dog Ruled all England
under the Hog. **1973** A. CHRISTIE *Postern of Fate* I. ii. *The cat, the rat and Lovell,
the dog, Rule all England under the hog.* . .
The hog was Richard the Third.
  [1] coat of arms.

The CAT would eat fish, but would not wet her feet

  [Med. Lat. *catus amat piscem, sed non
vult tingere plantas,* the cat loves a fish,
but does not wish to wet its feet.]
c **1225** in *Englische Studien* (1902) XXXI. 7
Cat lufat visch, ac he nele his feth wete.
c **1380** CHAUCER *House of Fame* III. 1783
For ye be lyke the sweynte[1] cat That
wolde have fissh; but wostow[2] what? He
wolde nothing wete his clowes. *c* **1549**
J. HEYWOOD *Dialogue of Proverbs* I. xi. B8[v]
But you lust not to do, that longeth therto.
The cat would eate fyshe, and wold not
wet her feete. **1605–6** SHAKESPEARE
Macbeth I. vii. 44 Letting 'I dare not' wait
upon 'I would', Like the poor cat i' th'
adage. **1732** T. FULLER *Gnomologia* no.
6130 Fain would the Cat Fish eat, But
she's loth her Feet to wet. **1928** *Sphere* 7
Jan. 36 'The cat would fain eat fish, but
would not wet his feet.' . . In modern

days one might paraphrase it into 'bad
sailors would fain enjoy the sun, but
would not cross the channel'.
  [1] tired.   [2] do you know.

cat: *see also* CARE killed the cat; CURIOSITY
killed the cat; KEEP no more cats than
will catch mice; WANTON kittens make
sober cats; there are more WAYS of killing
a cat than choking it with cream; *also*
CATS.

You cannot CATCH old birds with chaff

The wise are not easily fooled.
  1481 CAXTON *Reynard the Fox* (1880) xl.
Wenest[1] thou thus to deceyue. . . I am
no byrde to be locked ne take by chaf. I
know wel ynowh good corn. *c* **1590**
Timon (1842) IV. ii. Tis well.—An olde
birde is not caught with chaffe. **1670** J.
RAY *English Proverbs* 126 You can't catch
old birds with chaff. **1853** THACKERAY
Newcomes II. xv. They ogled him as they
sang . . with which chaff our noble bird
was by no means to be caught. **1936**
C. F. GREGG *Danger at Cliff House* v. Henry Prince was too old a bird to be caught
with such chaff.
  [1] *wenest,* do you think or expect.

catch: *see also* a CAT in gloves catches no
mice; a DROWNING man will clutch at
a straw; EAGLES don't catch flies; the
EARLY bird catches the worm; FIRST
catch your hare; HONEY catches more
flies than vinegar; KEEP no more cats
than will catch mice; if you RUN after two
hares you will catch neither; if the SKY
falls we shall catch larks; set a THIEF to
catch a thief.

All CATS are grey in the dark

The proverb, which turns on the concept that the night obscures all disting-
uishing features, is used in a variety of
contexts.
  c **1549** J. HEYWOOD *Dialogue of Proverbs*
I. v. A6[v] When all candels be out, all
cats be grey. All thyngs are then of one
colour. **1596** T. LODGE *Margarite of America* H2[v] All cattes are grey in the darke . .
and therefore (good madam) you doe
well to preferre the eie.[1] **1771** SMOLLETT

Humphry Clinker III. 67 He knew not which was which; and, as the saying is, all cats in the dark are gray. **1886** H. JAMES *Princess Casamassima* I. xiv. 'If she isn't, what becomes of your explanation?' . . 'Oh, it doesn't matter; at night all cats are grey.' **1941** M. SEELEY *Chuckling Fingers* iii. 'Cecile wouldn't know,' Jean said. 'In the dark all cats are grey.' **1981** J. ROSS *Dark Blue & Dangerous* xxvii. You forgot that all cats are grey in the dark and so are uniformed policemen.

[1] because *the* EYES *are the window of the soul.*

cause: *see* no one should be JUDGE in his own cause.

cease: *see* WONDERS will never cease.

certain: *see* CHILDREN are certain cares, but uncertain comforts; NOTHING is certain but death and taxes; NOTHING is certain but the unforeseen.

chaff: *see* you cannot CATCH old birds with chaff.

A CHAIN is no stronger than its weakest link

1856 C. KINGSLEY *Letter* 1 Dec. (1877) II. 499 The devil is very busy, and no one knows better than he, that 'nothing is stronger than its weakest part'. **1868** L. STEPHEN in *Cornhill Magazine* XVII. 295 A chain is no stronger than its weakest link; but if you show how admirably the last few are united . . half the world will forget to test the security of the . . parts which are kept out of sight. **1942** I. S. SHRIBER *Body for Bill* 298 'I've gathered that Stansfield was a pretty weak individual. . . He was threatening to give the whole thing away.' . . 'A chain is no stronger than its weakest link, remember?'

Don't CHANGE horses in mid-stream

The proverb is also used in the phrase *to change horses in mid-stream.*

1864 LINCOLN *Collected Works* (1953) VII. 384, I am reminded . . of a story of an old Dutch farmer, who remarked to a companion once that 'it was best not to swap horses when crossing streams'. **1889** W. F. BUTLER *C. G. Gordon* ii. Clothing and equipment were then undergoing a vigorous process of 'swopping' at the moment the animals were in the mid-stream of the siege of Sebastopol. **1929** R. GRAVES *Good-bye to All That* xxiii. 'If ours is the true religion why do you not become a Catholic?' . . 'Reverend father, we have a proverb in England never to swap horses while crossing a stream. **1967** RIDOUT & WITTING *English Proverbs Explained* 41 Don't change horses in mid-stream. . . If we think it necessary to make changes, we must choose the right moment to make them. **1979** D. MAY *Revenger's Comedy* ix. Changing horses, love? I should look before you leap.

A CHANGE is as good as a rest

1895 J. THOMAS *Randigal Rhymes* 59 Change of work is as good as touchpipe.[1] **1903** V. S. LEAN *Collectanea* III. 439 Change of work is rest. (Manx.) **1951** M. COLES *Now or Never* ii. On the principle that a change of work is a rest, we redecorated the bathroom. . . We came to the conclusion that the saying is a fallacy. **1967** O. MILLS *Death enters Lists* viii. There would be no fish-bits for Whiskers . . but she could buy him some fish-pieces; and a change was as good as a rest, she remembered.

[1] a short interval of rest.

change: *see also* (verb) the LEOPARD does not change his spots; TIMES change and we with time.

chapel: *see* where GOD builds a church, the Devil will build a chapel.

CHARITY begins at home

c **1383** in Wyclif *English Works* (EETS) 78 Charite schuld bigyne at hem-self. *a* **1625** BEAUMONT & FLETCHER *Wit without Money* v. ii. Charity and beating begins at home. **1659** T. FULLER *Appeal of Injured Innocence* I. 25 Charity begins, but doth not end, at home. . . My Church-History . . began with our own Domestick affairs. . . I intended . . to

have proceeded to forrain Churches.
1748 SMOLLETT *Roderick Random* I. vi.
The world would do nothing for her if
she should come to want—charity
begins at home. **1935** E. F. BENSON
Mapp & Lucia vii. Besides, the loss of
Foljambe *had* occurred to him first.
Comfort, like charity, began at home.

chase: *see* a STERN chase is a long chase.

It is as CHEAP sitting as standing
Commonly applied literally.

1666 G. TORRIANO *Italian Proverbs* 277
The English say, It is as cheap sitting as
standing, my Masters. **1858** R. S. SUR-
TEES *Ask Mamma* xlix. Let's get chairs,
and be snug; it's as cheap sitting as
standing. **1932** 'J. J. CONNINGTON'
Sweepstake Murders ix. He returned to
Tommie Redhill's car. 'Jump in, Inspec-
tor,' Tommie suggested, opening the
door at his side. 'It's as cheap sitting as
standing.'

cheap: *see also* why buy a cow when milk is
so cheap?

cheapest: *see* BUY in the cheapest market
and sell in the dearest.

CHEATS never prosper
[a **1612** HARINGTON *Epigrams* (1618) IV.
5 Treason doth neuer prosper, what's
the reason? For if it prosper, none dare
call it Treason.] **1805** R. PARKINSON *Tour
in America* II. xxix. It is a common saying in England, that 'Cheating never
thrives': but, in America, with honest
trading you cannot succeed. **1903** V. S.
LEAN *Collectanea* II. 38 'Cheating never
prospers.' A proverb frequently thrown
at each other by young people when
playing cards. **1935** R. CROMPTON *William—the Detective* vi. They avenged
themselves upon the newcomer . . by
shouting the time-honoured taunt
'Cheats never prosper'. **1966** P. O'DON-
NELL *Sabre-Tooth* v. Willie pulled off his
blindfold and knelt up, looking at her.
'Cheats never prosper,' he said with
pious severity.

cheeping: *see* MAY chickens come cheeping.

A CHERRY year, a merry year; a plum year, a dumb year
1678 J. RAY *English Proverbs* (ed. 2) 52
A cherry year a merry year: A plum year
a dumb year. This is a puerile and sence-
less rythme . . as far as I can see. **1869**
R. INWARDS *Weather Lore* 14 The progress of the seasons may be watched by
observing the punctuality of the vegetable world. . . A cherry year, a merry
year. A plum year, a dumb year. **1979**
V. CANNING *Satan Sampler* ix. Warboys
was studying an arrangement of cherry
blossom. . . The blossom was good this
year. A cherry year, a merry year.

chicken: *see* don't COUNT your chickens
before they are hatched; CURSES,
like chickens, come home to roost; MAY
chickens come cheeping.

Monday's CHILD is fair of face
Each line of the verse (quot. 1838) may
be used separately. Examples relating to
different days of the week are illustrated
here for convenience.

1838 A. E. BRAY *Traditions of Devon* II.
287 Monday's child is fair of face, Tuesday's child is full of grace, Wednesday's
child is full of woe, Thursday's child has
far to go, Friday's child is loving and
giving, Saturday's child works hard for
its living, And a child that's born on the
Christmas[1] day Is fair and wise and
good and gay. **1915** J. BUCHAN *Salute to
Adventurers* i. I was a Thursday's bairn,
and so, according to the old rhyme, 'had
far to go'. **1935** D. JONES *Journal* 12 Nov.
in R. Hague *Dai Greatcoat* (1980) ii. 81
Which day's child is 'loving and giving'
in the rhyme?. . Is it Wednesday's?
1957 V. BRITTAIN *Testament of Experience*
i. ii. From the outset Shirley sustained
the nursery adage which commends
'Sunday's child', for she put on weight
steadily and was the easiest of infants to
rear. **1980** A. WILSON *Setting World on
Fire* ii. iii. She showed her contrition by
stroking his hair. 'Saturday's child

works hard for his living,' she murmured.

[1] *Sabbath* is usually found instead of *Christmas.*

The CHILD is the father of the man
An assertion of the unity of character from youth to manhood.

[**1671** MILTON *Paradise Regained* IV. 220 The childhood shews the man, As morning shews the day.] **1807** WORDSWORTH *Poems* (1952) I. 226 My heart leaps up when I behold A rainbow in the sky: So was it when my life began. . . The Child is father of the Man. **1871** S. SMILES *Character* ii. The influences which contribute to form the character of the child . . endure through life. . . 'The child is father of the man.' **1907** E. GOSSE *Father & Son* xii. We are the victims of hallowed proverbs, and one of the most classic of these tells us that 'the child is the father of the man'. **1979** A. SANDERS *Victorian Historical Novel* v. The lonely, politically involved child, is father to the man who only slowly grasps the meaning of his experience.

child: *see also* a BURNT child dreads the fire; PRAISE the child, and you make love to the mother; SPARE the rod and spoil the child; it is a WISE child that knows its own father.

CHILDREN and fools tell the truth
[*c* **1425** in *Anglia* (1885) VIII. 154 Atte laste treuthe was tryed oute of a childe & dronken man.] **1537** in *Letters & Papers of Reign of Henry VIII* (1929) Addenda I. I. 437 It is 'an old saying that a child, a fool and a drunken man will ever show . . the truth'. **1591** LYLY *Endymion* IV. ii. Children must not see Endimion, because children & fooles speake true. **1652** J. TATHAM *Scots Figgaries* III. 23, I am a fool 'tis confest, but children and fooles tell truth sometimes; you know. **1805** SCOTT *Letter* Jan. (1932) I. 233 It is a proverb, that children and fools talk truth and I am mistaken if even the same valuable quality may not sometimes be extracted out of the tales made to entertain both. **1921** *Evening Standard* 21 Oct.

9 Solicitor. . 'Are you telling the truth in this case?' Witness.—Only children and fools tell the truth. [**1981** 'E. PETERS' *Saint Peter's Day* 180 Children and drunken men are the world's only innocents.]

CHILDREN are certain cares, but uncertain comforts
The sense is reversed in the last quotation.

1639 J. CLARKE *Parœmiologia Anglo-Latina* 240 Children are uncertaine comforts, but certaine cares. **1641** R. BRATHWAIT *English Gentleman* (ed. 3) 27 Children reflect constant cares, but uncertaine comforts. **1732** T. FULLER *Gnomologia* no. 1095 Children are certain Cares, but uncertain comforts. **1885** E. J. HARDY *How to be Happy though Married* xvi. Children are *not* 'certain sorrows and uncertain pleasures' when properly managed. **1915** J. WEBSTER *Dear Enemy* 203 My new little family has driven everything out of my mind. Bairns are certain joy, but nae sma' care.

CHILDREN should be seen and not heard
Originally applied specifically to (young) women.

c **1400** J. MIRK *Festial* (EETS) I. 230 Hyt ys an old Englysch sawe:[1] 'A mayde schuld be seen, but not herd.' **1560** T. BECON *Works* I. Bbb2 This also must honest maids provide, that they be not full of tongue. . . A maid should be seen, and not heard. **1773** R. GRAVES *Spiritual Quixote* I. III. xviii. It is a vulgar maxim, 'that a pretty woman should rather be seen than heard'. **1820** J. Q. ADAMS *Memoirs* (1875) V. xii. My dear mother's constant lesson in childhood, that children in company should be seen and not heard. **1959** M. BRADBURY *Eating People is Wrong* ii. 'You think that children should be seen and not heard then?' asked the novelist. **1980** L. LEWIS *Private Life of Country House* v. Two or three children . . supposed to be seen and not heard and not to speak unless spoken to.

[1] proverb, saying.

children: *see also* the DEVIL's children have the Devil's luck; HEAVEN protects children, sailors, and drunken men.

choice: *see* you PAYS your money and you takes your choice.

choke: *see* it is idle to SWALLOW the cow and choke on the tail.

choking: *see* there are more WAYS of killing a cat than choking it with cream; there are more WAYS of killing a dog than choking it with butter.

Never CHOOSE your women or your linen by candlelight

1573 J. SANFORDE *Garden of Pleasure* 51 Choose not a woman, nor linnen clothe by the candle. 1678 J. RAY *English Proverbs* (ed. 2) 64 Neither women nor linnen by candle-light. 1737 B. FRANKLIN *Poor Richard's Almanack* (May) Fine linnen, girls and gold so bright. Chuse not to take by candlelight. 1928 *Bystander* 17 Oct. 136 The title ['By Candle Light'] is explained by the grandmotherly aphorism: 'Choose neither women nor linen by candlelight.' 1980 *Woman's Journal* Dec. 105 'Never choose your women or your linen by candlelight,' they used to say: a testimony to the soft, flattering glow that candles always give.

choose: *see also* of two EVILS choose the less.

chooser: *see* BEGGARS can't be choosers.

Christmas: *see* a GREEN Yule makes a fat churchyard.

The CHURCH is an anvil which has worn out many hammers

1908 A. MACLAREN *Acts of Apostles* I. 136 The Church is an anvil which has worn out many hammers and the story of the first collision is, in essentials, the story of all. 1920 J. BUCHAN *Path of King* vii. 'From this day I am an exile from France so long as it pleases God to make His Church an anvil for the blows of His

enemies.' . . 'God's church is now an anvil, but remember . . it is an anvil which has worn out many hammers.'

church: *see also* the BLOOD of the martyrs is the seed of the Church; where GOD builds a church, the Devil will build a chapel; he is a GOOD dog who goes to church; the NEARER the church, the farther from God.

churchyard: *see* a GREEN Yule makes a fat churchyard.

CIRCUMSTANCES alter cases

1678 T. RYMER *Tragedies of Last Age* 177 There may be circumstances that alter the case, as when there is a sufficient ground of partiality. 1776 W. HEATH *Memoirs* (1798) 92 Our General reflected for a moment, that as circumstances alter cases, Gen. Washington . . might possibly wish for some aid. 1895 J. PAYN *In Market Overt* xxxix. Circumstances alter cases even with the best of us, as was shown in a day or two in the conduct of the Lord Bishop. 1938 A. CHRISTIE *Appointment with Death* xiii. It is undoubtedly true that circumstances alter cases. I do feel . . that in the present circumstances decisions may have to be reconsidered. 1960 J. WAIN *Nuncle* 116 Well, circumstances alter cases. . . With me sitting here . . you haven't got a free hand.

circus: *see* if you can't RIDE two horses at once, you shouldn't be in the circus.

cite: *see* the DEVIL can quote Scripture for his own ends.

city: *see* if every man would SWEEP his own doorstep the city would soon be clean.

CIVILITY costs nothing

Politeness sometimes replaces *civility.* Cf. *There is* NOTHING *lost by civility.*

1706 J. STEVENS *Spanish & English Dict.* s.v. *Cortesía,* Mouth civility is worth much and costs little. 1765 LADY M. W. MONTAGU *Letter* 30 May (1967) III. 107 Remember Civillity costs nothing, and buys every thing. **1765** H. TIMBERLAKE *Memoirs* 73 Politeness . . costs but little. **1841** S. WARREN *Ten Thousand a Year* I. iii. It may be as well . . to acknowledge the fellow's note. . . Civility costs nothing. **1873** W. ALLINGHAM *Rambles* xiv. Civility costs nothing, it is said—Nothing, that is, to him that shows it; but it often costs the world very dear. **1980** E. HARRIS *Medium for Murder* x. 'It made me hopping mad to hear you kow-towing to him.' . . 'Politeness costs nothing,' said Brooker.

civility: *see also* there is NOTHING lost by civility.

clay: *see* CANDLEMAS day, put beans in the clay, put candles and candlesticks away.

clean: *see* NEW brooms sweep clean; if every man would SWEEP his own door-step the city would soon be clean.

CLEANLINESS is next to godliness

[**1605** BACON *Advancement of Learning* II. 44 Cleannesse of bodie was euer esteemed to proceed from a due reuerence to God.] *a* **1791** WESLEY *Works* (1872) VII. 16 Slovenliness is no part of religion. . . 'Cleanliness is indeed next[1] to godliness.' **1876** F. G. BURNABY *Ride to Khiva* x. 'Cleanliness is next to Godliness.' The latter quality, as displayed in a Russian devotee, is more allied with dirt than anything else. **1935** J. T. FARRELL *Judgement Day* xvii. His sister Loretta . . called: Cleanliness is next to Godliness. **1979** C. EGLETON *Backfire* i. The hospital staff had a thing about personal cleanliness, next to godliness, you might say.

[1] i.e. immediately following, as in serial order.

CLERGYMEN's sons always turn out badly

1885 E. J. HARDY *How to be Happy though Married* xix. The Proverb says that 'Clergymen's sons always turn out badly' . . because the children are surfeited with severe religion, *not* with the true religion of Christ. **1922** W. R. INGE *Outspoken Essays* 2nd Ser. vii. An Eton boy . . when asked why the sons of Eli turned out badly, replied 'The sons of clergymen always turn out badly'.

client: *see* a man who is his own LAWYER has a fool for his client.

climb: *see* he that would EAT the fruit must climb the tree; the HIGHER the monkey climbs the more he shows his tail.

From CLOGS to clogs is only three generations

Said to be a Lancashire proverb. The clog, a shoe with a thick wooden sole, was commonly used by factory and other manual workers in the north of England. Cf. *From* SHIRTSLEEVES *to shirtsleeves in three generations.*

[Cf. **1700** DRYDEN *Wife of Bath* in *Fables Ancient & Modern* 493 Seldom three descents continue good. **1721** J. KELLY *Scottish Proverbs* 312 The Father buys, the Son biggs,[1] The Grandchild sells, and his Son thiggs.[2]] **1871** *Notes & Queries* 4th Ser. VII. 472 'From clogs to clogs is only three generations.' A Lancashire proverb, implying that, however rich a poor man may eventually become, his great-grandson will certainly fall back to poverty and 'clogs'. **1966** G. HAMILTON-EDWARDS *In Search of Ancestry* ix. Families . . easily fall in the social scale. There is a lot in that old Lancashire proverb 'There's nobbut[3] three generations between clog and clog.'

[1] builds.   [2] begs.   [3] only.

close: *see* when ONE door shuts, another opens.

cloth: *see* CUT your coat according to your cloth.

Every CLOUD has a silver lining

A poetic sentiment that even the gloomiest outlook contains some hopeful or consoling aspect.

[**1634** MILTON *Comus* I. 93 Was I deceiv'd, or did a sable cloud Turn forth

her silver lining on the night?] **1863**
D. R. LOCKE *Struggles of P. V. Nasby*
(1872) xxiii. Ther is a silver linin to evry
cloud. **1869** P. T. BARNUM *Struggles &
Triumphs* 406 'Every cloud', says the
proverb, 'has a silver lining.' **1939** E. F.
BENSON *Trouble for Lucia* xi. She always
discovered silver linings to the blackest
of clouds, but now, scrutinize them as
she might, she could detect in them
none but the most sombre hues. **1976**
T. SHARPE *Wilt* vi. At least that way he
would learn if it was a sensible method
of getting rid of Eva. There was that to
be said for it. Every cloud has. . .

clout: *see* ne'er CAST a clout till May be out.

clutch: *see* a DROWNING man will clutch at a
straw.

coat: *see* CUT your coat according to your
cloth.

Let the COBBLER stick to his last
Attributed to the Greek painter Apelles
(4th-cent. BC): see quot. 1721.

[PLINY *Natural History* XXXV. 85 *ne
supra crepidam sutor iudicaret*, the cobbler
should not judge beyond his shoe; ERAS-
MUS *Adages* I. vi. 16 *ne sutor ultra
crepidam*.] **1539** R. TAVERNER tr. *Erasmus' Adages* 17 Let not the shoemaker
go beyond hys shoe. **1616** J. WITHALS
Dict. (rev. ed.) 567 Cobler keepe your
last.[1] **1639** J. CLARKE *Parœmiologia
Anglo-Latina* 21 Cobler keepe to your
last. **1721** J. KELLY *Scottish Proverbs* 242
Let not the Cobler go beyond his last. . .
Taken from the famous Story of Apelles,
who could not bear that the Cobler
should correct any part of his Picture
beyond the Slipper. **1868** W. CLIFT *Tim
Bunker Papers* lix. I understood the use
of a plow . . better than the use of a pen
. . remembering the old saw 'Let the
cobbler stick to his last.' **1966** J. I. M.
STEWART *Aylwins* vi. I don't think I'd
care to write novels. . . The cobbler
should stick to his last, I suppose.

[1] a wooden or metal model on which a
shoemaker fashions shoes or boots.

**The COBBLER to his last and the gunner
to his linstock**
A fanciful variant of the preceding
proverb.

1748 SMOLLETT *Roderick Random* II.
xlii. I meddle with no body's affairs but
my own; The gunner to his linstock,[1]
and the steersman to the helm, as the
saying is. **1893** H. MAXWELL *Life of
W. H. Smith* II. v. He . . never showed
any disposition to trespass on the pro-
vince of science or literature . . . There is
sound sense in the adage, 'The cobbler
to his last and the gunner to his linstock.'

[1] a staff with a forked head to hold a lighted
match.

**Every COCK will crow upon his own
dunghill**
Everyone is confident or at ease when
on home ground.

[SENECA *Apocolocynthosis* vii. *Gallum
in suo sterquilinio plurimum posse*, the
cock is most powerful on his own dunghill; the work is a satire on Claudius'
deification at death: Seneca is punning
on Claudius' provincial origin and interests, as *gallus* means both a cock and a
Gaul.] *a* **1250** *Ancrene Wisse* (1952) 62
Coc is kene[1] on his owune mixerne.[2]
1387 J. TREVISA tr. *Higden's Polychronicon*
(1879) VIII. 5 As Seneca seith, a cok is
most myghty on his dongehille. **1546**
J. HEYWOOD *Dialogue of Proverbs* I. xi. D2
He was at home there, he myght speake
his will. Euery cocke is proude on his
owne dunghill. **1771** SMOLLETT *Humphry Clinker* II. 178 Insolence . . akin to
the arrogance of the village cock, who
never crows but upon his own
dunghill. **1935** D. L. SAYERS *Gaudy
Night* xix. 'I believe you're showing off.'
. . 'Every cock will crow upon his own
dunghill.' **1980** M. GILBERT *Death of
Favourite Girl* vii. Mariner seemed to be
easy enough. A cock on his own dunghill.

[1] bold.   [2] midden, dunghill.

cock: *see also* there's many a GOOD cock
come out of a tattered bag; the ROBIN
and the wren are God's cock and hen.

COLD hands, warm heart

1903 V. S. LEAN *Collectanea* III. 380 A cold hand and a warm heart. **1910** W. G. COLLINGWOOD *Dutch Agnes* 206, I did take her hand . . . Cold hand, warm heart! **1927** J. M. BARRIE *Shall We join Ladies* in *Plays* (1928) 840, I knew you would be on my side . . . Cold hand—warm heart. That is the saying, isn't it? **1962** E. LININGTON *Knave of Hearts* xv. A hot, humid night, but her hands cold. Cold hands, warm heart.

cold: *see also* (noun) as the DAY lengthens, so the cold strengthens; FEED a cold and starve a fever; (adjective) REVENGE is a dish that can be eaten cold.

colour: *see* a GOOD horse cannot be of a bad colour.

come: *see* ALL things come to those who wait; the BIGGER they are, the harder they fall; don't CROSS the bridge till you come to it; EASY come, easy go; FIRST come, first served; all is GRIST that comes to the mill; LIGHT come, light go; come LIVE with me and you'll know me; MARCH comes in like a lion, and goes out like a lamb; if the MOUNTAIN will not come to Mahomet, Mahomet must go to the mountain; when POVERTY comes in at the door, love flies out of the window; QUICKLY come, quickly go; TOMORROW never comes.

comfort: *see* CHILDREN are certain cares, but uncertain comforts.

COMING events cast their shadows before

1803 T. CAMPBELL *Poetical Works* (1907) 159 'Tis the sunset of life gives me mystical love, And coming events cast their shadows before. **1857** TROLLOPE *Barchester Towers* II. v. The coming event of Mr. Quiverful's transference to Barchester produced a delicious shadow in the shape of a new outfit for Mrs. Quiverful. **1979** D. LESSING *Shikasta* 231 'Coming events cast their shadows before.' This Shikastan[1] observation was of particular appropriateness during an epoch when the tempo of events was so speeded up.

[1] i.e. Earthly.

communication: *see* EVIL communications corrupt good manners.

A man is known by the COMPANY he keeps

Originally used as a moral maxim or exhortation in the context of (preparation for) marriage.

1541 M. COVERDALE tr. *H. Bullinger's Christian State of Matrimony* F6 So maye much be spyed also, by the company and pastyme that a body vseth. For a man is for the moost parte condicioned euen lyke vnto them that he kepeth company wythe all. **1591** H.SMITH *Preparative to Marriage* 42 If a man can be known by nothing els, then he maye bee known by his companions. **1672** W. WYCHERLEY *Love in Wood* I. i. There is a Proverb, Mrs. Joyner, You may know him by his Company. **1854** H. MELVILLE *Complete Short Stories* (1951) 182 'Surely he does not mean me.' . . 'One is known by the company he keeps.' **1912** 'SAKI' *Chronicles of Clovis* 286 (*heading*) A man is known by the company he keeps. **1942** W. O'FARRELL *Repeat Performance* xvi. I have always contended . . that a man is best known by the company he keeps. **1976** L. ALTHER *Kinflicks* ii. People knew a man by the company he kept, but they generally knew a woman by the man who kept her.

The COMPANY makes the feast

1653 I. WALTON *Compleat Angler* iii. Take this for a rule, you may pick out such times and such companies, that you may make yourselves merrier, . . for 'tis the company and not the charge[1] that makes the feast. **1911** F. W. HACKWOOD *Good Cheer* xxxii. Epicurus maintained that you should rather have regard to the company with whom you eat . . than to what you eat. . . This has been crystallised into the terse English proverb, 'The company makes the feast'. **1981** 'J. STURROCK' *Suicide most*

Foul vi. It is the company which makes the occasion, not the surroundings.

[1] expense.

company: *see also* MISERY loves company; TWO is company, but three is none.

COMPARISONS are odious

[Early 14th-cent. Fr. *comparisons sont haïneuses*, comparisons are odious.] *c* **1440** LYDGATE *Minor Poems* (EETS) 561 Odious of old been all comparisouns. **1456** *Gilbert of Hay's Prose MS*. (STS) 282, I will nocht here mak questioun . . quhy[1] that alwayis comparisoun is odious. *c* **1573** G. HARVEY *Letter-Book* (1884) 7 But thai wil sai, Comparisons ar odius: in deed, as it fals out, thai ar too odious. **1724** SWIFT *Drapier's Letters* X. 82 A Judge . . checked the Prisoner . . taxing him with 'reflecting on the court by such a Comparison, because Comparisons were odious'. **1872** H. JAMES *Letter* 15 Sept. (1920) I. 32 Nuremburg is excellent—and comparisons are odious; but I would give a thousand N[uremburg]'s for one ray of Verona. **1939** G. MITCHELL *Printer's Error* ii. 'I will study the psychology of pigs instead of that of . . refugees.' 'Comparisons are odious,' observed Carey. **1979** *Guardian* 25 June 4 If there is one message . . it is that comparisons are odious. Incomes and prices have to be related to local conditions.

[1] why.

He that COMPLIES against his will is of his own opinion still

1678 BUTLER *Hudibras* III. iii. He that complies against his Will, Is of his own Opinion still; Which he may adhere to, yet disown, For Reasons to himself best known. **1965** M. SPARK *Mandelbaum Gate* v. No one should submit their mind to another mind: He that complies against his will Is of his own opinion still—that's *my* motto. I won't be brainwashed.

CONFESS and be hanged

1589 'MISOPHONUS' *De Caede Gallorum Regis* A2[v] Confesse and be hangede

man In English some saie. **1604** SHAKESPEARE *Othello* IV. i. 37 Handkerchief—confessions—handkerchief! To confess, and be hanged for his labour. **1672** MARVELL *Rehearsal Transprosed* 74 After so ample a Confession as he hath made, must he now be hang'd too to make good the Proverb? **1821** SCOTT *Pirate* III. xii. At the gallows! . . Confess and be hanged is a most reverend proverb. **1951** M. C. BARNES *With all my Heart* vii. 'People who commit high treason get hanged.' . . 'Very well, confess and be hanged!'

confessed: *see* a FAULT confessed is half redressed.

Open CONFESSION is good for the soul

c **1641** in E. Beveridge *D. Fergusson's Scottish Proverbs* (1924) no. 159 Ane open confessione is good for the soul. **1721** J. KELLY *Scottish Proverbs* 270 Open Confession is good for the Soul. Spoken ironically, to them that boast of their ill Deeds. **1881** J. PAYN *Grape from Thorn* III. xxxix. Confession may be good for the soul; but it is doubtful whether the avowal of incapacity to the parties desirous of securing our services is quite judicious. **1942** R. A. J. WALLING *Corpse with Eerie Eye* v. That's open confession, but I don't know that it does my soul any good. **1980** M. GILBERT *Death of Favourite Girl* xii. Confession is good for the soul. I made a fool of myself.

CONSCIENCE makes cowards of us all

Quot. 1912 is a humorous perversion of the proverb.

[**1594** SHAKESPEARE *Richard III* I. iv. 133 Where's thy conscience now?—I'll not meddle with it—it makes a man a coward.] **1600–1** SHAKESPEARE *Hamlet* III. i. 83 Conscience does make cowards of us all. **1697** VANBRUGH *Provoked Wife* v. 75 It mayn't be amiss to deferr the Marriage till you are sure they [mortgages] are paid off . . . Guilty Consciences make Men Cowards. **1912** 'SAKI' *Chronicles of Clovis* 134 The English have a proverb, 'Conscience makes cowboys of us all.' **1941** H. G. WELLS *You can't be too Careful* viii. 'Why doesn't he face it out?' . . 'Conscience makes cowards of us all, Whittaker.'

conscience: *see* a GUILTY conscience needs no accuser.

consent: *see* SILENCE means consent.

CONSTANT dropping wears away a stone

Primarily used to mean that persistence will achieve a difficult or unlikely objective (but see also quots. 1874 and 1912). *Continual* frequently occurs instead of *constant* in the US.

[CHOERILUS OF SAMOS *Fragments* ix. πέτρην κοιλαίνει ῥανὶς ὕδατος ἐνδελεχείῃ, with persistence a drop of water hollows out the stone; OVID *Letters from Pontus* IV. x. *gutta cavat lapidem*, a drop hollows the stone.] *a* **1250** *Ancrene Wisse* (1962) 114 Lutle dropen thurleth[1] the flint the[2] ofte falleth theron. *c* **1477** CAXTON *Jason* (EETS) 26 The stone is myned and holowed by contynuell droppyng of water. **1591** SHAKESPEARE *Henry VI Pt. 3* III. ii. 50 He plies her hard; and much rain wears the marble. **1793** T. COKE *Extracts from Journals* III. ii. The Negroes of Barbadoes . . are much less prepared for the reception of genuine religion. But constant dropping, 'tis said, will wear out a stone. **1841** DICKENS *Old Curiosity Shop* I. vii. As to Nell, constant dropping will wear away a stone, you know you may trust me as far as she is concerned. **1874** G. J. WHYTE-MELVILLE *Uncle John* I. vi. Constant dropping wears away a stone; constant flirtation saps the character. **1912** D. H. LAWRENCE *Letter* 19 Dec. (1962) I. 169 She says a woman can only have one husband. . . Constant dropping will wear away a stone, as my mother used to say. **1931** C. WELLS *Horror House* ii. [They] were possessed of decided opinions and had no hesitancy in airing them. This produced more or less friction, and, like the continual dropping that wears away a stone, it had begun to react on the habitual goodness of the Baileys.

[1] pierce, penetrate, bore a hole in.
[2] that.

contempt: *see* FAMILIARITY breeds contempt.

continual: *see* CONSTANT dropping wears away a stone.

contrary: *see* DREAMS go by contraries.

cook: *see* GOD sends meat, but the Devil sends cooks; TOO many cooks spoil the broth.

CORPORATIONS have neither bodies to be punished nor souls to be damned

A large organization, unlike a private individual, can act unjustly or highhandedly without fear of being brought to account.

1658 E. BULSTRODE *Reports* II. 233 The opinion of Manwood, chief Baron [*c* 1580], was this, as touching Corporations, that they were invisible, immortall, and that they have no soule; and therefore no Subpœna lieth against them, because they have no Conscience nor soule. *c* **1820** J. POYNDER *Literary Extracts* (1844) I. 268 Lord Chancellor Thurlow said [*c* 1775] that the corporations have neither bodies to be punished nor souls to be damned. *a* **1845** S. SMITH in S. Holland *Memoir* (1855) I. xi. Why, you never expected justice from a company, did you? They have neither a soul to lose, nor a body to kick. **1932** ERNST & LINDEY *Hold your Tongue* xii. A corporation is just like any natural person, except that it has no pants to kick or soul to damn, and, by God, it ought to have both.

corrupt: *see* EVIL communications corrupt good manners; POWER corrupts.

cost: *see* CIVILITY costs nothing.

cough: *see* LOVE and a cough cannot be hid.

COUNCILS of war never fight
i.e. never reach the decision to fight,
which an individual would make.
1863 H. W. HALLECK *Telegram* 13 July
(1877) III. 148 Act upon your own judgment and make your Generals execute
your orders. Call no counsel [*sic*] of war.
It is proverbial that counsels of war
never fight. **1891** A. FORBES *Barracks,
Bivouacs & Battles* 191 Solomon's adage
that in the multitude of counsellors
there is wisdom does not apply to war.
'Councils of war never fight' has passed
into a proverb.

counsel: *see* a FOOL may give a wise man
counsel.

**Don't COUNT your chickens before they
are hatched**
An instruction not to make, or act upon,
an assumption (usually favourable)
which might turn out to be wrong. The
metaphorical phrase *to count one's chickens* is also used.
c **1570** T. HOWELL *New Sonnets* C2
Counte not thy Chickens that vnhat-
ched be. **1579** S. GOSSON *Ephemerides of
Phialo* 19, I woulde not haue him to
counte his Chickens so soone before
they be hatcht, nor tryumphe so long
before the victorie. **1664** BUTLER *Hudibras* II. iii. To swallow Gudgeons ere
th'are catch'd, And count their Chickens ere th'are hatch'd. **1829** SCOTT *Journal* 20 May (1946) 69, I see a fund . .
capable of extinguishing the debt . . in
ten years or earlier . . . But we must not
reckon our chickens before they are
hatchd. **1906** in Lady D. Nevill *Remini-
scences* xxii. A victory may be snatched,
But never count your little chicks, Before they're safely hatched. **1964** RID-
OUT & WITTING *English Proverbs Explained*
42 Mr. Smith hoped to be made manager before the end of the year . . . 'Don't
count your chickens before they are
hatched,' warned his wife.

**In the COUNTRY of the blind, the one-
eyed man is king**
A little wit, among foolish people, will
pass a man for a great genius: T. Field-

ing *Proverbs of all Nations* (1824) 23; also
used of ability as well as wit.
[ERASMUS *Adages* III. iv. *in regione
caecorum rex est luscus*, in the kingdom of
the blind the one-eyed man is king.]
1522 SKELTON *Works* (1843) II. 43 An one
eyed man is Well syghted when He is
amonge blynde men. **1640** G. HERBERT
Outlandish Proverbs no. 469 In the kingdome of blind men the one ey'd is king.
1830 J. L. BURCKHARDT *Arabic Proverbs*
34 The one-eyed person is a beauty in
the country of the blind. **1904** H. G.
WELLS in *Strand* Apr. 405 Through his
thoughts ran this old proverb . . 'In the
Country of the Blind, the One-Eyed
Man is king.' **1937** W. H. SAUMAREZ
SMITH *Letter* 7 Mar. in *Young Man's Country* (1977) ii. You exaggerate the alleged
compliment paid to me by the Bengal
Govt. in wanting to retain my services.
'In the country of the blind the one-eyed
man is king'. **1979** *Guardian* 3 Oct. 11 At
last among the blind the one-eyed man
was king . . . There are men much more
limited than David.

**Happy is the COUNTRY which has no
history**
[**1740** B. FRANKLIN *Poor Richard's Almanack* (Feb.) Happy that Nation,—fortunate that age, whose history is not
diverting.] **1807** T. JEFFERSON *Letter* 29
Mar. in *Writings* (1904) XI. 182 Blest
is that nation whose silent course of
happiness furnishes nothing for history
to say. **1860** G. ELIOT *Mill on Floss* VI. iii.
The happiest women, like the happiest
nations, have no history. **1864** CARLYLE
Frederick the Great IV. XVI. i. Happy the
people whose annals are blank in
history.[1] **1880** BLACKMORE *Mary Anerley*
vi. This land, like a happy country, has
escaped, for years and years, the afflic-
tion of much history. **1957** V. BRITTAIN
Testament of Experience I. iv. Quoting the
familiar dictum: 'Happy is the country
which has no history,' I remarked that I
belonged, like Edward VIII, to a generation which was still on the early side of
middle age but had already seen almost
more history than any generation could
bear. **1981** *Nature* 23 Apr. 698 An old

proverb . . tells us that 'happy is the nation that has no history.' . . DNA . . is the unhappiest of molecules, for it is the subject of innumerable biographies.

¹ attributed here to the French political philosopher, Montesquieu (1689–1755).

country: *see also* GOD made the country, and man made the town; OTHER times, other manners; a PROPHET is not without honour save in his own country.

couple: *see* a DEAF husband and a blind wife are always a happy couple.

The COURSE of true love never did run smooth

1595 SHAKESPEARE *Midsummer Night's Dream* I. i. 134 For aught that I could ever read . . The course of true love never did run smooth. **1836** M. SCOTT *Cruise of Midge* I. xi. 'The course of true love never did run smooth.' And the loves of Saunders Skelp and Jessy Miller were no exception to the rule. **1980** *Tablet* 26 Jan. 89 The course of true love could never run smooth with Sybylla's temperament.

course: *see also* HORSES for courses.

court: *see* HOME is home, as the Devil said when he found himself in the Court of Session.

cover: *see* you can't tell a BOOK by its cover.

coverlet: *see* everyone STRETCHES his legs according to the length of his coverlet.

Why buy a COW when milk is so cheap?

An argument for choosing the least troublesome alternative; frequently used as an argument against marriage. One of the few proverbs in the form of a rhetorical question; cf. *Why* KEEP *a dog, and bark yourself?*

1659 J. HOWELL *Proverbs* p.ii, It is better to buy a quart of Milk by the penny then keep a Cow. **1680** BUNYAN *Mr. Badman* 293 Who would keep a Cow of their own, that can have a quart of milk for a penny? Meaning, Who would be

at the charge to have a Wife, that can have a Whore when he listeth?¹ **1895** S. BUTLER *Note-Books* (1926) xvii. It was cheaper to buy the milk than to keep a cow. **1942** S. ACRE *Yellow Overcoat* v. 'He ain't marryin' . . any more! . . 'Why buy a cow when milk is so cheap, eh?' **1979** S. ALLAN *Mortal Affair* iv. 'Why don't you learn to drive?' 'Oh . . I'm too nervous and sensitive. Besides—' she halted herself in time. When you've already got a cow, why buy milk?²

¹ wishes. ² the form is reversed in this quotation.

cow: *see also* a BELLOWING cow soon forgets her calf; BETTER a good cow than a cow of a good kind; it is idle to SWALLOW the cow and choke on the tail; THREE things are not to be trusted.

coward: *see* a BULLY is always a coward; CONSCIENCE makes cowards of us all.

COWARDS die many times before their death

The popular form is a misquotation of Shakespeare.

[**1596** DRAYTON *Mortimeriados* S1 Every houre he dyes, which ever feares.] **1599** SHAKESPEARE *Julius Caesar* II. ii. 32 Cowards die many times before their deaths: The valiant never taste of death but once. **1800** M. EDGEWORTH *Castle Rackrent* p. xliv, In Ireland, not only cowards, but the brave 'die many times before their death'. **1927** *Sphere* 3 Dec. 414 It is true that cowards die many times before their death, and Noel Coward will come back again and again, and . . win his niche among the great dramatists.

The COWL does not make the monk

Do not judge men by what they appear to be.

[Med. L. *cucullus non facit monachum*, the cowl does not make the monk. *a* **1250** *Ancrene Wisse* (1962) 10 Her in is religiun, nawt i the wide hod ne i the blake cape.] **1387** USK *Testament of Love* in Chaucer *Complete Works* (1897) II. xi. For habit maketh no monk; ne weringe

of gilte spurres maketh no knight. **1588**
GREENE *Pandosto* IV. 289 Trueth quoth
Fawnia, but all that weare Cooles[1] are
not Monkes. **1613** SHAKESPEARE *Henry
VIII* III. i. 23 They should be good men,
their affairs as righteous; But all hoods
make not monks. **1820** SCOTT *Abbot* II.
xi. 'Call me not doctor . . since I have
laid aside my furred gown and bonnet.'
. . 'O, sir . . the cowl makes not the
monk.' **1891** G. B. SHAW *Music in London*
(1932) I. 217 Such impostures are sure of
support from the sort of people . . who
think that it is the cowl that makes the
monk.
  [1] cowls, hooded garments worn by monks.

cradle: *see* the HAND that rocks the cradle
rules the world.

A CREAKING door hangs longest
Usually said as a comfort to the infirm,
though sometimes implying that the
weak or faulty is a nuisance the longest.
Gate sometimes replaces *door*.
  1776 T. COGAN *John Buncle, Junior* I. vi.
They say a creaking gate goes the
longest upon its hinges; that's my comfort. **1888** F. HUME *Madame Midas* II. ii. It
is said that 'creaking doors hang the
longest'. Mrs. Pulchop . . was an excellent illustration of the truth of this
saying. **1944** A. CHRISTIE *Towards Zero*
62 But it seems I am one of these creaking gates—these perpetual invalids who
never die. **1970** L. DEIGHTON *Bomber* vi.
The Flight Engineer said, 'A creaking
door hangs longest.' Digby christened
her [an aeroplane] 'Creaking Door'.

cream: *see* there are more WAYS of killing a
cat than choking it with cream.

Give CREDIT where credit is due
The older form with *honour* (principally
in the sense 'obeisance, homage') is
now rare.
  [Cf. BIBLE (Rheims) *Romans* xiii. 7
Render therfore to al men their dew: . .
to whom honour, honour.] **1777** S.
ADAMS *Letter* 29 Oct. in *Collections of
Massachusetts Hist. Society* (1917) LXXII.
375 May Honor be given to whom Hon-

or may be due. **1834** M. FLOY *Diary* 17
Jan. (1941) 50 Loudon must be a man of
taste . . and disposed to give all credit
where any credit is due. **1894** *Girl's
Own Paper* 6 Jan. 228 The justice and
magnanimity which would show 'honour to whom honour is due' . . is not
always found equal to the occasion
when it involves the granting of a
degree. **1968** M. WOODHOUSE *Rock Baby*
xxii. You aren't half as daft as I thought.
. . Credit where credit's due. **1976**
T. SHARPE *Wilt* viii. 'Some maniac. . .'
'Come now, give credit where credit is
due,' interrupted Dr. Board.

crime: *see* POVERTY is not a crime.

crop: *see* good SEED makes a good crop.

Don't CROSS the bridge till you come to it
Do not concern yourself with difficulties
until they arise. Now also common as
the metaphorical phrase *to cross one's
bridges when one comes to them*.
  1850 LONGFELLOW *Journal* 29 Apr. in
Life (1886) II. 165 Remember the
proverb, 'Do not cross the bridge till you
come to it.' **1895** S. O. ADDY *Household
Tales* xiv. One who anticipates difficulty
is told not to cross the bridge till he gets
to it. **1967** T. STOPPARD *Rosencrantz &
Guildenstern are Dead* II. 43 We cross our
bridges when we come to them and
burn our bridges behind us, with nothing to show for our progress except a
memory of the smell of smoke, and a
presumption that once our eyes wa-
tered.

CROSSES are ladders that lead to heaven
There are two strands to the proverb: in
one, *cross* signifies the crucifix; in the
other, it means 'trouble, misfortune'.
  1616 T. DRAXE *Adages* 36 The Crosse is
the ladder of heauen. **1670** J. RAY *English Proverbs* 6 Crosses are ladders that
do lead to heaven. **1859** S. SMILES *Self-
Help* xi. If there be real worth in the
character . . it will give forth its finest
fragrance when pressed. 'Crosses' says
the old proverb, 'are ladders that lead to

heaven.' **1975** J. O'FAOLAIN *Women in Wall* iv. The cross, they say, is the ladder to heaven and so I have sent your lordship . . two.

crow: *see* (noun) on the FIRST of March, the crows begin to search; HAWKS will not pick out hawks' eyes; ONE for the mouse, one for the crow; (verb) every COCK will crow upon his own dunghill; *also* CROWING.

crowd: *see* TWO is company, but three is none.

crowing: *see* a WHISTLING woman and a crowing hen are neither fit for God nor men.

crown: *see* the END crowns the work.

cry: *see* MUCH cry and little wool.

It is no use CRYING over spilt milk
i.e. when it is too late to remedy the misfortune.
1659 J. HOWELL *Proverbs* (British) 40 No weeping for shed milk. **1738** SWIFT *Polite Conversation* I. 27 'I would cry my Eyes out.' . . 'Tis a Folly to cry for spilt Milk.' **1884** J. PAYN *Canon's Ward* I. xv. There would be a row . . but he would say, like a wise man, 'There's no use in crying over spilt milk.' **1936** M. DE LA ROCHE *Whiteoak Harvest* xxv. It's no use crying over spilt milk. The money's gone . . and that's that. **1979** R. CASSILIS *Arrow of God* IV. xvii. He couldn't be in two places at once. . . It was no use crying over spilt milk, was it, Nanny?

cup: *see* FULL cup, steady hand; the LAST drop makes the cup run over; there's MANY a slip 'twixt cup and lip.

He that will to CUPAR maun to Cupar
1721 J. KELLY *Scottish Proverbs* 141 He that will to Cowper,[1] will to Cowper. A Reflection upon obstinate Persons, that will not be reclaim'd. **1817** SCOTT *Rob Roy* III. i. The Hecate . . ejaculated, 'A wilfu' man will hae his way—them that will to Cupar maun[2] to Cupar!' **1893**

R. L. STEVENSON *Catriona* xiii. He stood part of a second . . , hesitating. 'He that will to Cupar, maun to Cupar,' said he, and . . was hauled into the skiff. **1958** J. CANNAN *And be Villain* v. 'I shall take the first plane to Paris.' . . 'Well, he who will to Cupar maun to Cupar, but I think it's very silly of you.'
[1] i.e. Cupar, a town in Fife. [2] must (Scottish).

cure: *see* PREVENTION is better than cure.

What can't be CURED must be endured
[**1377** LANGLAND *Piers Plowman* B. x. 439 When *must* comes forward, there is nothing for it but to *suffer*. *c* **1408** LYDGATE *Reason & Sensuality* (EETS) l. 4757 For thyng that may nat be eschiwed But of force mot be sywed.[1]] **1579** SPENSER *Shepherd's Calendar* (Sept.) 88 And cleanly couer, that cannot be cured. Such il, as is forced, mought nedes be endured. **1763** C. CHURCHILL *Prophecy of Famine* 18 Patience is sorrow's salve; what can't be cur'd, so Donald right areeds,[2] must be endur'd. **1870** KINGSLEY *Madam How* i. That stupid resignation which some folks preach . . is merely saying—what can't be cured, must be endured. **1936** W. HOLTBY *South Riding* VI. i. We all have our bad turns. What can't be cured must be endured, you know. **1979** J. MASTERS *Now God be Thanked* iii. What can't be cured must be endured . . and then I don't think he'll beat you any more.
[1] followed (said of a misfortune). [2] areads, conjures (archaic).

CURIOSITY killed the cat
Cf. CARE killed the cat.
1921 E. O'NEILL *Diff'rent* II. 252 'What'd you ask 'em, for instance?'. . 'Curiosity killed a cat! Ask me no questions and I'll tell you no lies.' **1939** L. I. WILDER *By Shores of Silver Lake* xvi. 'Whatever are you making, Pa? . . [There was no reply.] Curiosity killed a cat, Pa.' Tantalizing, he sat there whittling. **1973** A. CHRISTIE *Postern of Fate* I. iv. 'A curiosity death,' said Tommy. 'Curiosity killed the cat.'

curried (combed): *see* a SHORT horse is soon curried.

CURSES, like chickens, come home to roost

c **1390** CHAUCER *Parson's Tale* l. 620 And ofte tyme swich cursynge wrongfully retorneth agayn to hym that curseth, as a bryd that retorneth agayn to his owene nest. **1592** *Arden of Feversham* G4 For curses are like arrowes shot upright, Which falling down light on the suters[1] head. **1810** SOUTHEY *Kehama* (title-page), Curses are like young chicken; they always come home to roost. **1880** S. SMILES *Duty* iv. Their injustice will return upon them. Curses, like chickens, come home to roost. **1932** S. GIBBONS *Cold Comfort Farm* vii. Curses, like rookses, flies home to nest in bosomses and barnses.

[1] shooter's.

The CUSTOMER is always right

1928 C. SANDBURG *Good Morning, America* 17 Behold the proverbs of a nation. . . Let one hand wash the other. The customer is always right. **1941** D. LODGE *Death & Taxes* ii. 'I'm drunk.'. . 'You shouldn't do it, George.' 'Business,' he said solemnly. 'The customer is always right.' **1973** 'J. HERRIOT' *Let Sleeping Vets Lie* xxv. You've heard the old saying, 'The customer is always right.' Well I think it's a good working axiom. **1980** *Times* 30 Sept. 9 That the customer is always right is a theory attributed to John Wanamaker, the American retail prince who founded the stores which bear his name.

Don't CUT off your nose to spite your face

A warning against spiteful revenge which results in one's own hurt or loss. The metaphorical phrase *to cut off one's nose to spite*[1] *one's face* is very frequently found.

[Med. L. *male ulciscitur dedecus sibi illatum, qui amputat nasum suum*, he who cuts off his nose takes poor revenge for a shame inflicted on him.] *c* **1560** *Deceit of Women* I1 He that byteth hys nose of, shameth hys face. **1788** GROSE *Dict. Vulgar Tongue* (ed. 2) U3^v He cut off his nose to be revenged of his face, said of one who, to be revenged of his neighbour, has materially injured himself. **1889** R. L. STEVENSON *Master of Ballantrae* x. He was in that humour when a man—in the words of the old adage—will cut off his nose to spite his face. **1964** RIDOUT & WITTING *English Proverbs Explained* 43 Don't cut off your nose to spite your face. **1974** W. FOLEY *Child in Forest* II. vi. After my previous jobs I thought perhaps I had cut off my nose to spite my face, but I felt a kind of glory in my rebellion.

[1] be revenged upon.

CUT your coat according to your cloth

Actions should suit circumstances or resources. Also common as the metaphorical phrase *to cut one's coat according to one's cloth*.

1546 J. HEYWOOD *Dialogue of Proverbs* I. viii. C1, I shall Cut my cote after my cloth. **1580** LYLY *Euphues* II. 188 Be neither prodigall to spende all, nor couetous to keepe all, cut thy coat according to thy cloth. **1778** G. WASHINGTON *Writings* (1936) XIII. 79 General McIntosh . . must . . yield to necessity; that is, to use a vulgar phraze, 'shape his Coat according to his Cloth'. **1873** TROLLOPE *Phineas Redux* II. xxxv. An unselfish, friendly, wise man, who by no means wanted other men to cut their coats according to his pattern. **1951** 'P. WENTWORTH' *Miss Silver comes to Stay* xxxvii. 'You must cut your coat according to your cloth.'. . 'My trouble is that I do like the most expensive cloth.' **1974** T. SHARPE *Porterhouse Blue* iii. I'm afraid the . . exigencies of our financial position do impose certain restraints. . . A case of cutting our coats to suit our cloth.

cut: *see also* (participial adjective) a SLICE off a cut loaf isn't missed; (verb) DIAMOND cuts diamond.

D

daisy: *see* it is not SPRING until you can plant your foot upon twelve daisies.

damned: *see* CORPORATIONS have neither bodies to be punished nor souls to be damned.

They that DANCE must pay the fiddler
Cf. *He who* PAYS *the piper calls the tune*, where the emphasis is reversed. *To pay the piper (fiddler,* etc.) means 'to bear the cost (of an enterprise)'. The proverb is now predominantly found in US use.

 1638 J. TAYLOR *Taylor's Feast* in *Works* (1876) 94 One of the Fidlers said, Gentlemen, I pray you to remember the Musicke,[1] you have given us nothing yet. . . Alwayes those that dance must pay the Musicke. **1837** A. LINCOLN *Speech* 11 Jan. in *Works* (1953) I. 64, I am decidedly opposed to the people's money being used to pay the fiddler. It is an old maxim and a very sound one, that he that dances should always pay the fiddler. *a* **1957** L. I. WILDER *First Four Years* (1971) i. Laura was going to have a baby. . . She remembered a saying of her mother's: 'They that dance must pay the fiddler.'
 [1] company of musicians.

dance: *see also* he that LIVES in hope dances to an ill tune.

danger: *see* the post of HONOUR is the post of danger; OUT of debt, out of danger.

dangerous: *see* DELAYS are dangerous; a LITTLE knowledge is a dangerous thing.

dark: *see* all CATS are grey in the dark.

The DARKEST hour is just before the dawn
 1650 T. FULLER *Pisgah Sight* II. xi. It is always darkest just before the Day dawneth. **1760** in J. Wesley *Journal* (1913) IV. 498 It is usually darkest before day break. You shall shortly find pardon. **1897** J. McCARTHY *Hist. our Own Times* V. iii. Ayoob Khan now laid siege to Candahar. . . As so often happens in the story of England's struggles in India, the darkest hour proved to be that just before the dawn. **1979** R. CASSILIS *Arrow of God* IV. x. I know how you feel. . . It's probably just the darkest-hour-before-the-dawn syndrome.

darling: *see* BETTER be an old man's darling, than a young man's slave.

daughter: *see* like FATHER, like son; like MOTHER, like daughter; my SON is my son till he gets him a wife, but my daughter's my daughter all the days of her life.

dawn: *see* the DARKEST hour is just before the dawn.

As the DAY lengthens, so the cold strengthens
 1631 E. PELLHAM *God's Power* 27 The New Year now begun, as the Days began to lengthen, so the Cold began to strengthen. **1639** J. CLARKE *Parœmiologia Anglo-Latina* 18 As the day lengthens so the cold strengthens. **1721** J. KELLY *Scottish Proverbs* 52 As the Day lengthens the Cold strengthens. It is often found that February and March are much more cold and piercing than December or January. **1899** A. WEST *Recollections* II. xxi. The weather at this time was bearing out the old adage and the cold strengthened as the days lengthened. **1933** L. I. WILDER *Farmer Boy* x. The days were growing longer, but the cold was more intense. Father said: When the days begin to lengthen The cold begins to strengthen. **1978** R. WHITLOCK *Calendar of Country Customs* iii. As the day lengthens, So the cold strengthens, is still a well-known country proverb, applicable to January and early February.

day: *see also* an APPLE a day keeps the doctor away; BARNABY bright, Barnaby bright, the longest day and the shortest night; the BETTER the day, the better the deed; every DOG has his day; FAIR and softly goes far in a day; FISH and guests stink after three days; OTHER times, other manners; ROME was not built in a day; my SON is my son till he gets him a wife, but my daughter's my daughter all the days of her life; SUFFICIENT unto the day is the evil thereof; TOMORROW is another day.

Let the DEAD bury the dead
[MATTHEW viii. 22 Iesus said vnto him, Follow me, & let the dead, bury their dead.] **1815** L. DOW *Hist. Cosmopolite* (1859) 340 A religious bigot made a motion to mob me; but none would second it. A worldling replied to him, 'Let the dead bury their dead.' **1931** J. S. HUXLEY *What dare I Think?* vi. Let, then, the dead bury the dead. The task for us is to rejuvenate ourselves and our subject. **1974** J. O'FAOLAIN *Man in Cellar* 101 That stuff's out of date. . . Our economy is linked to England's. Let the dead bury the dead!

DEAD men don't bite
[Gr. νεϰϱὸς οὐ δάϰνει, a dead man does not bite; ERASMUS *Adages* III. vi. *mortui non mordent*, the dead do not bite.] *a* **1547** E. HALL *Chronicle* (1548) Hen. VI 92ᵛ A prouerbe . . saith, a dead man doth no harme: Sir John Mortimer . . was attainted[1] of treason and put to execucion. *a* **1593** *Jack Straw* F2 I trow[2] they cannot bite when they be dead. **1655** T. FULLER *Church Hist. Britain* IX. iv. The dead did not bite; and, being dispatch'd out of the way, are forgotten. **1883** R. L. STEVENSON *Treasure Island* xi. 'What are we to do with 'em anyway? . . Cut 'em down like that much pork?'. . 'Dead men don't bite,' says he. **1902** A. LANG *Hist. Scotland* II. xii. The story that Gray 'whispered in Elizabeth's ear, The dead don't bite', is found in Camden.

¹ convicted. ² believe.

DEAD men tell no tales
[**1560** T. BECON *Works* II. 97 He that hath his body loden with meat & drinke is no more mete to prai vnto god then a dead man is to tel a tale.] **1664** J. WILSON *Andronicus Comnenius* I. iv. 'Twere best To knock 'um i' th' head. . . The dead can tell no tales. **1702** G. FARQUHAR *Inconstant* v. 76 Ay, ay, Dead Men tell no Tales. **1850** KINGSLEY *Alton Locke* I. iv. Where are the stories of those who have . . ended in desperation? . . Dead men tell no tales. **1979** J. LEASOR *Love & Land Beyond* viii. Dead men told no tales, and the living were not always overcommunicative.

Blessed are the DEAD that the rain rains on
Cf. *Happy is the* BRIDE *that the sun shines on.*
1607 *Puritan* I. i. If, Blessed bee the coarse[1] the raine raynes vpon, he had it, powring downe. **1787** F. GROSE *Provincial Glossary* (Superstitions) 61 It is . . esteemed a good sign if it rains whilst a corpse is burying: . . Happy is the corpse that the rain rains on. **1925** F. S. FITZGERALD *Great Gatsby* 210, I could only remember, without resentment, that Daisy hadn't sent a message or a flower. Dimly I heard someone murmur, 'Blessed are the dead that the rain falls on.'
¹ corpse.

dead: *see also* the only GOOD Indian is a dead Indian; it's ILL waiting for dead men's shoes; a LIVE dog is better than a dead lion; never SPEAK ill of the dead; STONE-dead hath no fellow; THREE may keep a secret, if two of them are dead.

deadly: *see* the FEMALE of the species is more deadly than the male.

There's none so DEAF as those who will not hear
Cf. *There's none so* BLIND *as those who will not see.*
1546 J. HEYWOOD *Dialogue of Proverbs* II. ix. K4 Who is so deafe, or so blynde, as is hee, That wilfully will nother here

nor see? *c* **1570** T. INGELEND *Disobedient Child* C2ᵛ I perceyve by thys geare, That none is so deaf, as who wyll not heare. **1766** in B. Franklin *Papers* (1969) XIII. 18, I have not interfered in this Trial one word, only in my Applications to you and Mr. Foxcraft, both of which turn a deaf Ear: for none so deaf as those who will not hear. **1824** J. BENTHAM *Book of Fallacies* I. V. None are so completely deaf as those who will not hear. **1943** 'P. WENTWORTH' *Chinese Shawl* xvii. You can't put won't for don't, Miss Lucy. There's none so deaf as those who won't hear.

A DEAF husband and a blind wife are always a happy couple

1578 J. FLORIO *First Fruits* 26 There neuer shal be chiding in that house, where the man is blynd, and the wife deafe. **1637** T. HEYWOOD *Pleasant Dialogues* VI. 334 Then marriage may be said to be past in all quietnesse, When the wife is blind, and the husband deafe. **1710** S. PALMER *Proverbs* 338 The Husband must not See, and the Wife must be blind. **1940** H. W. THOMPSON *Body, Boots & Britches* xix. When the wooing is o'er and the maid wed . . the neighbours will observe . . 'A deaf husband and a blind wife are always a happy couple.'

dear: *see* EXPERIENCE keeps a dear school; GOLD may be bought too dear.

dear-bought: *see* FAR-FETCHED and dear-bought is good for ladies.

dearest: *see* BUY in the cheapest market and sell in the dearest.

DEATH is the great leveller

[CLAUDIAN *De Raptu Proserpinae* II. 302 *omnia mors aequat*, death levels all things.] **1732** T. FULLER *Gnomologia* no. 1250 Death is the grand leveller. **1755** YOUNG *Centaur* ii. Is diversion grown a leveller, like death? **1961** M. DICKENS *Heart of London* I. 101 'All this is going to be a great leveller.'. . 'It is death which is the great leveller.'

DEATH pays all debts

[**1597–8** SHAKESPEARE *Henry IV Pt. 1* III. ii. 157 The end of life cancels all bands.[1]] **1611** —— *Tempest* III. ii. 126 He that dies pays all debts. **1827** SCOTT *Two Drovers* in *Chronicles of Canongate* I. xiv. 'It must be sorely answered.'. . 'Never you mind that—Death pays all debts; it will pay that too.' **1979** K. BONFIGLIONI *After You* xvi. I have no particular objection to death as such; it pays all bills.

[1] bonds, obligations.

death: *see also* COWARDS die many. times before their death; NOTHING is certain but death and taxes; there is a REMEDY for everything except death.

debt: *see* DEATH pays all debts; OUT of debt, out of danger; SPEAK not of my debts unless you mean to pay them.

deceptive: *see* APPEARANCES are deceptive.

deed: *see* the BETTER the day, the better the deed.

deep: *see* STILL waters run deep.

defence: *see* ATTACK is the best form of defence.

deferred: *see* HOPE deferred makes the heart sick.

defiled: *see* he that TOUCHES pitch shall be defiled.

DELAYS are dangerous

[*c* **1300** *Havelok* (1915) l. 1352 Dwelling haueth ofte scathe[1] wrouht.] **1578** LYLY *Euphues* I. 212 Delayes breed daungers, nothing so perillous as procrastination. **1655** J. SHIRLEY *Gentlemen of Venice* v. 62 Shall we go presently,[2] delaies are dangerous. **1824** J. FAIRFIELD *Letters* (1922) p.xxxi, I have always found on all subjects that 'delays are dangerous'. . . It is expedient that we marry young. **1930** B. FLYNN *Murder en Route* xxxiii. What a pity Master

Hector left it too late. . . Delays are
proverbially dangerous.
  [1] harm, damage. [2] immediately (archaic).

delved: *see* when ADAM delved and Eve
span, who was then the gentleman?

Derbyshire: *see* YORKSHIRE born and Yorkshire bred, strong in the arm and weak
in the head.

deserve: *see* none but the BRAVE deserve
the fair; one GOOD turn deserves
another.

**DESPERATE diseases must have desper-
ate remedies**
The proverb is found in many variant
forms.
  [L. *extremis malis extrema remedia*, extreme remedies for extreme ills.] **1539**
R. TAVERNER tr. *Erasmus' Adages* 4 A
stronge disease requyreth a stronge
medicine. **1600–1** SHAKESPEARE *Hamlet*
IV. iii. 9 Diseases desperate grown By
desperate appliance are reliev'd, Or not
at all. **1639** J. CLARKE *Parœmiologia
Anglo-Latina* 200 Desperate cuts must
have desperate cures. **1659** J. RUSH-
WORTH *Hist. Collections* I. 120 According
to the usual Proverb, A desperate disease must have a desperate remedy.
1748 RICHARDSON *Clarissa* VI. 292, I
must . . have an interview with the
charmer of my Soul: For desperate diseases must have desperate remedies.
1935 'A. WYNNE' *Toll House Murder* ix.
These circumstances are wholly exceptional. Desperate diseases, they say, call
for desperate remedies. **1961** 'A. GIL-
BERT' *She shall Die* xi. She'd have sold
the roof over her head sooner than have
you know. Desperate situations require
desperate remedies.

destroy: *see* whom the GODS would destroy, they first make mad.

**The DEVIL can quote Scripture for his
own ends**
The proverb alludes to the temptation of
Christ by the Devil (MATTHEW iv.).

1596 SHAKESPEARE *Merchant of Venice*
I. iii. 93 The devil can cite Scripture for
his purpose. An evil soul producing
holy witness Is like a villain with a smiling cheek. **1761** C. CHURCHILL *Apology*
15 Thus Candour's maxims flow from
Rancour's throat, As devils, to serve
their purpose, Scripture quote. **1843**
DICKENS *Martin Chuzzlewit* xi. Is any one
surprised at Mr. Jonas making such a
reference to such a book for such a purpose? Does any one doubt the old saw
that the Devil (being a layman) quotes
Scripture for his own ends.' **1937** 'C.
DICKSON' *Ten Teacups* xiii. The versatile
personage in our popular proverbs,
who . . quotes Scripture for his own
ends. **1955** G. GREENE *Loser takes All* I.
54 His use of a biblical phrase gave her
a touch of shivers, of *diablerie*—the
devil at his old game of quoting
scripture.

**The DEVIL finds work for idle hands to
do**
  [ST. JEROME *Letters* (Migne) no. 125
*fac et aliquid operis, ut semper te diabolus
inveniat occupatum*, do something, so
that the devil may always find you busy;
cf. *c* **1386** CHAUCER *Tale of Melibee* l.
2785 Therfore seith Seint Jerome:
'Dooth somme goode dedes that the de-
vel, which is oure enemy, ne fynde yow
nat unoccupied.'] **1715** I. WATTS *Divine
Songs* 29 In Works of Labour or of Skill I
would be busy too: For Satan finds some
mischief still for idle Hands to do. **1721**
J. KELLY *Scottish Proverbs* 221 If the Devil
find a Man idle, he'll set him on Work.
1792 M. WOLLSTONECRAFT *Vindication of
Rights of Women* ix. There is a homely
proverb, which speaks a shrewd truth,
that whoever the devil finds idle he will
employ. . . What but habitual idleness
can hereditary wealth and titles produce? **1941** A. UPDEGRAFF *Hills look
Down* iv. Better keep busy, and the devil won't find so much for your idle
hands to do. **1980** *Times* 30 June 15
There is risk that youngsters will leave
school and college to find themselves
unwanted, with the devil finding work
for idle hands to do.

Why should the DEVIL have all the best tunes?

Many hymns are sung to popular secular melodies; this practice was especially favoured by the Methodists. The expression is commonly attributed to the English evangelist Rowland Hill (1744–1833).

1859 W. CHAPPELL *Popular Music* II. 748 The Primitive Methodists . . acting upon the principle of 'Why should the devil have all the pretty tunes?' collect the airs which are sung at pot and public houses, and write their hymns to them. **1879** J. E. HOPKINS *Work amongst Working Men* vi. If Wesley could not see why the devil should have all the good tunes, still less should we be able to see why he should have all the good amusements. **1979** *Listener* 31 May 748 *Kids* (Yorkshire Television) makes sure the devil doesn't have the best tunes; he has hardly a note.

The DEVIL is not so black as he is painted

a **1535** MORE *Dialogue of Comfort* (1553) III. xxii. Some saye in sporte, and thinke in earnest: The devill is not so blacke as he is painted. **1642** J. HOWELL *Instructions for Foreign Travel* xiv. The Devill is not so black as he is painted, no more are these Noble Nations and Townes as they are tainted. **1834** MARRYAT *Peter Simple* II. x. Fear kills more people than the yellow fever. . . The devil's not half so black as he's painted. **1928** S. HUDDLESTON *Bohemian Life in Paris* xi. Even if we admit that the devil is as black as he is painted, we cannot admit that everybody is a devil. **1953** A. CHRISTIE *Pocket full of Rye* xxiii. Lance patted her on the arm. 'You didn't believe the devil was as black as he was painted? Well, perhaps he wasn't.'

The DEVIL looks after his own

1606 J. DAY *Isle of Gulls* D4ᵛ You were worse then the devil els, for they say hee helps his Servants. **1721** J. KELLY *Scottish Proverbs* 310 The Dee'ls ay good to his own. . . Spoken when they whom we affect not, thrive and prosper in the World; as if they had their Prosperity from the Devil. **1837** F. CHAMIER *Arethusa* II. i. Weazel was the only midshipman saved besides myself: the devil always takes care of his own. **1940** R. A. J. WALLING *Why did Trethewy Die?* vii. 'The devil looks after his own,' said Pierce. 'Yes, doesn't he? But even he's not so clever, either.' **1979** L. MEYNELL *Hooky & Villainous Chauffeur* xii. I managed; as the saying goes, the devil looks after his own.

The DEVIL's children have the Devil's luck

Cf. **The DEVIL looks after his own.**

1678 J. RAY *English Proverbs* (ed. 2) 126 The Devils child the Devils luck. **1721** J. KELLY *Scottish Proverbs* 333 The Dee'ls Bairns have Dee'ls luck. Spoken enviously when ill People prosper. **1798** LD. NELSON *Letter* 20 July (1845) III. 42 It is an old saying, 'the Devil's children have the Devil's luck.' I cannot find . . where the French Fleet are gone to. **1841** F. CHAMIER *Tom Bowling* II. x. The luck of the fellow! . . Not a leg or an arm missing. . . The devil's children have the devil's luck. **1980** G. RICHARDS *Red Kill* viii. The Devil's son has the Devil's luck. We're going to need that kind of luck.

DEVIL take the hindmost

Cf. **EVERY** *man for himself, and the Devil take the hindmost.*

[HORACE *Ars Poetica* 417 *occupet extremum scabies,* may the itch take the one who is last.] **1620** BEAUMONT & FLETCHER *Philaster* v. i. What if . . they run all away, and cry the Devil take the hindmost. **1725** DEFOE *Everybody's Business* 29 In a few years the navigation . . will be entirely obstructed. . . Every one of these gentlemen-watermen hopes it will last his time, and so they all cry, the Devil take the hindmost. **1824** *Tales of American Landlord* I. ix. The troops . . hurried away . . with a precipitation which seemed to say 'De'il tak the hindmost.' **1953** P. GALLICO *Foolish Immortals* vii. Hannah grew up in . . a land of unlimited resources and opportunity for acquiring them and let the devil take the hindmost. **1980** S. KING *Firestarter* 67 His motto . . had been full speed ahead and devil take the hindmost.

The DEVIL was sick, the Devil a saint would be; the Devil was well, the devil a saint was he!
Promises made in adversity may not be kept in prosperity.
[Med. Lat. *aegrotavit daemon, monachus tunc esse volebat; Daemon convaluit, daemon ut ante fuit,* when the Devil was ill, he wished to be a monk; when the Devil recovered, he was a Devil just as before. **1586** J. WITHALS *Dict.* (rev. ed.) K8 The diuell was sicke and crasie; Good woulde the monke bee that was lasie.] **1629** T. ADAMS *Works* 634 God had need to take what deuotion he can get at our hands in our misery; for when prosperity returnes, wee forget our vowes. . . The Deuill was sicke, the deuill a Monke would be, The Deuill was well, the deuill[1] of [*sic*] Monke was he. **1881** D. C. MURRAY *Joseph's Coat* II. xvii. A prisoner's penitence is a thing the quality of which it is very difficult to judge until you see it . . tried outside. 'The devil was sick.' **1913** H. JAMES *Small Boy* xxviii. The old, the irrepressible adage . . was to live again between them: 'When the devil was sick the devil a saint would be; when the devil was well the devil a saint was he!'

[1] a colloquial use of *devil* expressing strong negation. The last section means that when the Devil was well, he was nothing like the saint he had previously promised to be.

Devil: *see also* set a BEGGAR on horseback, and he'll ride to the Devil; BETTER the devil you know than the devil you don't know; EVERY man for himself, and the Devil take the hindmost; GIVE a thing, and take a thing, to wear the Devil's gold ring; GIVE the Devil his due; where GOD builds a church, the Devil will build a chapel; GOD sends meat, but the Devil sends cooks; what is GOT over the Devil's back is spent under his belly; HASTE is from the Devil; HOME is home, as the Devil said when he found himself in the Court of Session; an IDLE brain is the

Devil's workshop; NEEDS must when the Devil drives; PARSLEY seed goes nine times to the Devil; it is easier to RAISE the Devil than to lay him; he who SUPS with the Devil should have a long spoon; TALK of the Devil, and he is bound to appear; TELL the truth and shame the Devil.

DIAMOND cuts diamond
Used of persons well matched in wit or cunning. Also frequently found as a descriptive phrase *diamond cut diamond*.
[**1593** NASHE *Christ's Tears* II. 9 An easie matter is it for anie man to cutte me (like a Diamond) with mine own dust.] **1604** MARSTON *Malcontent* IV. i. None cuttes a diamond but a diamond. **1629** J. FORD *Lover's Melancholy* I. 18. We're caught in our own toyles. Diamonds cut Diamonds. **1863** C. READE *Hard Cash* II. xi. You might say I robbed you . . . It is diamond cut diamond. **1958** M. STEWART *Nine Coaches Waiting* xi. I'll always have prospects. Diamond cuts diamond. **1979** *Guardian* 19 Apr. 26 *When the boat comes in*: Diamond cut diamond. James Bolam as the rough one turned smoothie.

die: *see* COWARDS die many times before their death; we must EAT a peck of dirt before we die; whom the GODS love die young; the GOOD die young; call no man HAPPY till he dies; he who LIVES by the sword dies by the sword; OLD habits die hard; OLD soldiers never die; YOUNG men may die, but old men must die; *also* DYING.

diet: *see* the best DOCTORS are Dr Diet, Dr Quiet, and Dr Merryman.

differ: *see* TASTES differ.

The DIFFICULT is done at once; the impossible takes a little longer
1873 TROLLOPE *Phineas Redux* II. xxix. What was it the French Minister said. If it is simply difficult it is done. If it is impossible, it shall be done. **1967** H. HARRISON *Technicolor Time Machine* iv. The impossible may take a while,

but we do it, you know the routine. **1981** P. McCUTCHAN *Shard calls Tune* iv. A well-worn precept of the British Navy was that the difficult was done at once; the impossible took a little longer.

difficult: *see also* it is the FIRST step that is difficult.

difficulty: *see* ENGLAND's difficulty is Ireland's opportunity.

DILIGENCE is the mother of good luck
1591 W. STEPNEY *Spanish Schoolmaster* L2ᵛ Diligence is the mother of good fortune. *La diligencia es madre de la buena ventura.* **1736** B. FRANKLIN *Poor Richard's Almanack* (Feb.) Diligence is the mother of good Luck. **1875** S. SMILES *Thrift* ix. Diligence is the mother of good luck. . . A man's success in life will be proportionate to his efforts. **1972** B. EMECHETA *In Ditch* vi. Where do people get a system that allows a man to be better off when out of work? . . People . . used to say that diligence was the mother of fortune.

dinner: *see* AFTER dinner rest a while, after supper walk a mile; BETTER a dinner of herbs than a stalled ox where hate is.

Throw DIRT enough, and some will stick
Persistent slander will eventually pass for truth.
[L. *calumniare fortiter, et aliquid adhaerebit,* slander strongly and some will stick.] **1656** *Trepan* 34 She will say before company, Have you never had the French Pox? speak as in the sight of God: let them Reply what they will, some dirt will stick. **1678** B.R. *Letter to Popish Friends* 7 'Tis a blessed Line in Matchiavel—If durt enough be thrown, some will stick. **1705** E. WARD *Hudibras Redivivus* II. 11 Scurrility's a useful Trick, Approv'd by the most Politic; Fling Dirt enough, And some will stick. **1857** T. HUGHES *Tom Brown's Schooldays* I. viii. Whatever harm a . . venomous tongue could do them, he took care should be done. Only throw dirt enough and some of it is sure to

stick. **1979** C. DEXTER *Service of all Dead* xli. If I tell everybody you're having an affair . . you'd suddenly find yourself trying like hell to prove you weren't. . . Like they say, you throw enough mud and some of it'll stick.

dirt: *see also* we must EAT a peck of dirt before we die.

DIRTY water will quench fire
In quots. 1546 and 1945, used to mean that a man's lust can be satisfied by any woman, however loose or ugly.
1546 J. HEYWOOD *Dialogue of Proverbs* I. v. B2 As this prouerbe saieth, for quenchyng hot desire, Foul water as soone as fayre, wyl quenche hot fire. **1796** W. COBBETT *Political Censor* Sept. 62 That I have made use . . of the British Corporal for a good purpose, I have little doubt—Dirty water will quench fire. **1945** O. ONIONS *Ragged Robin* vi. It's flocks and straw for us. . . Well, dirty water's good enough to quench a fire with.

dirty: *see also* it's an ILL bird that fouls its own nest; don't THROW out your dirty water until you get in fresh; one does not WASH one's dirty linen in public.

disappointed: *see* BLESSED is he who expects nothing, for he shall never be disappointed.

DISCRETION is the better part of valour
[Cf. EURIPIDES *Suppliants* l. 510 καὶ τοῦτ' ἐμοὶ τἀνδρεῖον ἡ προμηθία, and bravery consists in foresight. *c* **1477** CAXTON *Jason* (EETS) 23 Than as wyse and discrete he withdrewe him sayng that more is worth a good retrayte than a folisshe abydinge.] **1597-8** SHAKESPEARE *Henry IV Pt.* 1 V. iv. 121 The better part of valour is discretion; in the which better part, I have saved my life. **1885** C. LOWE *Prince Bismarck* I. v. Napoleon . . had vowed that he would free Italy 'from the Alps to the Adriatic', but . . he acted on the maxim that discretion is the better part of valour. **1961** M. SPARK *Prime of Miss Jean Brodie* iii.

'The apples do not come under my own jurisdiction, but discretion is . . discretion is . . Sandy?' 'The better part of valour, Miss Brodie.'

disease: *see* DESPERATE diseases must have desperate remedies.

disgrace: *see* POVERTY is no disgrace, but it is a great inconvenience.

dish: *see* REVENGE is a dish that can be eaten cold.

dismount: *see* he who RIDES a tiger is afraid to dismount.

dispose: *see* MAN proposes, God disposes.

DISTANCE lends enchantment to the view

　　1799 T. CAMPBELL *Pleasure of Hope* I. 3 Why do those cliffs of shadowy tint appear More sweet than all the landscape smiling near?—'Tis distance lends enchantment to the view, And robes the mountain in its azure hue. **1827** T. HOOD *Poems* (1906) 78 What black Mont Blancs arose, Crested with soot and not with snows. . . I fear the distance did not 'lend enchantment to the view'. **1901** C. FITCH *Captain Jinks* II. 118 'I wish you'd taike me hout of the second row and put me in the front.' . . 'You forget the old adage, . . "Distance lends enchantment."' **1974** T. SHARPE *Porterhouse Blue* xviii. As ever with Lady Mary's affections, distance lent enchantment to the view, and . . she was herself the intimate patroness of this idol of the media.

ditch: *see* when the BLIND lead the blind, both shall fall into the ditch.

DIVIDE and rule

　　A political axiom that government is more easily maintained if factions are set against each other, and not allowed to unite against the ruler.

　　[L. *divide et impera*, divide and rule. **1588** tr. M. Hurault's *Discourse upon Present State of France* 44 It hath been

alwaies her [Catherine de Medici's] custome, to set in France, one against an other, that in the meane while shee might rule in these diuisions.] **1605** J. HALL *Meditations* I. 109 For a Prince . . is a sure axiome, Diuide and rule. **1732** SWIFT *Poems* III. 805 As Machiavel[1] taught 'em, divide and ye govern. **1907** *Spectator* 20 Apr. 605 The cynical maxim of 'Divide and rule' has never clouded our relations with the daughter–States. **1979** D. WILLIAMS *Genesis & Exodus* ii. Matters concerning the estate were put in the hands of a secretary and a steward who were responsible not to Benson but to the Governors. But 'divide and rule' was not in his nature.

　　[1] the Italian political philosopher Niccolo Machiavelli (1469–1527) in fact denounced this principle.

divided: *see* UNITED we stand, divided we fall.

divine: *see* to ERR is human (to forgive divine); on SAINT Thomas the Divine kill all turkeys, geese, and swine.

DO as I say, not as I do

　　[MATTHEW xxiii. 3 Doe not ye after their workes: for they say, and doe not.] *a* **1100** in N. R. Ker *Anglo-Saxons* (1959) 277 Ac theah ic wyrs do thonne ic the lære ne do thu na swa swa ic do, ac do swa ic the lære gyf ic the wel lære [Although I do worse than I teach you, do not do as I do, but do as I teach you if I teach you well]. **1546** J. HEYWOOD *Dialogue of Proverbs* II. v. H4ᵛ It is as folke dooe, and not as folke say. **1689** J. SELDON *Table-Talk* 45 Preachers say, Do as I say, not as I do. **1911** *Spectator* 24 June 957 It has always been considered allowable to say . . to children, 'Do as I say, rather than as I do.' **1979** D. CLARK *Heberden's Seat* v. I saw you spooning sugar into coffee. . . Do as I say, not as I do.

DO as you would be done by

　　See DO *unto others as you would they should do unto you*. The forms are sometimes mixed.

c **1596** A. MUNDAY et al. *Sir Thomas More* 9 A[1] saies trewe: letts do as we may be doon by. **1747** CHESTERFIELD *Letter* 16 Oct. (1932) III. 1035 'Do as you would be done by,' is the surest method that I know of pleasing. **1863** KINGSLEY *Water Babies* v. I shall grow as handsome as my sister . . the loveliest fairy in the world; . . her name is Mrs. Doasyouwouldbedoneby. **1928** 'J. J. CONNINGTON' *Mystery at Lynden Sands* viii. 'Do unto others as you'd be done by' is my motto.

[1] he.

DO right and fear no man

c **1450** *Proverbs of Good Counsel* in *Book of Precedence* (EETS) 68 The beste wysdom that I Can,[1] Ys to doe well, and drede no man. **1721** J. KELLY *Scottish Proverbs* 89 Do well and doubt[2] no Man. But rest satisfied in the Testimony of a good Conscience. **1979** *Guardian* 31 Mar. 10 It used to be, 'Do right and fear no man. Don't write and fear no women.'

[1] know.   [2] fear.

DO unto others as you would they should do unto you

Cf. DO *as you would be done by* above.

[LUKE vi. 31 As yee would that men should doe to you, doe yee also to them likewise.] *a* **901** *Laws of Alfred* in F. Liebermann *Gesetze Angelsachsen* (1903) I. 44 Þæt ge willen, þæt oþre men eow ne don, ne doth ge þæt othrum monnum [What you do not wish others to do to you, do not to other men]. **1477** A. WYDEVILLE *Dicts of Philosophers* 62 Do to other as thou woldest they should do to the, and do to noon other but as thou woldest be doon to. **1790** W. HAZLITT *Letter* 9 July (1979) 48 He wished to have him out, merely because 'he would do to others as he would be done to'. **1880** TROLLOPE *Life of Cicero* II. xii. The lesson which had governed his [Cicero's] life: 'I will do unto others as I would they should do unto me.' **1964** P. VANSITTART *Lost Lands* I. 22 They could perhaps be freed . . *Do unto*

others—said Hillel, the courteous Pharisee.

do: *see also* the KING can do no wrong; whatever MAN has done, man may do; when in ROME, do as the Romans do; if you WANT a thing done well, do it yourself.

doctor: *see* an APPLE a day keeps the doctor away; *also* DOES, DOING, DONE.

The best DOCTORS are Dr Diet, Dr Quiet, and Dr Merryman

[*a* **1449** LYDGATE *Minor Poems* (EETS) 704 Thre lechees[1] consarue a mannys myht, First *a glad hert . . Temperat* diet . . And best of all, for no thyng *take no thouht.*] **1558** W. BULLEIN *Government of Health* 50[v], I should not staye my selfe vpon the opinion of any one phisicion, but rather vpon three. . . The first was called doctor diet, the seconde doctor quiet, the thirde doctor mery man. **1738** SWIFT *Polite Conversation* II. 154 The best Doctors in the World, are Doctor Dyet, Doctor Quiet, and Doctor Merryman. **1909** *Spectator* 30 Jan. 175 A proverb prescribes for sickness Dr. Diet, Dr. Quiet, and Dr. Merryman. The merry heart goes all the way in all but the worst sicknesses.

[1] doctors.

doer: *see* EVIL doers are evil dreaders.

does: *see* he who CAN does, he who cannot teaches; it's DOGGED as does it; EASY does it.

Give a DOG a bad name and hang him

The principle is that a person's plight is hopeless once his reputation has been blackened. Cf. *He that has an* ILL *name is half hanged.*

1706 J. STEVENS *Spanish & English Dict.* s.v. Perro, We say, Give a Dog an ill name and his work is done. **1721** J. KELLY *Scottish Proverbs* 124 Give a Dog an ill Name, and he'll soon be hanged. Spoken of those who raise an ill Name on a Man on purpose to prevent his Advancement. **1803** *Norfolk* (Virginia)

Herald 14 Apr. 3 It is an old saying, 'give a dog a bad name and hang him'. **1928** G. B. SHAW *Intelligent Woman's Guide to Socialism* lvii. The Liberal impulse is almost always to give a dog a bad name and hang him: that is, to denounce the menaced proprietors as enemies of mankind, and ruin them in a transport of virtuous indignation. **1979** E. KYLE *Summer Scandal* xiii. 'A man is innocent till he's proved guilty. That's English law, isn't it? 'This is Scotland. . . Give a dog a bad name—.'

DOG does not eat dog

[VARRO *De Lingua Latina* VII. 32. *canis caninam non est*, a dog does not eat dog's flesh.] **1543** W. TURNER *Hunting of Romish Fox* A2ᵛ The prouerb . . on dog will not eat of an other dogges fleshe. **1602** SHAKESPEARE *Troilus & Cressida* v. vii. 19 One bear will not bite another, and wherefore should one bastard? **1790** 'P. PINDAR' *Epistle to Bruce* 31 Dog should not prey on dog, the proverb says: Allow then brother-trav'lers crumbs of praise. **1866** KINGSLEY *Hereward the Wake* II. xi. Dog does not eat dog and it is hard to be robbed by an Englishman, after being robbed a dozen times by the French. **1933** F. D. GRIERSON *Empty House* viii. Dog doesn't eat dog, my dear fellow. To put it more politely, the physician attends his brother practitioner without charge. **1979** *Guardian* 26 May 8 Mr Donnet's assault on Mr Fisher breaks an honoured union tradition in which dog is not expected to eat dog.

Every DOG has his day

1545 R. TAVERNER tr. *Erasmus' Adages* (ed. 2) 63 A dogge hath a day. **1600–1** SHAKESPEARE *Hamlet* v. i. 286 Let Hercules himself do what he may, The cat will mew, and dog will have his day. **1611** R. COTGRAVE *Dict. French & English* s.v. Fevrier, Euerie dog hath his day. **1726** POPE *Odyssey* V. xxii. Dogs, ye have had your day; ye fear'd no more Ulysses vengeful from the Trojan shore. **1837** CARLYLE *French Revolution* III. I. i. How changed for Marat, lifted

from his dark cellar! . . All dogs have their day; even rabid dogs. **1863** KINGSLEY *Water Babies* ii. Young blood must have its course, lad, And every dog his day. **1978** 'M. CRAIG' *Were He Stranger* x. 'She could be his sister.' 'No way— not with a face like that.' 'Well, every dog deserves his day.'

Every DOG is allowed one bite

The proverb is based on the common law rule (dating at least from the seventeenth century) by which the keeper of a domestic animal was not liable for harm done by it unless he knew of its vicious propensities. The first time a dog bit someone might not be evidence of its wildness.

1902 V. S. LEAN *Collectanea* I. 439 Every dog is allowed his first bite i.e. is not punished. **1913** *Spectator* 15 Mar. 440 Every dog is allowed by the law one free bite. After the dog has once bitten a person it is presumed that its owner knows it to be 'savage'. **1968** P. FOOT *Politics of Harold Wilson* x. In March 1967 . . Wilson rounded on the Left at a Parliamentary Party meeting, warning them that 'a dog is only allowed one bite' and threatening them with a General Election unless they came to heel. **1980** 'A. BLAISDELL' *Consequence of Crime* (1981) ii. She got arrested. . . They say every dog allowed one bite. . . But it was a vice thing. . . I told her to get out.

The DOG returns to its vomit

The expression is frequently found in various metaphorical and allusive forms, such as *to return like a dog to his vomit*, as illustrated below. Before 1534 (see quot.), the proverb is used in more or less similar forms in earlier versions of, and commentaries upon, the Bible. The concept enjoyed wide popularity in the Middle Ages.

[PROVERBS xxvi. 11 As a dogge returneth to his vomite: so a foole returneth to his folly; 2 PETER ii. 22 (see below).] c **1390** CHAUCER *Parson's Tale* l. 138 Ye trespassed so ofte tyme as dooth the hound that retourneth to eten his spewyng. **1534** W. TYNDALE tr. *Bible* 2

Peter ii. 22 It is happened vnto them accordinge to the true proverbe: The dogge is turned to his vomet agayne. **1832** S. WARREN *Diary of Late Physician* II. vi. His infatuated wife betook herself—'like a dog to his vomit' . . —to her former . . extravagance and dissipation. **1981** P. McCUTCHAN *Shard calls Tune* xvi. The old saying that the dog returns to his vomit, the criminal to the scene of his crime.

A DOG that will fetch a bone will carry a bone
A gossip carries talk both ways.

1830 R. FORBY *Vocabulary of East Anglia* 429 'The dog that fetches will carry.'— i.e. A talebearer will tell tales *of* you, as well as *to* you. **1941** L. I. WILDER *Little Town on Prairie* xv. So Nellie twisted what you said and told it to Miss Wilder . . . 'A dog that will fetch a bone, will carry a bone.'

dog: *see also* a BARKING dog never bites; BRAG is a good dog, but Holdfast is better; the CAT, the rat, and Lovell the dog, rule all England under the hog; he is a GOOD dog who goes to church; why KEEP a dog and bark yourself?; if you LIE down with dogs, you will get up with fleas; a LIVE dog is better than a dead lion; LOVE me, love my dog; let SLEEPING dogs lie; it is easy to find a STICK to beat a dog; you cannot TEACH an old dog new tricks; THREE things are not to be trusted; there are more WAYS of killing a dog than choking it with butter; there are more WAYS of killing a dog than hanging it; a WOMAN, a dog, and a walnut tree, the more you beat them the better they be; *also* DOGS.

It's DOGGED as does it
Similar in form to the expression EASY *does it.*

1864 M. B. CHESNUT *Diary* 6 Aug. (1949) 429 'It's dogged as does it,' says Isabella. **1867** TROLLOPE *Last Chronicle of Barset* lxi. There ain't nowt a man can't bear if he'll only be dogged. . . It's dogged as does it. It's not thinking about it. **1916** J. BUCHAN *Greenmantle* i. We've

got the measure of the old Boche now, and it's dogged as does it. **1942** N. MARSH *Death & Dancing Footman* x. 'If we stick . . they can damn well produce a farm animal to lug us out.' . . 'It's dogged as does it,' said Chloris.

DOGS bark, but the caravan goes on
1924 C. K. SCOTT MONCRIEFF tr. Proust's *Within Budding Grove* I. 45 In the words of a fine Arab proverb, 'The dogs may bark; the caravan[1] goes on!' . . Its effect was great, the proverb being familiar to us already. It had taken the place, that year, among people who 'really counted', of 'He who sows the wind shall reap the whirlwind.' **1930** *Time* 4 July 17, I was struggling to explain the situation to an old Moor. . . After thinking it over he murmured: 'Dogs bark but the caravan goes on.' **1979** A. FOX *Threat Warning Red* ii. Just a few tantrums in the Commons, dogs barking while the caravan rolled on.

[1] a company of people travelling through the desert.

doing: *see* if a THING's worth doing, it's worth doing well.

What's DONE cannot be undone
Also found in the more casual form *what's done is done.* Cf. *Things* PAST *cannot be recalled.*

[Cf. SOPHOCLES *Ajax* l. 378 οὐ γὰρ γένοιτ' ἂν ταῦθ' ὅπως οὐχ ὧδ' ἔχειν, things could not now be otherwise.] *c* **1450** *King Ponthus* in *Publications of Modern Language Association of America* (1897) XII. 107 The thynges that be doone may not be undoone. **1546** J. HEYWOOD *Dialogue of Proverbs* I. x. C4 Things done, can not be undoone. **1605–6** SHAKESPEARE *Macbeth* III. ii. 12 Things without all remedy Should be without regard. What's done is done. *Ibid.* v. i. 65 What's done cannot be undone. **1791** G. WASHINGTON *Letter* 1 Dec. in *Writings* (1939) XXXI. 433 What has been done cannot be undone, and it would be unfortunate . . if disputes amongst the friends of the federal City should Arm the enemies of it with weapons to wound it. **1818** S. FERRIER *Marriage* III. xxi. I hope you will think twice about it. Second thoughts are best. What's done cannot be undone. **1967** H. HARRISON *Technicolor Time Machine* vii. What's done is done . . I'll see you don't suffer for it. **1981** J. STUBBS *Ironmaster* xxii. What is done cannot be undone . . but you must take responsibility for it.

done: *see also* DO as you would be done by; whatever MAN has done, man may do; NOTHING should be done in haste but gripping a flea; WELL begun is half done.

A DOOR must either be shut or open

Said of two mutually exclusive alternatives.

[Cf. Fr. *il faut qu'une porte soit ouverte ou fermée*, it is necessary that a door be open or shut.] **1762** GOLDSMITH *Citizen of World* I. xlix. There are but the two ways; the door must either be shut, or it must be open. **1896** G. SAINTSBURY *Hist. Nineteenth-Century Literature* vii. Fiction . . pleads in vain for detailed treatment. For all doors must be shut or open; and this door must now be shut.

door: *see also* a CREAKING door hangs longest; a GOLDEN key can open any door; when ONE door shuts, another opens; OPPORTUNITY never knocks twice at any man's door; a POSTERN door makes a thief; when POVERTY comes in at the door, love flies out of the window; it is too late to shut the STABLE-door after the horse has bolted.

doorstep: *see* if every man would SWEEP his own doorstep the city would soon be clean.

When in DOUBT, do nowt

1884 G. WEATHERLY *'Little Folks' Proverb Painting Book* 64 Err ever on the side that's safe, And when in doubt, abstain. **1917** J. C. BRIDGE *Cheshire Proverbs* 155 When in doubt, do nowt.[1] This shows the cautious Cheshireman at his best. **1972** E. GRIERSON *Confessions of Country Magistrate* vii. 'When in doubt say nowt' is a precept enshrined over most magistrates' courts. **1981** E. AGRY *Assault Force* i. What to do? . . 'When in doubt, do nowt,' had always been my grandfather's advice.

[1] i.e. nought; nothing.

Whosoever DRAWS his sword against the prince must throw the scabbard away

Whoever seeks to assassinate or depose a monarch must remain constantly prepared to defend himself, and his sword will never be able to return to its scabbard.

1604 R. DALLINGTON *View of France* F3[v] His King, against whom when yee drawe the sword, ye must throw the scabberd into the riuer. **1659** J. HOWELL *Proverbs* (English) 17 Who draweth his sword against his Prince, must throw away the scabbard. **1962** S. E. FINER *Man on Horseback* viii. [The Military] must still fear the results of a fall from power. . . 'Whosoever draws his sword against the prince must throw the scabbard away' . . pithily express[es] the logic of the situation.

dread: *see* a BURNT child dreads the fire.

dreader: *see* EVIL doers are evil dreaders.

DREAM of a funeral and you hear of a marriage

The proverb is also found in the reverse form.

1639 J. CLARKE *Parœmiologia Anglo-Latina* 236 After a dreame of weddings comes a corse.[1] **1766** GOLDSMITH *Vicar of Wakefield* x. My wife had the most lucky dreams in the world. . . It was one night a coffin and cross-bones, the signs of an approaching wedding. **1883** C. S. BURNE *Shropshire Folklore* xx. We have the sayings . . 'Dream of a funeral, hear of a wedding' . . and vice versa. **1909** *British Weekly* 8 July 331 'Dream of a funeral and you hear of a marriage' . . has probably been verified many times in the experience of ordinary people.

[1] corpse.

dream: *see also* MORNING dreams come true.

DREAMS go by contraries

c **1400** *Beryn* (EETS) l. 108 Comynly of these swevenys[1] the contrary man shul fynde. **1584** LYLY *Sappho & Phao* IV. iii. I dreamed last night, but I hope dreams are contrary, that . . all my hair blazed on a bright flame. **1673** W. WYCHERLEY *Gentleman Dancing-Master* IV. 64 Ne're fear it, dreams go by the contraries. **1731** FIELDING *Grub-Street Opera* I. xi. Oh! the perjury of men! I find dreams do not always go by contraries. **1860** T. C. HALIBURTON *Season-Ticket* 30 The events of life, like dreams, appear in the words of the old proverb, 'to go by contraries'. **1932** J. H. WALLIS *Capital City Murder* iv. There was no sign . . of Lester Armande. 'Dreams go by contraries,' said Lily. **1973** 'P. SIMPLE' *Stretchford Chronicles* (1980) 198 They say dreams go by opposites. . . Perhaps you'll dream about that AA man again.

[1] dreams.

drink: *see* you can take a HORSE to the water, but you can't make him drink.

A DRIPPING June sets all in tune

1742 *Agreeable Companion* 35 A dripping June Brings all Things in Tune. **1883** W. ROPER *Weather Sayings* 22 A dry May and a dripping June brings all things in tune. **1912** *Spectator* 28 Dec. 1094 'A dripping June sets all in tune,' and on sandy soils not only farm crops but garden flowers do best in a wet summer.

DRIVE gently over the stones

1711 SWIFT *Letter* 30 June in *Journal to Stella* (1948) I. 301 A gallop: sit fast, sirrah, and don't ride hard upon the stones. **1844** DICKENS *Martin Chuzzlewit* xxix. Gently over the stones, Poll. Go a-tiptoe over the pimples! Poll . . scraped the lather off again with particular care. **1885** E. J. HARDY *How to be Happy though Married* xi. Drive gently over the stones! This piece of advice . . given to inexperienced whips, may be

suggested metaphorically to the newly-married.

You can DRIVE out Nature with a pitchfork, but she keeps on coming back

[HORACE *Epistles* I. x. *naturam expelles furca, tamen usque recurret,* you may drive out nature with a pitchfork, but she will always return.] **1539** R. TAVERNER tr. *Erasmus' Adages* 44 Thurst out nature wyth a croche,[1] yet woll she styll runne backe agayne. **1831** T. L. PEACOCK *Crotchet Castle* i. Mr. Crotchet . . seemed . . to settle down . . into an English country gentleman . . But as, though you expel nature with a pitchfork, she will always come back. **1867** J. A. FROUDE *Short Studies* II. 252 Drive out nature with a fork, she ever comes running back. **1927** 'D. YATES' *Blind Corner* vi. There's a proverb which is rather in point. 'You can drive Nature out with a pitchfork, *but she'll always come back.*' I know you're using a bucket instead. But . . the result is the same. **1980** C. GAVIN *How sleep Brave* xiv. There was feminine logic for you! . . 'You can drive out Nature with a pitchfork, . . but she keeps on coming back.'

[1] crook, staff.

drive: *see also* BAD money drives out good; HUNGER drives the wolf out of the wood; NEEDS must when the Devil drives; ONE nail drives out another.

drop: *see* the LAST drop makes the cup run over.

dropping: *see* CONSTANT dropping wears away a stone.

drowned: *see* if you're BORN to be hanged then you'll never be drowned.

A DROWNING man will clutch at a straw

One grabs at the slightest chance when all hope is slipping away. *To clutch at a straw* (or *straws*) is frequently used as a metaphorical phrase.

1534 MORE *Dialogue of Comfort* (1553) iii. Lyke a man that in peril of drowning catcheth whatsoeuer cometh next to hand . . be it neuer so simple a sticke. **1583** J. PRIME *Fruitful & Brief Discourse* I. 30 We do not as men redie to be drowned, catch at euery straw. **1623** J. HALL *Contemplations* VII. XIX. 252 The drowning man snatches at every twig. . . The messengers of Benhadad catch hastily at that stile of grace, and hold it fast. **1748** RICHARDSON *Clarissa* VII. i. A drowning man will catch at a straw, the Proverb well says. **1829** P. EGAN *Boxiana* 2nd Ser. II. 721 Drowning men will catch at straws, but it is a pity Crosbie practised his stale trick of crying 'Foul'. **1915** CONRAD *Victory* IV. viii. Wang seemed to think my insistence . . very stupid and tactless. But a drowning man clutches at straws. **1967** RIDOUT & WITTING *English Proverbs Explained* 49 A drowning man will clutch at a straw. **1967** T. STOPPARD *Rosencrantz & Guildenstern are Dead* III. 80 We drift down time, clutching at straws. But what good's a brick to a drowning man?

drunken: *see* CHILDREN and fools tell the truth; HEAVEN protects children, sailors, and drunken men.

druv (drove, driven): *see* SUSSEX won't be druv.

dry: *see* you never MISS the water till the well runs dry; put your TRUST in God and keep your powder dry.

due: *see* give CREDIT where credit is due; GIVE the Devil his due.

dumb: *see* a CHERRY year a merry year.

dunghill: *see* every COCK will crow upon his own dunghill.

dust: *see* a PECK of March dust is worth a king's ransom.

duty: *see* the FIRST duty of a soldier is obedience.

dying: *see* you cannot SHIFT an old tree without it dying.

E

eagle: *see* where the CARCASE is, there shall the eagles be gathered together.

EAGLES don't catch flies

Great or important people do not concern themselves with trifling matters or insignificant people.

[ERASMUS *Adages* III. ii. *aquila non captat muscas*, the eagle does not catch flies.] **1563** *Mirror for Magistrates* (1938) 405 The iolly Egles catche not little flees. **1581** G. PETTIE tr. *S. Guazzo's Civil Conversation* II. 48ᵛ That is the right act of a Prince, and therefore it is well saide, That the Egle catcheth not flies. **1786** H. L. PIOZZI *Anecdotes of Johnson* 185 With regard to slight insults . . 'They sting one (says he) but as a fly stings a horse; and the eagle will not catch flies.' **1942** H. C. BAILEY *Nobody's Vineyard* i. 'Eagles don't catch flies.' 'What do you mean?' 'Inspectors of Police don't trail urchins.' **1980** A. T. ELLIS *Birds of Air* 93 'Aquila non captat muscas,' she told him in a reassuring whisper. 'Eagles don't catch flies.'

ear: *see* FIELDS have eyes, and woods have ears; LITTLE pitchers have large ears; you can't make a SILK purse out of a sow's ear; WALLS have ears.

The EARLY bird catches the worm

1636 W. CAMDEN *Remains concerning Britain* (ed. 5) 307 The early bird catcheth the worme. **1859** H. KINGSLEY *Geoffrey Hamlyn* II. xiv. Having worked . . all the week . . a man comes into your room at half-past seven . . and informs you that the 'early bird gets the worm'. **1935** 'N. BLAKE' *Question of Proof* xii. You're very skittish this morning, superintendent. I shall always know now what the early bird looks like when it has caught the worm. **1979** G. WATSON *Black Jack* vii. The early bird catches the worm, Dr. Ellis. We have been looking for you for a long time.

The EARLY man never borrows from the late man

1659 J. HOWELL *Proverbs* (English) 17 The rath[1] sower never borroweth of the late. **1732** T. FULLER *Gnomologia* no. 4492 The early Sower never borrows of the Late. **1978** R. WHITLOCK *Calendar of Country Customs* iii. Oats, too, benefit from early sowing. . . Another agricultural proverb . . declares that, 'the early man never borrows from the late man'.

[1] early.

EARLY to bed and early to rise, makes a man healthy, wealthy, and wise

1496 *Treatise of Fishing with Angle* H1 As the olde englysshe prouerbe sayth in this wyse. Who soo woll ryse erly shall be holy helthy and zely.[1] **1523** J. FITZHERBERT *Husbandry* (1530 52ᵛ At gramer scole I lerned a verse, . . erly rysynge maketh a man hole in body, holer in soule, and rycher in goodes. **1639** J. CLARKE *Parœmiologia Anglo-Latina* 91 Earley to bed and earely to rise, makes a man healthy, wealthy, and wise. **1853** R. S. SURTEES *Sponge's Sporting Tour* ix. Early to bed and early to rise being among Mr. Sponge's maxims, he was enjoying the view . . shortly after daylight. **1979** T. WISEMAN *Game of Secrets* iv. 'I wake you?' 'You did, Walter.' 'Early to bed, early to rise—.'

[1] happy, fortunate.

earned: *see* a PENNY saved is a penny earned.

easier: *see* it is easier to PULL down than to build up; it is easier to RAISE the Devil than to lay him; *also* EASY.

EAST, west, home's best

Cf. *There's no* PLACE *like home.*

1859 W. K. KELLY *Proverbs of all Nations* 36 'East and west, at home the best' (German). . . *Ost und West, daheim das Best.* **1869** C. H. SPURGEON *John Plough-*

man's Talk xiii. East and west, Home is best. **1920** E. V. LUCAS *Verena in Midst* cxiii. None the less I don't envy the traveller. 'East, west, home's best.' **1979** *Times* 21 Dec. 9 All the most prosaic and conservative morals: East, West, home's best; . . nothing succeeds like success.

east: *see also* when the WIND is in the east, 'tis neither good for man nor beast.

EASY come, easy go
Cf. LIGHT *come, light go;* QUICKLY *come, quickly go.*
   1650 A. BRADSTREET *Tenth Muse* 126 That which easily comes, as freely goes. **1832** S. WARREN *Diary of Late Physician* II. xi. 'Easy come, easy go' is . . characteristic of rapidly acquired commercial fortunes. **1960** I. JEFFERIES *Dignity & Purity* ii. She's your only daughter, isn't she. . . Well, easy come, easy go.

EASY does it
Also *gently does it.* Similar in form to *It's* DOGGED *as does it.*
   1863 T. TAYLOR *Ticket-of-Leave Man* IV. i. Easy does it, Bob. Hands off, and let's take things pleasantly. **1928** J. P. McEVOY *Showgirl* 21 No high pressure stuff, sis. Easy does it with Dick. **1955** L. P. HARTLEY *Perfect Woman* xxx. I won't ask her yet what's the matter he decided. Easy does it. **1972** A. PRICE *Colonel Butler's Wolf* iii. Easy does it—the nails are big, but they are old and brittle. **1981** S. RUSHDIE *Midnight's Children* I. 103 Important to build bridges . . between the faiths. Gently does it.

easy: *see* it is easy to find a STICK to beat a dog; it is easy to be WISE after the event; *also* EASIER.

We must EAT a peck of dirt before we die
A consolatory expression, frequently used in literal contexts.
   [**1603** H. CHETTLE et al. *Patient Grisel* A3ᵛ I thinke I shall not eate a pecke[1] of salt: I shall not liue long sure.] **1738** SWIFT *Polite Conversation* I. 48 'Why

then, here's some Dirt in my Tea-cup.' . . 'Poh! you must eat a Peck of Dirt before you die.' **1819** J. KEATS *Letter* 19 Mar. (1939) 339 This is the second black eye I have had since leaving school. . . We must eat a peck before we die. **1939** F. THOMPSON *Lark Rise* vi. The children . . were told: 'Us've all got to eat a peck o' dust before we dies, an' it'll slip down easy in this good yarb[2] beer.' **1979** M. BABSON *Twelve Deaths of Christmas* xxix. She tried to rinse off the ice cubes. 'Never mind. . . They say, we all have to eat a peck of dirt before we die!'
   [1] a dry measure of two gallons.   [2] herb.

He that would EAT the fruit must climb the tree
   [**1577** J. GRANGE *Golden Aphroditis* M1 Who will the fruyte that haruest yeeldes, must take the payne.] **1721** J. KELLY *Scottish Proverbs* 141 He that would eat the Fruit must climb the Tree. **1843** 'R. CARLTON' *New Purchase* I. xxiv. It is a proverb, 'He that would eat the fruit must first climb the tree and get it': but when that fruit is honey, he that wants it must first cut it down. **1970** V. CANNING *Great Affair* xiv. 'Son, are you one of those who like to eat the fruit and then walk away from the tree?' 'I want to marry her.'

EAT to live, not live to eat
   [Attributed to SOCRATES (see Diogenes Laertius *Socrates* ii.) ἔλεγέ τε τοὺς μέν ἄλλους ἀνθρώπους ζῆν ἐσθίοιεν; αὐτός δὲ ἐσθίειν ἵνα ζώη, and he said that other men live to eat, but he eats to live; CICERO *Rhetoricorum* IV. vii. *edere oportet ut vivas, non vivas ut edas,* one must eat to live, not live to eat.] **1387** J. TREVISA tr. *Higden's Polychronicon* (1871) III. 281 Socrates seide that meny men wil leve forto ete and drynke, and that they wolde ete and drynke . . forto lyve. *c* **1410** in *Secreta Secretorum* (1898) 67, I will ete so that y leue, and noght lyf that y ete. **1672** T. SHADWELL *Miser* 46 Eat to live, not live to eat; as the Proverb says. **1912** A. W. PINERO *Preserving Mr. Panmure* II. 85, I shall eat sufficient. . . But I eat to live; I don't live to eat.

eat: *see also* the CAT would eat fish, but would not wet her feet; DOG does not eat dog; you cannot HAVE your cake and eat it; if you won't WORK you shan't eat.

eaten: *see* REVENGE is a dish that can be eaten cold.

eating: *see* APPETITE comes with eating; the PROOF of the pudding is in the eating.

egg: *see* you cannot make an OMELETTE without breaking eggs; there is REASON in the roasting of eggs; don't TEACH your grandmother to suck eggs.

Don't put all your EGGS in one basket
Don't chance everything on a single venture, but spread your risk. *To put all one's eggs in one basket* is commonly used as a metaphorical phrase.

   1662 G. TORRIANO *Italian Proverbial Phrases* 125 To put all ones Eggs in a Paniard, *viz.* to hazard all in one bottom.[1] **1710** S. PALMER *Proverbs* 344 Don't venture all your Eggs in One Basket. **1894** J. LUBBOCK *Use of Life* iii. Do not put too many eggs in one basket. However well you may be advised . . something may occur to upset all calculations. **1967** RIDOUT & WITTING *English Proverbs Explained* 46 Don't put all your eggs in one basket. **1979** J. BARNETT *Backfire is Hostile* x. When you are an older and wiser man . . you'll learn not to put all your eggs in one basket.
   [1] a ship's hull; a boat.

eight: *see* SIX hours' sleep for a man, seven for a woman, and eight for a fool.

eleven: *see* POSSESSION is nine points of the law; RAIN before seven, fine before eleven.

Every ELM has its man
   [**1906** KIPLING *Puck of Pook's Hill* 32 Ellum she hateth mankind, and waiteth Till every gust be laid To drop a limb on the head of him That any way trusts her shade.] **1928** *Times* 29 Nov. 10 Owing to the frequency with which this tree sheds its branches, or is uprooted in a

storm, it has earned for itself a sinister reputation. 'Every elm has its man' is an old country saying.

EMPTY sacks will never stand upright
Extreme need makes survival impossible. Also used as an argument for taking food or drink (see last quotation).

   1642 G. TORRIANO *Select Italian Proverbs* 90 *Sacco vuoto non può star in piedi.* An emptie sack cannot stand upright: nota, Applied to such as either pinch themselves, or are pincht by hard fortune. **1758** B. FRANKLIN *Poor Richard's Almanack* (Introduction) Poverty often deprives a Man of all Spirit and Virtue; 'Tis hard for an empty Bag to stand upright. **1860** G. ELIOT *Mill on Floss* I. i. viii. There's folks as things 'ull allays go awk'ard with: empty sacks 'ull never stand upright. *a* **1895** F. LOCKER-LAMPSON *My Confidences* (1896) 395 Gibbs . . by this artifice . . made a hundred per cent. . . Gibbs was a needy man, and . . would often say, 'It's hard for an empty sack to stand upright.' **1978** J. McGAHERN *Getting Through* 99 'Give this man something.' . . 'A cup of tea will do fine,' he had protested. . . 'Nonsense. . . Empty bags can't stand.'

EMPTY vessels make the most sound
Foolish or witless persons are the most talkative or noisy; also used of achievement (see quot. 1932).

   a **1430** LYDGATE *Pilgrimage of Man* (EETS) l. 15933 A voyde vessel[1] . . maketh outward a gret soun, Mor than . . what yt was ful. **1547** W. BALDWIN *Treatise of Moral Philosophy* IV. Q4 As emptye vesselles make the lowdest sounde: so they that haue least wyt, are the greatest babblers. **1599** SHAKESPEARE *Henry V* IV. iv. 64, I did never know so full a voice issue from so empty a heart: but the saying is true— The empty vessel makes the greatest sound. **1707** SWIFT *Essay on Faculties of Mind* I. 249 Empty Vessels sound loudest. **1932** W. McFEE *Harbourmaster* ii. 'You think we don't show up too well in the test?' . . 'Oh, dear, no. It is the empty vessel that makes the most noise.' **1967** RIDOUT & WITTING *English Proverbs Explained* 51 Empty vessels make the most sound.

¹ i.e. a receptacle for liquid, etc., as a drinking-vessel.

empty: *see* don't THROW the baby out with the bathwater.

enchantment: *see* DISTANCE lends enchantment to the view.

The END crowns the work

[L. *finis coronat opus,* the end crowns the work; 15th-cent. Fr. *la fin loe l'oeuvre,* the end praises the work.] **1509** H. WATSON *Ship of Fools* Dd1 For the ende crowneth. **1592** G. DELAMOTHE *French Alphabet* II. 29 The end doth crowne the worke. . . *La fin couronne l'œuure.* **1602** SHAKESPEARE *Troilus & Cressida* IV. v. 223 The end crowns all; And that old common arbitrator, Time, Will one day end it. **1820** SCOTT *Abbot* I. xiii. As the end crowns the work, it also forms the rule by which it must be ultimately judged. **1870** DICKENS *Edwin Drood* xviii. Proof, sir, proof, must be built up stone by stone. . . As I say, the end crowns the work. **1942** T. JOB *Uncle Harry* I. 5 But the end crowns the work, Mr. Jenkins. Murderers, like artists, must be hung to be appreciated.

The END justifies the means

The negative of this is often asserted.

[OVID *Heroides* ii. *exitus acta probat,* the outcome justifies the deeds.] **1583** G. BABINGTON *Exposition of Commandments* 260 The ende good, doeth not by and by make the meanes good. **1718** M. PRIOR *Literary Works* (1971) I. 186 The End must justify the Means: He only Sins who Ill intends. **1941** 'H. BAILEY' *Smiling Corpse* 238 'The police don't like to have their bodies moved.' . . 'In this case the end justifies the means.' **1979** O. SELA *Petrograd Consignment* 65 Ulyanov has demonstrated that the end justifies the means.

end: *see also* (noun) ALL good things must come to an end; the DEVIL can quote Scripture for his own ends; EVERYTHING has an end; he who WILLS the end, wills the means; (verb) all's WELL that ends well.

ending: *see* a GOOD beginning makes a good ending.

endured: *see* what can't be CURED must be endured.

enemy: *see* the BEST is the enemy of the good; it is good to make a BRIDGE of gold to a flying enemy; the GOOD is the enemy of the best; there is no LITTLE enemy; SAVE us from our friends.

ENGLAND's difficulty is Ireland's opportunity

1856 *Tribune* 19 Jan. 188 Some few years ago, we followed O'Connell, and when he declared that 'England's difficulty is Ireland's opportunity', we threw our hats in the air. **1916** G. B. SHAW in *New York Times* (Magazine) 9 Apr. 2 The cry that 'England's Difficulty Is Ireland's Opportunity' is raised in the old senseless, spiteful way as a recommendation to stab England in the back when she is fighting some one else. **1969** T. PAKENHAM *Year of Liberty* i. Successive plantations—of Scottish Presbyterians in Ulster . . did not secure Ireland. The Catholics' watchword remained: 'England's difficulty is Ireland's opportunity.'

England: *see also* the CAT, the rat, and Lovell the dog, rule all England under the hog; what MANCHESTER says today, the rest of England says tomorrow; TURKEY, heresy, hops, and beer came into England all in one year.

The ENGLISH are a nation of shopkeepers

Attributed to Napoleon.

[**1766** J. TUCKER *Letter from Merchant* 46 A Shop-keeper will never get the more Custom by beating his Customers; and what is true of a Shop-keeper, is true of a Shop-keeping Nation. **1776** A. SMITH *Wealth of Nations* II. IV. vii. To found a great empire for the sole purpose of raising up a people of customers, may at first sight appear a project fit only for a nation of shopkeepers.] **1831** DISRAELI *Young Duke* I. xi. Hast thou brought this, too, about that ladies' hearts should be won . . over a counter. . . We are indeed a nation of shopkeepers. **1911** *Times Weekly* 17 Feb. 132 Napoleon . . described the English as a nation of shopkeepers. Uttered in a sneering spirit, it embodied . . the profound truth that our prosperity is based upon our trade. **1981** R. RENDELL *Put on by Cunning* xiv. Americans . . are a nation of salesmen just as the English are a nation of small shopkeepers.

One ENGLISHMAN can beat three Frenchmen

Now also used of other persons, and in different proportions.

1599 SHAKESPEARE *Henry V* III. vi. 144 When they were in health . . I thought upon one pair of English legs Did march three Frenchmen. **1745** H. WALPOLE *Letter* 13 July (1941) IX. 17 We, who formerly . . could any one of us beat three Frenchmen, are now so degenerated that three Frenchmen can evidently beat one Englishman. **1834** MARRYAT *Peter Simple* III. viii. My men . . there are three privateers. . . It's just a fair match for you—one Englishman can always beat three Frenchmen. **1851** G. BORROW *Lavengro* I. xxvi. In the days of pugilism it was no vain boast to say, that one Englishman was a match for two of t'other race [the French]. **1913** A. LUNN *Harrovians* i. Peter knew that an Englishman can tackle three foreigners, and forgot that the inventor of this theory took care to oppose three Englishmen to one foreigner as often as possible. **1981** *London Review of Books* 16 July–5 Aug. 5 Spain's conquest of Mexico 'gave Europeans a new and potent myth', the conviction of one European as equal to twenty others.

An ENGLISHMAN's home is his castle

1581 R. MULCASTER *Positions* xl. He is the appointer of his owne circumstance,

and his house is his castle. **1642** T. FULLER *Fast Sermon* 28 It was wont to be said *A mans house is his Castle* but if this Castle of late hath proved unable to secure any, let them make their conscience their castle. **1791** BOSWELL *Life of Johnson* II. 284 In London . . a man's own house is truly his *castle*, in which he can be in perfect safety from intrusion. **1837** DICKENS *Pickwick Papers* xxiv. Some people maintains that an Englishman's house is his castle. That's gammon.[1] **1965** *Evening Standard* 14 July 15 Mr. Ashling claimed 150 guineas compensation. . . 'The Englishman's home is his castle and this [hedge-clipping] upset my amenities.'

[1] nonsense.

An ENGLISHMAN's word is his bond

c **1500** *Lancelot of Lake* (STS) l. 1671 O kingis word shuld be o kingis bonde. *a* **1606** *Nobody & Somebody* C2ᵛ Nobodies worde is as good as his bond. **1642** T. FULLER *Holy State* v. xiii. He hath this property of an honest man, that his word is as good as his band. **1754** RICHARDSON *Grandison* I. Letter ix. I am no flincher. . . The word of Sir Rowland Meredith is as good as his bond. **1841** DICKENS *Old Curiosity Shop* lviii. 'Marchioness,' said Mr. Shriveller, rising, 'the word of a gentleman is as good as his bond—sometimes better, as in the present case.' **1924** G. BENHAM *Book of Quotations* (rev. ed.) 735 (Proverb) An Englishman's word is his bond. **1981** A. GRAHAM-YOOLL *Forgotten Colony* xvi. If a verbal promise is made the native, to seal the contract, usually says *palabra de inglés*, . . meaning that he will act as an Englishman, whose word is his bond.

ENOUGH is as good as a feast

c **1375** J. BARBOUR *Bruce* (EETS) xiv. 363 He maid thame na gud fest, perfay,[1] And nocht-for-thi[2] yneuch had thai. *c* **1470** MALORY *Morte d'Arthur* (1967) I. 246 Inowghe is as good as a feste. **1546** J. HEYWOOD *Dialogue of Proverbs* II. xi. M1 Here is enough, I am satisfied (sayd he) . . For folke say, enough is as good

as a feast. **1833** LAMB *Elia's Last Essays*
vi. That enough is as good as a feast.
Not a man, woman, or child in ten miles
round Guildhall, who really believes
this saying. **1928** D. H. LAWRENCE
Woman who rode Away 165 I'll *live* with
another woman but I'll never *marry*
another. Enough is as good as a feast.
1976 *New Yorker* 26 Apr. 77 If he's so
smart he'll amount to something in
good time—enough is as good as a
feast: that's an old saying and a true
one.

¹ truly.   ² *nought for thi*, nevertheless.

enough: *see also* throw DIRT enough, and
some will stick; give a man ROPE enough
and he will hang himself; a WORD to the
wise is enough.

envied: *see* BETTER be envied than pitied.

To ERR is human (to forgive divine)
   [L. *humanum est errare*, it is human to
err. Cf. **1539** R. MORISON tr. *J. L. Vives'*
Introduction to Wisdom D7 It is naturally
gyuen to al men, to erre, but to no man
to perseuer . . therin.] **1578** H. WOTTON
tr. *J. Yver's Courtly Controversy* E3 To
offend is humaine, to repent diuine,
and to perseuere diuelish. **1659** J.
HOWELL *Proverbs* (French) 12 To erre is
humane, to repent is divine, to perse-
vere is Diabolicall. **1711** POPE *Essay on*
Criticism l. 525 Good-Nature and Good-
Sense must ever join; To Err is Humane;
to Forgive, Divine. **1908** *Times Literary*
Supplement 27 Mar. 1 The modern
moralist pardons everything, because
he is not certain of anything, except that
to err is human. **1980** 'C. AIRD' *Passing*
Strange xiv. After all 'to err was human'
and—dedicated dominie¹ that he was—
Burton knew all about mistakes.

¹ schoolmaster, pedagogue (chiefly Scottish).

escape: *see* LITTLE thieves are hanged, but
great ones escape.

eternal: *see* HOPE springs eternal.

Eve: *see* when ADAM delved and Eve span,
who was then the gentleman?

even: *see* (adjective) never give a SUCKER an
even break; (adverb) even a WORM will
turn.

event: *see* COMING events cast their shadows before; it is easy to be WISE after
the event.

EVERY little helps
   [**1590** G. MEURIER *Deviz Familiers* A6
peu ayde, disçoit le formy, pissant en mer en
plein midy, every little helps, said the
ant, pissing into the sea at midday.]
1602 P. GAWDY *Letters* (1906) 118 The
wrenn sayde all helpte when she . . in
the sea. **1623** W. CAMDEN *Remains con-*
cerning Britain (ed. 3) 268 Euery thing
helpes, quoth the Wren when she pist i'
the sea. **1787** E. HAZARD in *Collections of*
Massachusetts Hist. Society (1877) 5th Ser.
II. 477 A guinea is a guinea, and every
little helps. **1840** MARRYAT *Poor Jack* xiii.
It's a very old saying, that every little
helps. . . Almost all the men were on
the larboard side. **1980** S. T. HAYMON
Death & Pregnant Virgin xxii. 'I'll see if
we can't put it off a day or two.' 'Every
little helps.'

EVERY man for himself
See also the later expanded forms in the
next two proverbs.
   c **1390** CHAUCER *Knight's Tale* l. 1182
At the kynges court, my brother, Ech
man for hymself, ther is noon oother.
1478 J. WHETLEY *Letter* 20 May in *Paston*
Letters (1976) II. 427 Your moder . . hath
made her wyll, the wyche ye shall
understond more when I com, for ther is
every man for hym selff. **1678** J. RAY
English Proverbs (ed. 2) (Scottish) 366
Every man for himself (quoth the
Merteine). **1795** D. YANCEY *Letter* 6 June
in *Virginia Magazine of Hist. & Biography*
(1922) XXX. 224 The old adage might
well be applied in many cases. Every
man for himself. **1974** A. PRICE *Other*
Paths to Glory II. vi. It was pretty much
every man for himself. But I was hit
quite early on.

EVERY man for himself, and God for us all

1546 J. HEYWOOD *Dialogue of Proverbs* II. ix. L2 Praie and shifte eche one for hym selfe, as he can. Every man for hym selfe, and god for us all. **1615** T. ADAMS *White Devil* (ed. 2) 83 That by-word, Euery man for himselfe, and God for vs all, is vncharitable, vngodly. **1830** MARRYAT *King's Own* III. xiii. The captain . . ordered the sailor to leave the boat. 'Every man for himself, and God for us all!' was the cool answer of the refractory seaman. **1979** *Times* 29 Dec. 12 Each for himself, and God for us all, as the elephant said when he danced among the chickens.

EVERY man for himself, and the Devil take the hindmost

The two earliest examples are more closely aligned in form to the preceding proverb. Cf. also DEVIL *take the hindmost*.

1530 A. BARCLAY *Eclogues* (EETS) I. 1009 Eche man for him selfe, and the fiende for all. **1573** J. SANFORDE 108ᵛ Euery man for him selfe; and the Deuill for all. **1858** D. M. MULOCK *Women's Thoughts about Women* ii. The world is hard enough, for two-thirds of it are struggling for the dear life—'each for himself, and de'il tak the hindmost.' **1939** L. I. WILDER *By Shores of Silver Lake* xxv. There wasn't any standing in line. . . It was each fellow for himself and devil take the hindmost. **1957** tr. G. Guareschi's *Don Camillo & Devil* iv. 'I'm paying for what I eat.' . . 'Every man must pay for himself and the devil take the hindmost.'

EVERY man has his price

1734 W. WYNDHAM in *Bee* VIII. 97 'It is an old Maxim, that every Man has his Price,' if you can but come up to it. **1798** W. COXE *Memoirs of Sir Robert Walpole* I. lxiv. The political axiom generally attributed to him [Walpole] . . was perverted by leaving out the word *those*. . . He ascribed to the interested views of themselves . . the declaration of pretended patriots, of whom he said, 'All those men have their price.' **1845**

G. P. R. JAMES *Smuggler* I. x. 'You can do nothing with Mowle. He never took a penny in his life.' 'Oh, every man has his price.' **1949** N. MAILER *Naked & Dead* II. xi. It was the sort of deal his father might have pulled. 'Every man has his price.' **1976** J. I. M. STEWART *Memorial Service* xvi. What was reputable could be left with some confidence to Edward Pococke. Every man has his price.

EVERY man is the architect of his own fortune

[PSEUDO-SALLUST *Ad Caesarem Senem Oratio* i. *sed res docuit id verum esse, quod in carminibus Appius ait, fabrum esse suae quemque fortunae*, but experience had shown what Appius said in his verses to be true, that each man is the architect of his own fortune.] **1533** N. UDALL *Flowers for Latin Speaking* (1560) 24 A prouerbiall spekyng. . . Euery man . . is causer of his own fortune. **1649** MILTON *Eikonoklastes* III. 542 They in whomsoeuer these vertues dwell . . are the architects of thir own happiness. **1707** J. DUNTON *Athenian Sport* 454 We are . . architects of our own fortune. **1818** S. FERRIER *Marriage* III. vi. As every man is said to be the artificer of his own fortune, so every one . . had best be the artificer of their own friendship. **1873** *Notes & Queries* 4th Ser. XII. 514 We have not a commoner saying among us than 'Every man is the architect of his own fortune,' and we have very few much older.

EVERY man to his taste

[Cf. Fr. *chacun à son goût*, each to his taste.] **1580** LYLY *Euphues* II. 161 Betweene them it was not determined, but euery one as he lyketh. *a* **1640** MIDDLETON et al. *Old Law* II. ii. Every one to their liking. But I say An honest man's worth all. **1760** STERNE *Tristram Shandy* I. vii. I own I never could envy Didius in these kinds of fancies of his:—But every man to his own taste. **1849** BULWER-LYTTON *Caxtons* III. XVII. i. 'Sheep are dull things to look at after a bull-hunt.' . . 'Every man to his taste in the Bush.' **1929** E. LINKLATER *Poet's Pub* xxvi. 'I like fairy tales,' said the professor. . . 'Every man to his taste,' agreed the landlord.

EVERY man to his trade

[I CORINTHIANS vii. 20 Let euery man abide in the same calling wherein he was called. Cf. **1539** R. TAVERNER tr. *Erasmus' Adages* E1 Let euerye man exercise hym selfe in the facultie that he knoweth. **1590–1** SHAKESPEARE *Henry VI Pt. 2* IV. ii. 15 And yet it is said 'Labour in thy vocation'; which is as much to say as 'Let the magistrates be labouring men'; and therefore should we be magistrates.] **1597–8** SHAKESPEARE *Henry IV Pt. 1* II. ii. 75 Every man to his business. **1605** MARSTON *Dutch Courtesan* I. i. Every man must follow his trade, and every woman her occupation. *a* **1721** M. PRIOR *Dialogues of Dead* (1907) 221 Every man to his trade, Charles, you should have challenged me at long pike or broad sword. **1930** C. BUSH *Murder at Fenwold* viii. 'I dabble in mathematics but . . I'd rather have your Latinity.' 'Every man to his trade.' **1978** O. WHITE *Silent Reach* xvi. I understand your distrust of theories. . . Every man to his trade.

every: *see also* every CLOUD has a silver lining; every COCK will crow upon his own dunghill; every DOG has his day; every DOG is allowed one bite; every ELM has its man; there is an EXCEPTION to every rule; every HERRING must hang by its own gill; every JACK has his Jill; every LAND has its own law; every PICTURE tells a story; if every man would SWEEP his own doorstep the city would soon be clean; every TUB must stand on its own bottom; there are TWO sides to every question.

EVERYBODY loves a lord

1869 F. J. FURNIVALL in *Queen Elizabeth's Academy* (EETS) p. xii, The second tract . . is printed, not mainly because 'John Bull loves a Lord' . . but because the question of Precedence was so important a one in old social arrangements. **1908** *Spectator* 3 July 9 It is always said that an Englishman loves a lord. It would be more exact to say that he is in love with lordliness. **1980** M. NICHOLLS *Importance of being Oscar* (1981) 58 If pressed, he would probably have admitted that he was no exception to the adage that 'Everybody Loves a Lord'.

What EVERYBODY says must be true

a **1400** *Legends of Saints* (STS) III. 105 For I fynd suthfastnes,[1] that al men sais, is nocht les.[2] *c* **1475** in *Modern Philology* (1940) XXXVIII. 118 Hit ys cominly truye that all men sayth. *c* **1518** A. BARCLAY tr. *Mancinus' Mirror of Good Manners* F1[v] It nedes muste be trewe which euery man doth say. **1748** RICHARDSON *Clarissa* IV. 74 The most accomplished of women, as every one says; and what every one says must be true. **1905** A. MACLAREN *Gospel according to St. Matthew* II. 246 'What everybody says must be true' is a cowardly proverb. . . What most people say is usually false.
[1] truth.   [2] i.e. lies.

EVERYBODY's business is nobody's business

[ARISTOTLE *Politics* II. i. ἥκιστα γὰρ ἐπιμελείας τυγχάνει τὸ πλείστων κοινόν, a matter common to most men receives least attention.] **1611** R. COTGRAVE *Dict. French & English* s.v. Ouvrage, Euerie bodies worke is no bodies worke. **1655** I. WALTON *Compleat Angler* (ed. 2) I. ii. A wise friend of mine did usually say, That which is every bodies businesse, is no bodies businesse. **1725** DEFOE *(title)* Every-Body's Business, is No-Body's Business; . . exemplified in the Pride of our Woman-Servants. **1834** J. H. NEWMAN *Parochial Sermons* I. xix. It is commonly said, that what is every one's business is practically no one's. **1914** G. B. SHAW *Misalliance* 10 'The danger of public business is that it never ends.' . . 'What I say is that everybody's business is nobody's business.'

everyone: *see* you can't PLEASE everyone; everyone SPEAKS well of the bridge which carries him over; everyone

STRETCHES his legs according to the length of his coverlet.

EVERYTHING has an end
Cf. ALL *good things must come to an end.*
c **1385** CHAUCER *Troilus & Criseyde* III. 615 At the laste, as every thyng hath ende, She took hire leve. **1616** N. BRETON *Crossing of Proverbs* A6 'Euery thing hath an end.' . . 'Not so, a Ring hath none.' **1841** DICKENS *Barnaby Rudge* xx. Everything has an end. Even young ladies in love cannot read their letters for ever. **1980** D. FRANCIS *Reflex* viii. Can't go on for ever, more's the pity. Everything ends, doesn't it.

everything: *see also* ALL things come to those who wait; MONEY isn't everything; a PLACE for everything, and everything in its place; there is a REMEDY for everything except death; there is a TIME and place for everything; there is a TIME for everything.

EVIL communications corrupt good manners
Proper conduct is harmfully influenced by false information or knowledge; also used to assert the deleterious effect of bad example.
[I CORINTHIANS xv. 33 Bee not deceiued: euill communications corrupt good manners.] *c* **1425** J. ARDERNE *Treatises of Fistula* (EETS) 5 Shrewed speche corrumpith gode maners. **1533** MORE *Debellation¹ of Salem* xiv. (As saynt Poule speketh of such heresyes) euyl communicacion corrupteth good maners. **1749** FIELDING *Tom Jones* IV. XII. iii. I heartily wish you would . . not think of going among them.—Evil Communication corrupts good Manners. **1829** COBBETT *Advice to Young Men* V. cccviii. Evil communications corrupt good manners. **1874** TROLLOPE *Phineas Redux* I. xvi. [The horse] would have taken the fence . . but Dandolo had baulked . . and evil communications will corrupt good manners. **1939** W. S. MAUGHAM *Christmas Holiday* ii. A disposition of such sweetness that no evil communication could corrupt his good manners. **1955**

R. CROSSMAN *Journal* 28 Feb. (1981) 392 We are accused of misleading Attlee. . . I have no doubt that this is written to warn Mr Attlee that evil communications will corrupt good manners, as indeed they will.
¹ conquest, subjugation.

EVIL doers are evil dreaders
a **1568** R. ASCHAM *Schoolmaster* (1570) I. 27 Ill doinges, breed ill thinkinges. And corrupted manners, spryng perverted iudgements. **1721** J. KELLY *Scottish Proverbs* 176 Ill doers, ill deemers. **1737** A. RAMSAY *Scots Proverbs* xix. Ill doers are ay ill dreaders. **1828** SCOTT *Fair Maid of Perth* II. v. Put me not to quote the old saw, that evil doers are evil dreaders.—It is your suspicion, not your knowledge, which speaks. **1886** R. L. STEVENSON *Kidnapped* xxvii. If you were more trustful, it would better befit your time of life. . . We have a proverb . . that evil doers are aye evil-dreaders.

evil: *see also* a GREAT book is a great evil; IDLENESS is the root of all evil; MONEY is the root of all evil; SEE no evil, hear no evil, speak no evil; SUFFICIENT unto the day is the evil thereof.

Of two EVILS choose the less
[ARISTOTLE *Nicomachean Ethics* II. ix. φασί, πλοῦν τὰ ἐλάχιστα ληπτέον τῶν κακῶν, it is said that the best course is to choose the least of the evils; CICERO *De Officiis* III. xxix. *minima de malis,* of evils choose the least.] *c* **1385** CHAUCER *Troilus & Criseyde* II. 470 Of harmes two, the lesse is for to chese. *c* **1440** *Gesta Romanorum* (EETS) 10 Of too Evelis the lasse Evill is to be chosyn. **1546** J. HEYWOOD *Dialogue of Proverbs* I. v. B2 Of two yls, chose the least while choyse lyth in lot. **1785** BOSWELL *Journal of Tour to Hebrides* 464 'O ho! Sir, (said I), you are flying to me for refuge!' . . 'It is of two evils choosing the least.' **1891** A. FORBES *Barracks, Bivouacs & Battles* 187 Either the Turks would make a prisoner of me . . or I must . . take my chance of the Russian fire. . . 'Of two evils choose the less,' says the wise proverb.

EXAMPLE is better than precept

c **1400** J. MIRK *Festial* (EETS) 216 Then saythe Seynt Austeyn that an ensampull yn doyng ys mor commendabull then ys techyng other[1] prechyng. *a* **1568** R. ASCHAM *Schoolmaster* (1570) I. 20 One example, is more valiable . . than xx. preceptes written in bookes. **1708** M. PRIOR *Literary Works* (1971) I. 535 Example draws where Precept fails, And Sermons are less read than Tales. **1828** D. M. MOIR *Mansie Wauch* xix. Example is better than precept, as James Batter observes. **1894** J. LUBBOCK *Use of Life* xix. Men can be more easily led than driven: example is better than precept. **1930** E. M. BRENT-DYER *Chalet Girls in Camp* v. Example is better than precept, my child. Likewise, practise what you preach. **1981** P. O'DONNELL *Xanadu Talisman* ii. Example is always better than precept, remember.

[1] or.

The EXCEPTION proves the rule

'The very fact of an exception proves there must be a rule' (Brewer); now frequently misunderstood and used to justify inconsistency.

[L. *exceptio probat regulam in casibus non exceptis*, the exception confirms the rule in cases not excepted.] **1640** G. WATTS *Bacon's Advancement of Learning* VIII. iii. Exception strengthens the force of a Law in Cases not excepted. **1664** J. WILSON *Cheats* A2[v] I think I have sufficiently justifi'd the Brave man, even by this Reason, That the Exception proves the Rule. **1765** JOHNSON *Shakespeare* Preface C2[v] There are a few passages which may pass for imitations, but so few that the exception only confirms the rule. **1907** H. W. FOWLER *Si Mihi* 80 It is one of those cryptic sayings, like 'The exception proves the rule', which always puzzle me. **1979** *Daily Telegraph* 9 Mar. 15 The better the play, the worse the film. This rule I find to prevail, with the occasional exception 'to prove it'.

There is an EXCEPTION to every rule

Cf. the preceding entry.

1579 T.F. *News from North* D1[v] There is no rule so generall, that it admitteth not exception, albeit I dout not . . that honors chaunge maners. **1608** T. HEYWOOD *Rape of Lucrece* V. 169 A general concourse of wise men. . . Tarquin, if the general rule have no exceptions, thou wilt have an empty consistory.[1] **1773** R. GRAVES *Spiritual Quixote* III. ix. xviii. The rules of Grammar cannot, in any language, be reduced to a strict analogy; but all general rules have some exceptions. **1836** MARRYAT *Midshipman Easy* I. xii. I have little reason to speak in its favour . . but there must be exceptions in every rule. **1981** *Listener* 21 May 683 'There is still something awe-inspiring about a duke,' we are informed. Only those who share such values will want to read Heirs and Graces. Even they should remember that there is an exception to every rule.

[1] council-chamber.

A fair EXCHANGE is no robbery

1546 J. HEYWOOD *Dialogue of Proverbs* II. iv. G4 Chaunge be no robbry for the changed case. *c* **1590** *John of Bordeaux* (1936) l. 213 Exchaung is no roberie. *a* **1628** in M. L. Anderson *Proverbs in Scots* (1957) no. 540 Fair shifts[1] na robberie. **1721** J. KELLY *Scottish Proverbs* 105 Fair Exchange is no Rob'ry. Spoken when we take up one Thing, and lay down another. **1748** SMOLLETT *Roderick Random* II. xli. Casting an eye at my hat and wig . . he took them off, and clapping his own on my head, declared, that a fair exchange was no robbery. **1930** M. C. KEATOR *Eyes through Trees* i. A fair exchange was no robbery. She might gain a new insight into the art of living as I also might gain a fuller comprehension of the heart of things. **1960** N. MITFORD *Don't tell Alfred* xx. 'So it was you who took away the Harar frescoes?' 'Took away? We exchanged them. . . A good exchange is no robbery, I believe?'

[1] exchange is.

excuse: *see* a BAD excuse is better than none; IGNORANCE of the law is no excuse for breaking it.

expand: *see* WORK expands so as to fill the time available.

What can you EXPECT from a pig but a grunt?

Used rhetorically with reference to coarse or boorish behaviour.

1731 *Poor Robin's Almanack* C6 If we petition a Hog, what can we expect but a grunt. **1827** SCOTT *Journal* 10 Apr. (1941) 41 They refuse a draught of £20, because, in mistake, it was £8 overdrawn. But what can be expected of a *sow* but a *grumph?* **1910** P. W. JOYCE *English as We speak it in Ireland* x. Of a coarse, ill-mannered man, who uses unmannerly language: 'What could you expect from a pig but a grunt.' **1921** B. DUFFY *Special Pleading* 7 What can one expect from a pig but the grunt! What could the likes of you know about fairies anyway?

expect: *see also* BLESSED is he who expects nothing, for he shall never be disappointed.

EXPERIENCE is the best teacher

The Latin tag *experientia docet*, experience teaches, gave rise to a large number of proverbs. There is no standard form, and the sayings given below illustrate the themes that 'one learns (also, fools learn) by experience' and that 'experience is a hard teacher'. See also EXPERIENCE *keeps a dear school.*

a **1568** R. ASCHAM *Schoolmaster* (1570) I. 19 Erasmus . . saide wiselie that experience is the common scholehouse of foles. **1618** N. BRETON *Court & Country* B4 Let ignorance be an enemy to wit, and experience be the Mistris of fools. **1670** J. RAY *English Proverbs* 86 Experience is the mistress of fools. *Experientia stultorum magistra.* Wise men learn by others harms, fools by their own. **1732** T. FULLER *Gnomologia* no. 1484 Experience teacheth Fools; and he is a great one, that will not learn by it. **1782** J. TURNBULL *M'Fingal* IV. 97 Alas, great Malcolm cried, experience Might teach you not to trust appearance. **1803** M. L. WEEMS *Letter* 12 Nov. (1929) 278

Experience, the best of teachers. **1814** R. BLAND *Proverbs* I. 280 'Experience', we say, 'makes even fools wise.' **1856** F. M. WHITCHER *Widow Bedott Papers* xxix. I . . dident know how to do anything as well as I do now. . . Experience is the best teacher, after all. **1874** G. J. WHYTE-MELVILLE *Uncle John* I. x. Experience does *not* make fools wise. . . Most proverbs are fallacious. None greater than that which says it does. **1962** *Infantry* Nov.—Dec. 26 Experience is a hard teacher, and we cannot afford to learn on the battlefield what should be taught during normal training. **1978** L. BLOCK *Burglar in Closet* i. Experience is as effective a teacher as she is because one does tend to remember her lessons.

EXPERIENCE is the father of wisdom

[ALCMAN *Fragments* lxiii. πεῖρά τοι μαθήσιος ἀρχά, experience is the beginning of knowledge.] **1539** R. TAVERNER *Garden of Wisdom* II. 24ᵛ This be comonly true, for experience is mother of prudence, yet suche prudence & wysedom cost the comon weale moch. *a* **1547** E. HALL *Chronicle* (1548) Rich. III 31 He by the longe and often alternate proof . . had gotten by great experience the very mother and mastres[1] of wisedome. **1581** G. PETTIE tr. *S. Guazzo's Civil Conversation* I. 11 Experience is the father of wisedom, and memorie the mother. **1788** *American Museum* III. 183 If it be true, that experience is the mother of wisdom, history must be an improving teacher. **1981** P. O'DONNELL *Xanadu Talisman* ii. Experience is the father of wisdom, remember.

[1] mistress.

EXPERIENCE keeps a dear school

1743 B. FRANKLIN *Poor Richard's Almanack* (Dec.) Experience keeps a dear school, but Fools will learn in no other. **1897** C. C. KING *Story of British Army* vii. But the British leaders were to learn the fact, they might have foreseen, in the 'only school fools learn in, that of experience'. **1938** E. O. LORIMER tr. *W. Frischauer's Twilight in Vienna* vii. The various Governments had . . to learn their lesson in blood and tears, for 'experience keeps a dear school'.

EXTREMES meet
Opposite extremes have much in common.
[*a* 1662 PASCAL *Pensées* (new ed., 1835) I. iv. *les extrèmes se touchent*, extremes meet.] 1762 J. WATTS in *Collections of New York Hist. Society* (1928) LXII. 48 But as extremes meet we may possibly the sooner have a peace for it. 1822 SCOTT *Nigel* III. iii. This Olifaunt is a Puritan?—not the less like to be a Papist . . for extremities meet. 1836 E. HOWARD *Rattlin the Reefer* I. xiv. Let us place at least one 'barring out' [i.e. action of schoolboys barricading themselves in a room] upon record, in order to let the radicals see, and seeing hope, when they find how nearly extremes meet, what a slight step there is from absolute despotism to absolute disorganization. 1905 J. B. CABELL *Line of Love* vi. It is a venerable saying that extremes meet. 1931 H. READ *Meaning of Art* II. lxi. It is a case of extremes meeting, but it may be suggested that extremes meet always on common ground.

extremity: *see* MAN's extremity is God's opportunity.

What the EYE doesn't see, the heart doesn't grieve over
[ST. BERNARD *Sermon* V. *vulgo dicitur: Quod non videt oculus cor non dolet*, it is commonly said: what the eye sees not, the heart does not grieve at.] 1545 R. TAVERNER tr. *Erasmus' Adages* (ed. 2) 13 That the eye seeth not, the hart rueth not. 1576 G. PETTIE *Petit Palace* 145 As the sence of seeinge is most sharp, so is that paine most pinching, to see the thing one seeketh, and can not possesse it. . . And as the common saying is, that which the eye seeth, the hart greeueth. 1721 J. KELLY *Scottish Proverbs* 341 What the Eye sees not, the Heart rues not. Men may have Losses, but if they be unknown to them they give them no Trouble. 1830 J. L. BURCKHARDT *Arabic Proverbs* 109 When the eye

does not see, the heart does not grieve. 1883 C. S. BURNE *Shropshire Folklore* xxxvi. These . . seem to be popular legal maxims. . . What the eye doesn't see, the heart doesn't grieve. 1939 G. HEYER *No Wind of Blame* iii. Anyone knows what men are, and what the eye doesn't see the heart won't grieve over. 1979 M. BABSON *So soon done For* xiii. 'The Norrises don't know about it yet.' 'There you are, "What the eye does not see"—' He broke off.

The EYE of a master does more work than both his hands
1744 B. FRANKLIN *Poor Richard's Almanack* (Oct.) The eye of a Master, will do more Work than his Hand. 1755—*Ibid.* (Sept.) The Master's Eye will do more Work than both his Hands. 1876 I. BANKS *Manchester Man* I. xiv. She was wont to say, 'The eye of a master does more work than both his hands,' accordingly in house or warehouse her active supervision kept other hands from idling.

eye: *see also* BEAUTY is in the eye of the beholder; the BUYER has need of a hundred eyes, the seller of but one; FIELDS have eyes, and woods have ears; FOUR eyes see more than two; HAWKS will not pick out hawks' eyes; PLEASE your eye and plague your heart.

The EYES are the window of the soul
The variant form *The face is the index of the mind* is among those also exemplified below.
[CICERO *Orator* lx. *ut imago est animi voltus sic indices oculi*, the face is a picture of the mind as the eyes are its interpreter; L. *vultus est index animi* (also *oculus animi index*), the face (also, eye) is the index of the mind.] 1545 T. PHAER *Regiment of Life* 14 The eyes . . are the wyndowes of the mynde, for both ioye & anger . . are seene . . through them. *a* 1575 J. PILKINGTON *Nehemiah* (1585) i. The affections of the minde declare them selues openlie in the face. 1601 JONSON *Cynthia's Revels* D3[v] I can refell[1] that Paradox . . of those, which hold the

face to be the Index of the minde. **1781** A. ADAMS in L. H. Butterfield et al. *Adams Family Correspondence* (1973) IV. 215, I did not study the Eye that best Index to the mind. **1864** MRS. H. WOOD *Trevlyn Hold* I. i. You have not to learn that the face is the outward index of the mind within. **1940** G. SEAVER *Scott of Antarctic* II. 48 The eye, which is the reflector of the external world, is also the mirror of the soul within. **1979** J. GERSON *Omega Factor* iii. If the old saying, the eyes are the window of the soul, were true then this young girl had misplaced her soul.

[1] refute.

F

face: *see* don't CUT off your nose to spite your face; the EYES are the window of the soul.

FACT is stranger than fiction
An alliterative version of TRUTH *is stranger than fiction.*

1853 T. C. HALIBURTON *Sam Slick's Wise Saws* 5 Facts are stranger than fiction, for things happen sometimes that never entered into the mind of man to imagine or invent. **1881** A. JESSOPP *Arcady for Better or Worse* iii. I have no desire to convince the world that . . in this . . case fact is stranger than fiction. But the following instance of Mr. Chowne's 'cunning' may be verified. **1910** A. M. FAIRBAIRN *Studies in Religion & Theology* II. v. Forgetting the fact which is stranger than fiction, that the sagest man in the theory of State may be the unwisest man in statecraft. **1929** E. J. MILLWARD *Copper Bottle* 64 Facts may be stranger than fiction, . . but fiction is generally truer than facts.

FACTS are stubborn things
1732 E. BUDGELL *Liberty & Progress* ii. Plain matters of fact are terrible stubborn things. **1749** J. ELIOT *Continuation of Essay on Field Husbandry* 20 Facts are stubborn things. **1866** BLACKMORE *Cradock Nowell* III. vi. Facts, however, are stubborn things, and will not even make a bow to the sweetest of young ladies. **1942** L. THAYER *Murder is Out* xxvii. You're . . too intelligent to think that suggestion would have any weight with a jury. . . Facts are stubborn things.

FAINT heart never won fair lady
[c **1390** GOWER *Confessio Amantis* V. 6573 Bot as men sein, wher herte is failed, Ther schal no castell ben assailed.] **1545** R. TAVERNER tr. *Erasmus' Adages* (ed. 2) 10 A coward verely neuer obteyned the loue of a faire lady. **1580** LYLY *Euphues* II. 131 Faint[1] hart Philautus neither winneth Castell nor Lady:

therfore endure all thinges that shall happen with patience. **1614** W. CAMDEN *Remains concerning Britain* (ed.2) 306 Faint heart neuer wonne faire Lady. **1754** RICHARDSON *Grandison* I. xvi. Then, madam, we will *not* take your denial. . . Have I not heard it said, that faint heart never won fair lady. **1899** G. GISSING *Crown of Life* xiii. Could he leave England, this time, without confessing himself to her? Faint heart—he mused over the proverb. **1972** J. I. M. STEWART *Palace of Art* xix. 'You proceed as planned.' 'Faint heart never won fair lady?'

[1] cowardly, timorous.

FAIR and softly goes far in a day
c **1350** *Douce MS* 52 no. 50 Fayre and softe me[1] ferre gose. **1670** J. RAY *English Proverbs* 87 Fair and softly goes far in a day. . . He that spurs on too fast at first setting out, tires before he comes to his journeys end. *Festina lente.* **1818** SCOTT *Heart of Midlothian* IV. viii. Reuben Butler isna the man I take him to be, if he disna learn the Captain to fuff[2] his pipe some other gate[3] than in God's house, or[4] the quarter be ower. 'Fair and softly gangs far,' said Meiklehose. **1914** K. F. PURDON *Folk of Furry Farm* ii. Maybe I'm like the singed cat, better than I look! I'm slow, but fair and easy goes far in a day.

[1] one. [2] puff. [3] place. [4] before.

All's FAIR in love and war
[**1578** LYLY *Euphues* I. 236 Anye impietie may lawfully be committed in loue, which is lawlesse.] **1620** T. SHELTON tr. *Cervantes' Don Quixote* II. xxi. Love and warre are all one. . . It is lawfull to use sleights and stratagems to . . attaine the wished end. **1845** G. P. R. JAMES *Smuggler* II. iv. In love and war, every stratagem is fair, they say. **1850** F. E. SMEDLEY *Frank Fairlegh* xlix. 'You opened the letter!' . . 'How was I to read it if I hadn't? All's . . fair in love and war, you know.' **1972** J. I. M. STEWART *Palace of Art* xii. 'Do you really suppose I

would tell?' he demanded coldly. 'Might do. All's fair in——.'

FAIR play's a jewel

1809 W. IRVING *Hist. New York* II. VI. vii. The furious Risingh, in despight of that noble maxim . . that 'fair play is a jewel', hastened to take advantage of the hero's fall. **1823** J. F. COOPER *Pioneers* II. v. Well, fair play's a jewel. But I've got the lead of you, old fellow. **1935** E. F. BENSON *Lucia's Progress* viii. There's been a lil' mistake. . . I want my lil' rubber of Bridge. Fair play's a jewel. **1948** L. A. G. STRONG *Trevannion* iv. 'It ain't good to win crooked.' 'Good for you, Stan. I agree. Fair play's a jewel.'

fair: *see also* none but the BRAVE deserve the fair; a fair EXCHANGE is no robbery; FAINT heart never won fair lady; GIVE and take is fair play; if SAINT Paul's day be fair and clear, it will betide a happy year; SAINT Swithun's day if thou be fair for forty days it will remain; TURN about is fair play.

FAITH will move mountains

[Cf. MATTHEW xvii. 20 If yee haue faith as a graine of mustard seed, yee shall say vnto yonder mountaine; Remoue hence to yonder place; and it shall remoue.] **1897** 'S. GRAND' *Beth Book* xvi. If mountains can be moved by faith, you can surely move your own legs! **1933** J. BETJEMAN *Ghastly Good Taste* iii. As faith can move mountains, so nothing was impossible to Holy Church. **1948** B. STEVENSON *Home Book of Proverbs* (rev. ed.) 745 Faith will move mountains. **1980** C. FREMLIN *With no Crying* xix. Faith moves mountains, they say: and Hope lights up our darkness.

fall: *see* (noun) PRIDE goes before a fall; (verb) the APPLE never falls far from the tree; BETWEEN two stools one falls to the ground; the BIGGER they are, the harder they fall; when the BLIND lead the blind, both shall fall into the ditch; the BREAD never falls but on its buttered side; when all FRUIT falls, welcome haws; a REED before the wind lives on, while

mighty oaks do fall; if the SKY falls we shall catch larks; as a TREE falls, so shall it lie; UNITED we stand, divided we fall.

FAMILIARITY breeds contempt

[PUBLILIUS SYRUS *Sententiae* 463 *parit contemptum nimia familiaritas*, too much familiarity breeds contempt.] *c* **1390** CHAUCER *Tale of Melibee* l. 2876 Men seyn that 'over-greet hoomlynesse[1] engendreth dispreisynge'. **1539** R. TAVERNER *Garden of Wisdom* II. 4^v Hys specyall frendes counsailled him to beware, least his ouermuche familiaritie myght breade him contempte. **1654** T. FULLER *Comment on Ruth* 176 With base and sordid natures familiarity breeds contempt. **1869** TROLLOPE *He knew He was Right* II. lvi. Perhaps, if I heard Tennyson talking every day, I shouldn't read Tennyson. Familiarity does breed contempt. **1928** D. H. LAWRENCE *Phoenix II* (1968) 598 We say . . Familiarity breeds contempt. . . That is only partly true. It has taken some races of men thousands of years to become contemptuous of the moon. **1979** E. H. GOMBRICH *Sense of Order* ii. Familiarity, it is said, breeds contempt and Jones' treatment of the Chinese tradition is not very far from contempt.

[1] familiarity.

The FAMILY that prays together stays together

The saying was invented by Al Scalpone, a professional commercial-writer, and was used as the slogan of the Roman Catholic Family Rosary Crusade by Father Patrick Peyton (P. Peyton, *All for Her*, 1967). The crusade began in 1942 and the slogan was apparently first broadcast on 6 Mar. 1947 during the radio programme *Family Theater of the Air*. The Crusade in Britain started in 1952, and the expression has come to have many (often humorous) variant forms.

1948 *St. Joseph Magazine* (Oregon) Apr. 3 'More things are wrought by prayer than this world dreams of', and 'The family that prays together stays together.' Such religious themes are hardly what one would expect to hear propounded over the air waves of our modern radio. **1949** *Catholic Digest* June 98 'The family that prays together stays together.' That is what Father Peyton has made it his business to remind you of every week. **1954** *Parents' Magazine* Feb. 119 The family that plays together stays together. **1980** *Times* 28 Nov. 8 One [Texas] hoarding said . . 'A family that prays together stays together.' **1980** R. HILL *Spy's Wife* xxi. The family that spies together, sties together. Old Cockney Russian proverb.

family: *see also* ACCIDENTS will happen (in the best-regulated families).

far: *see* BLUE are the hills that are far away; FAIR and softly goes far in a day.

FAR-FETCHED and dear-bought is good for ladies

Expensive or exotic articles are suitable for women.

c **1350** *Douce MS* 52 no. 7 Ferre ifet and dere i-bowght is goode for ladys. **1583** B. MELBANCKE *Philotimus* 18 Pallas . . is . . hard to be found, but easy to bee intreated, to be farre fetchte & deare boughte, but that we say is good for Ladies. **1616** JONSON *Epicœne* V. 163 When his cates[1] are all in brought, Though there be none far fet, there will dear-bought, Be fit for ladies. **1738** SWIFT *Polite Conversation* I. 60 But you know, far-fetch'd and dear-bought is fit for Ladies. I warrant, this cost your Father Two pence half-penny. **1876** I. BANKS *Manchester Man* III. x. 'Where did these beautiful things come from?' . . 'India . . They are "far-fetched and dear-bought", and so must be good for you, my lady.'
[1] dainties, delicacies.

fashion: *see* BETTER be out of the world than out of the fashion; when the GORSE is out of bloom, kissing's out of fashion.

fast: *see* BAD news travels fast; a MONEYLESS man goes fast through the market.

fastest: *see* he TRAVELS fastest who travels alone.

fasting: *see* it's ill speaking between a FULL man and a fasting.

fat: *see* a GREEN Yule makes a fat churchyard; the OPERA isn't over till the fat lady sings.

Like FATHER, like son

The variant form *Like father, like daughter* also occurs; cf. *Like* MOTHER, *like daughter.*

[L. *qualis pater talis filius*, as is the father, so is the son.] c **1340** R. ROLLE *Psalter* (1884) 342 Ill sunnys folous ill fadirs. **1509** A. BARCLAY *Ship of Fools* 98 An olde prouerbe hath longe agone be sayde That oft the sone in maners lyke wyll be Vnto the Father. **1616** T. DRAXE *Adages* 149 Like father like sonne. **1709** O. DYKES *English Proverbs* 30 Like Father, like Son. . . How many Sons inherit their Fathers Failings, as well as Estates? **1841** S. WARREN *Ten Thousand a Year* II. xiii. Two such bitter Tories . . for, like father, like son. **1936** W. HOLTBY *South Riding* v. i. Perhaps Lydia might do it once too often. . . Like father, like daughter. **1977** *Time* 22 Aug. 41 Like father, like son—usually perhaps, but not in the Hunt family.

father: *see also* the CHILD is father of the man; EXPERIENCE is the father of wisdom; it is a WISE child that knows its own father; the WISH is father to the thought.

A FAULT confessed is half redressed

1558 *Interlude of Wealth & Health* D2[v] Yf thou haue doone amisse, and be sory therfore, Then helfe a mendes is made. **1592** *Arden of Feversham* H1[v] A fault confessed is more than half amends, but men of such ill spirite as your selfe Worke crosses[1] and debates twixt man and wife. **1732** T. FULLER *Gnomologia* no. 1140 Confession of a Fault makes half amends. **1822** SCOTT *Nigel* III. v. Come, my Lord, remember your promise to confess; and indeed, to confess

is, in this case, in some slight sort to redress. **1855** H. G. BOHN *Handbook of Proverbs* 285 A fault confessed is half redressed. **1981** P. O'DONNELL *Xanadu Talisman* x. A fault confessed is half re-dressed, so I hope he will forgive us.
¹ troubles, arguments.

favour: *see* (noun) KISSING goes by favour; (verb) FORTUNE favours fools; FORTUNE favours the brave.

FEAR the Greeks bearing gifts
The saying is often used allusively.
[VIRGIL *Aeneid* II. 49 *timeo Danaos, et dona ferentes*, I fear the Greeks, even when bringing gifts (said by Laocoön as a warning to the Trojans not to admit the wooden horse); cf. **1777** S. JOHNSON *Letter* 3 May (1952) II. 515 Tell Mrs. Boswell that I shall taste her marmalade cautiously at first. *Timeo Danaos et dona ferentes.* Beware, says the Italian proverb, of a reconciled enemy.] **1873** TROLLOPE *Phineas Redux* I. xxxiii. The right honourable gentleman had prided himself on his generosity as a Greek. He would remind the right honourable gentleman that presents from Greeks had ever been considered dangerous. **1929** *Times* 26 Oct. 13 Mr. Moses . . must now be reflecting on the wisdom of the advice to 'fear the Greeks even when they bring gifts'. **1943** E. S. GARDNER *Case of Drowsy Mosquito* vi. 'It wasn't a trap, I tell you.' Nell Sims said . . 'Fear the Greeks when they bear olive branches.' **1980** J. GERSON *Assassination Run* iv. Fear the Greeks bearing gifts was the maxim to be drummed into every novice in the department.

fear: *see also* DO right and fear no man; FOOLS rush in where angels fear to tread.

feast: *see* the COMPANY makes the feast; ENOUGH is as good as a feast.

feather: *see* BIRDS of a feather flock together; FINE feathers make fine birds.

If in FEBRUARY there be no rain, 'tis neither good for hay nor grain
[**1670** J. RAY *English Proverbs* 40 All the moneths in the year curse a fair Februeer.] **1706** J. STEVENS *Spanish & English Dict.* s.v. Febrero, When it does not rain in February, there's neither good Grass nor good Rye. **1906** E. HOLDEN *Country Diary of Edwardian Lady* (1977) 13 If February bring no rain 'Tis neither good for grass nor grain. **1978** R. WHITLOCK *Calendar of Country Customs* iii. One farming adage asserts that 'If in February there be no rain Tis neither good for hay nor grain'.

FEED a cold and starve a fever
Probably intended as two separate admonitions, but the alternative interpretation is explained in quot. 1939.
[**1574** J. WITHALS *Dict.* 66 Fasting is a great remedie in feuers.] **1852** E. FITZGERALD *Polonius* p. ix. 'Stuff a cold and starve a fever,' has been grievously misconstrued, so as to bring on the fever it was meant to prevent. **1867** 'M. TWAIN' *Celebrated Jumping Frog* 69 It was policy to 'feed a cold and starve a fever'. **1910** A. BENNETT *Clayhanger* I. xvii. Edwin's cold was now fully developed; and Maggie had told him to feed it. **1939** C. MORLEY *Kitty Foyle* xxxi. I said I better go downstairs and eat a square meal, 'feed a cold and starve a fever.' . . 'You misunderstand that,' he says. 'It means if you feed a cold you'll have to starve a fever later.' **1961** L. PAYNE *Nose on my Face* iii. 'You ought to have something to eat, sir.' 'What about you?' 'I'll give it a miss.' . . 'Feed a cold, starve a fever.'

feel: *see* a MAN is as old as he feels, and a woman as old as she looks; PRIDE feels no pain.

feet: *see* the CAT would eat fish, but would not wet her feet; *also* FOOT.

fell: *see* LITTLE strokes fell great oaks.

fellow: *see* STONE-dead hath no fellow.

The FEMALE of the species is more deadly than the male
The phrase *the female of the species* is often used with allusion to the proverb.

1911 KIPLING in *Morning Post* 20 Oct. 7 The she-bear thus accosted rends the peasant tooth and nail, For the female of the species is more deadly than the male. **1922** WODEHOUSE *Clicking of Cuthbert* ix. The Bingley-Perkins combination, owing to some inspired work by the female of the species, managed to keep their lead. **1979** *Guardian* 28 Apr. 12 We know phrases about the female of the species being more deadly than the male, but the suffragettes . . seemed to have gone into . . abeyance.

fence: *see* GOOD fences make good neighbours; the GRASS is always greener on the other side of the fence.

fetch: *see* a DOG that will fetch a bone will carry a bone.

fever: *see* FEED a cold and starve a fever.

few: *see* you WIN a few, you lose a few.

fiction: *see* FACT is stranger than fiction; TRUTH is stranger than fiction.

fiddle: *see* there's many a GOOD tune played on an old fiddle.

fiddler: *see* they that DANCE must pay the fiddler.

FIELDS have eyes, and woods have ears
Cf. WALLS have ears.

c **1225** in *Englische Studien* (1902) XXXI. 8 Veld haued hege[=eye], and wude haued heare—*Campus habet lumen et habet nemus auris acumen.* *c* **1390** CHAUCER *Knight's Tale* l. 1522 But sooth is seyd, go sithen many yeres, That 'feeld' hath eyen and the wode hath eres'. **1640** J. HOWELL *Dodona's Grove* A4[v] Hedges have eares, the rurall Proverb sayes. **1738** SWIFT *Polite Conversation* III. 199 'O, Miss; 'tis nothing what we say among ourselves.' . . 'Ay Madam; but they say Hedges have Eyes, and Walls have Ears.' **1905** S. J. WEYMAN *Starvecrow Farm* xxviii. Heedful of the old saying, that fields have eyes and woods have ears, she looked

carefully round her before she laid her hand on the gate.

FIGHT fire with fire
An injunction to counter like with like.
[**1608** SHAKESPEARE *Coriolanus* IV. vii. 54 One fire drives out one fire; one nail, one nail.] **1846** J. F. COOPER *Redskins* III. i. If 'Fire will fight fire', 'Indian' ought to be a match for 'Injin' any day. **1869** P. T. BARNUM *Struggles & Triumphs* xl. I write to ask what your intentions are. . . Do you intend to fight fire with fire? **1980** C. SMITH *Cut-out* xi. 'You think the other Palestinians have hired some heavies as well?' 'Why not? Fight fire with fire.'

fight: *see also* COUNCILS of war never fight.

He who FIGHTS and runs away, may live to fight another day
The phrase *to live to fight another day* is also used allusively.
[Gr. ἀνὴρ ὁ φεύγων καὶ πάλιν μαχήσεται, a man who flees will fight again (attributed to Demosthenes); cf. *a* **1250** *Owl & Nightingale* (1960) l. 176 'Wel fight that wel flight,' seth the wise.] **1542** N. UDALL *Erasmus' Apophthegms* II. 335[v] That same manne, that renneth awaye, Maye again fight, an other daye. **1678** BUTLER *Hudibras* III. iii. For, those that fly, may fight againe, Which he can never do that's slain. **1747** J. RAY *Complete Hist. Rebellion* 61 The Dragoons . . thought proper . . a sudden Retreat; as knowing that, He that fights and runs away, May turn and fight another Day; But he that is in Battle slain, Will never rise to fight again. **1853** S. A. HAMMETT *Stray Yankee* i. I beat an immediate retreat . . having always had implicit faith in the old proverb touching 'fighting and running away', in the hope of 'living to fight another day'. **1876** J. A. AULLS *Sparks & Cinders* 5 For be it known he kept in view That ancient adage, trite but true, That 'He who fights and runs away, May live to fight another day.' **1981** *Daily Telegraph* 10 June 2 (*caption*) He who fights and runs away . . lives to fight another day!

fill: *see* WORK expands so as to fill the time available.

find: *see* the DEVIL finds work for idle hands to do; those who HIDE can find; LOVE will find a way; SAFE bind, safe find; SCRATCH a Russian and you find a Tartar; SEEK and ye shall find; it is easy to find a STICK to beat a dog.

FINDERS keepers (losers weepers)
A colloquial variant of the succeeding proverb.

    1825 J. T. BROCKETT *Glossary of North Country Words* 89 No halfers—findee keepee, lossee seekee. **1856** C. READE *Never too Late* III. xiii. We have a proverb—'Losers seekers finders keepers.' **1874** E. EGGLESTON *Circuit Rider* xv. If I could find the right owner of this money, I'd give it to him; but I take it he's buried. . . 'Finders, keepers,' you know. **1969** *Daily Express* 17 Mar. 9 Where I come from it's finders keepers, losers weepers.

FINDINGS keepings
See also the preceding proverb.

    [Cf. PLAUTUS *Trinummus* l. 63 *habeas ut nanctu's*, he may keep that finds. **1595** A. COOKE *Country Errors* in *Harley MS 5247* 108ᵛ That a man finds is his own, and he may keep it.] **1863** J. H. SPEKE *Discovery of Source of Nile* v. The scoundrels said, 'Findings are keepings, by the laws of our country; and as we found your cows, so we will keep them.' **1904** *Daily Chronicle* 27 Sept. 1 Harsh sentences of imprisonment for 'findings-keepings' offences. **1963** G. GREENE *Sense of Reality* 38 'I found them in the passage.' . . 'Finding's[1] not keeping here,' he said, 'whatever it may be up there.'

    [1] i.e. the action or fact of finding, rather than the objects found, as in earlier examples.

FINE feathers make fine birds
Beautiful clothes confer beauty on the wearer.

    [**1583** J. SPONDANUS in tr. *Odyssey* VI. 81 *apud meos Vascones . . hac parœmia . . : speciosae plumae avem speciosam consti-*

tuunt, this is a proverb amongst my fellow Gascons: beautiful feathers make a beautiful bird.] **1592** G. DELAMOTHE *French Alphabet* II. 29 The faire feathers, makes a faire foule. **1658** E. PHILLIPS *Mysteries of Love & Eloquence* 162 Fine feathers make fine birds. As you may see in Hide Park. **1858** R. S. SURTEES *Ask Mamma* x. Mrs. Joe . . essayed to pick her to pieces, intimating that she was much indebted to her dress—that fine feathers made fine birds. **1952** H. F. M. PRESCOTT *Man on Donkey* I. 69 'Well, if fine feathers'll make a fine bird—' he said, and brushed her cheek with the back of his hand.

FINE words butter no parsnips
Nothing is ever achieved by fine words alone. Parsnips are traditionally garnished with butter before serving.

    1639 J. CLARKE *Parœmiologia Anglo–Latina* 169 Faire words butter no parsnips. **1692** R. L'ESTRANGE *Fables of Aesop* cccxl. Relations, Friendships, are but Empty Names of Things, and Words Butter No Parsnips. **1763** A. MURPHY *Citizen* I. ii. What becomes of his Greek and Latin now? Fine words butter no parsnips. **1848** THACKERAY *Vanity Fair* xix. Who . . said that 'fine words butter no parsnips'? Half the parsnips of society are served and rendered palatable with no other sauce. **1947** 'M. WESTMACOTT' *Rose & Yew Tree* xviii. Fine words butter no parsnips. Who will help me to reopen my shop? Who will build me a house? **1978** R. PARKER *Men of Dunwich* xii. Fine words, they say, butter no parsnips. They certainly build no sea-walls.

fine: *see also* RAIN before seven, fine before eleven.

FINGERS were made before forks
Commonly used as a polite excuse for eating with one's hands at table.

    1567 *Loseley MSS* (1836) 212 As God made hands before knives, So God send a good lot to the cutler's wives. **1738** SWIFT *Polite Conversation* II. 136 (Colonel takes them [some fritters] out with his

Hand.) *Col.* Here, Miss, they say, Fingers were made before Forks, and Hands before Knives. **1857** TROLLOPE *Barchester Towers* II. iii. Miss Thorne . . was always glad to revert to anything and . . would doubtless in time have reflected that fingers were made before forks, and have reverted accordingly. **1934** A. CHRISTIE *Why didn't They ask Evans?* xxviii. As they truly used to say to me in my youth . . fingers were made before forks—and teeth were used before fingers.

fire: *see* a BURNT child dreads the fire; DIRTY water will quench fire; FIGHT fire with fire; you should KNOW a man seven years before you stir his fire; if you PLAY with fire you get burnt; no SMOKE without fire; THREE removals are as bad as a fire.

FIRST catch your hare
Commonly thought to originate in the recipe for hare soup in Mrs Glasse's *Art of Cookery* (1747) or in Mrs Beeton's *Book of Household Management* (1851), but not found there (see quot. 1896).
[Cf. *c* **1300** BRACTON *De Legibus Angliae* IV. xxi. *vulgariter dicitur, quod primo opportet cervum capere, & postea cum captus fuerit illum excoriare,* it is commonly said that one must first catch the deer, and afterwards, when he has been caught, skin him.] **1801** *Spirit of Farmers' Museum* 55 How to dress a dolphin, first catch a dolphin. **1855** THACKERAY *Rose & Ring* xiv. 'To seize wherever I should light upon him—' 'First catch your hare!' . . exclaimed his Royal Highness. **1896** *Daily News* 20 July 8 The familiar words, 'First catch your hare,' were never to be found in Mrs. Glasse's famous volume. What she really said was, 'Take your hare when it is cased.'[1] *a* **1961** A. CHRISTIE *Miss Marple's Final Cases* (1979) 39, I haven't begun yet. 'First catch your hare.'
[1] skinned.

FIRST come, first served
[Cf. late 13th-cent. Fr. *qui ainçois vient au molin ainçois doit moldre,* he who

comes first to the mill may grind first.] *c* **1390** CHAUCER *Wife of Bath's Prologue* l. 389 Whoso that first to mille comth, first grynt. **1548** H. BRINKELOW *Complaint of Roderick Mors* xvii. Ye haue a parciall lawe in making of tachmentes,[1] first come first serued. **1608** R. ARMIN *Nest of Ninnies* D1 He found Sexton . . making nine graues . . and who so dyes next, first comes, first seru'd. **1763** SMOLLETT *Letter* 19 Oct. in *Travels* (1766) I. viii. I pulled out the post-book, and began to read . . the article which orders that the traveller who comes first shall be first served. **1819** SCOTT *Montrose* in *Tales of My Landlord* 3rd Ser. IV. xii. All must . . take their place as soldiers should, upon the principle of, first come, first served. **1979** M. A. SHARP *Sunflower* v. In London they formed polite queues . . a nice orderly version of first come, first served.
[1] attachments, i.e. judicial seizure or apprehension of one's person or goods (obsolete).

The FIRST duty of a soldier is obedience
The last quotation presents a variation of the proverb.
1847 J. GRANT *Romance of War* IV. xv. 'What do the wiseacres at head-quarters mean, in sending a detachment there?' 'I suppose they scarcely know themselves. But obedience—We all know the adage.' **1872** G. J. WHYTE-MELVILLE *Satanella* II. xxiv. 'The first duty of a soldier is obedience,' he answered in great glee. **1915** F. M. HUEFFER *Good Soldier* III. iii. She had been handed over to him, like some patient mediæval virgin; she had been taught all her life that the first duty of a woman is to obey.

FIRST impressions are the most lasting
1700 CONGREVE *Way of World* IV. i. How shall I receive him? . . There is a great deal in the first Impression. **1791** H. JACKSON in *Publications of Colonial Society of Massachusetts* (1954) XXXVI. 112, I am affraid it is too late . . and you know that first impressions are the most lasting. **1844** DICKENS *Martin Chuzzlewit*

v. I didn't like to run the chance of being found drinking it . . for first impressions, you know, often go a long way, and last a long time. **1926** R. M. OGDEN *Psychology & Education* xii. Primacy is popularly expressed by the statement that 'first impressions are lasting'. **1946** J. B. PRIESTLEY *Bright Day* ii. He ought to look neat and tidy . . . It's half the battle . . making a good first impression.

On the FIRST of March, the crows begin to search

1846 M. A. DENHAM *Proverbs relating to Seasons, &c.* 39 On the first of March, the crows begin to search. Crows are supposed to begin pairing on this day. **1847** R. CHAMBERS *Popular Rhymes of Scotland* (ed. 2) 165 On the first of March, The craws begin to search; By the first o'April, They are sitting still. **1906** E. HOLDEN *Country Diary of Edwardian Lady* (1977) 132 By the 1.st of March the crows begin to search, By the 1.st of April they are sitting still, By the 1.st of May they are flown away, Creeping greedy back again With October wind and rain.

It is the FIRST step that is difficult

The proverb has various forms. The story to which the French saying alludes is referred to in the last quotation.

[Cf. Fr. *ce n'est que le premier pas qui coûte,* it is only the first step that costs.] *c* **1596** A. MUNDAY et al. *Sir Thomas More* 11 Would I were so far on my journey. The first stretch is the worst methinks. **1616** J. WITHALS *Dict.* (rev. ed.) 576 The first step is as good as halfe ouer. **1876** A. B. MEACHAM *Wi-Ne-Ma & her People* iv. He had fortified himself against the charms of the Indian maiden, as he thought, but consented to visit her. Ah! my man, have you not learned that when the first step is taken the next follows easily. **1979** J. BARNETT *Backfire is Hostile* xi. St. Denis was executed . . . Afterwards he picked up his head and walked for six miles . . . The wise man said, 'The distance . . is not important. It was the first step that was difficult.'

FIRST things first

1894 G. JACKSON (*title*) First things first; addresses to young men. **1920** W. RILEY *Yorkshire Suburb* 136 The dear lady was . . incapable . . of putting first things first. **1979** 'L. BLACK' *Penny Murders* iv. They dropped the talking; first things first, as Kate always felt about a pleasant meal.

first: *see also* every DOG is allowed one bite; whom the GODS would destroy, they first make mad; SELF-preservation is the first law of nature; if at first you don't SUCCEED, try, try, try, again; THINK first and speak afterwards; he that will THRIVE must first ask his wife.

The FISH always stinks from the head downwards

The freshness of a dead fish can be judged from the condition of its head. Thus, when the responsible part (as the leaders of a country, etc.) is rotten, the rest will soon follow.

[Gr. ἰχθὺς ἐκ τῆς κεφαλῆς ὄζειν ἄρχεται, a fish begins to stink from the head.] **1581** G. PETTIE tr. S. *Guazzo's Civil Conversation* III. 51 If the prouerbe be true, . . that a fishe beginneth first to smell at the head, . . the faultes of our seruantes will be layed vppon vs. **1611** R. COTGRAVE *Dict. French & English* s.v. Teste, Fish euer begins to taint at the head; the first thing that's deprau'd in man's his wit. **1674** J. JOSSELYN *Account of Two Voyages to New England* 9 The first part that begins to taint in a fish is the head. **1915** W. S. CHURCHILL *Letter* 3 Dec. in M. Gilbert *Winston S. Churchill* (1972) III. Compan. II. 1309 The guilt of criminality attaches to those responsible. 'Well,' said the Aga Khan, 'fish goes rotten by the head.' **1981** *Sunday Telegraph* 3 May 16 'The fish', as the saying goes, 'always stinks from the head downwards.' Last Sunday we deplored Mr. Michael Foot's liking for the street politics of marches and 'demos'. Since then, a hundred Labour MPs . . have followed their leader's example.

FISH and guests stink after three days

[Cf. PLAUTUS *Miles Gloriosus* l. 741 *nam hospes nullus tam in amici hospitium devorti potest, quin, ubi triduom continuom fuerit, iam odiosus siet*, no host can be hospitable enough to prevent a friend who has descended on him from becoming tiresome after three days.] **1580** LYLY *Euphues* II. 81 As we say in Athens, fishe and gestes in three dayes are stale. **1648** HERRICK *Hesperides* 169 Two dayes y'ave larded here; a third yee know, Makes guests and fish smell strong; pray go. **1736** B. FRANKLIN *Poor Richard's Almanack* (Jan.) Fish and visitors smell in three days. **1869** *Notes & Queries* 4th Ser. IV. 272 'See that you wear not out your welcome.' This is an elegant rendering of the vulgar saying that 'Fish and company stink in three days'. **1981** *Times* 18 June 14 The bedside ashtray read: Guests, like fish, stink after three days.

There are as good FISH in the sea as ever came out of it

Now often used as a consolation to rejected lovers, 'there are plenty more fish in the sea'.

c **1573** G. HARVEY *Letter-Book* (1884) 126 In the mayne sea theres good stoare of fishe, And in delicate gardens . . Theres always greate varietye of desirable flowers. **1816** T. L. PEACOCK *Headlong Hall* xiv. There never was a fish taken out of the sea, but left another as good behind. **1822** SCOTT *Nigel* III. x. Ye need not sigh sae deeply . . . There are as gude fish in the sea as ever came out of it. **1944** W. S. MAUGHAM *Razor's Edge* iii. I'm a philosopher and I know there are as good fish in the sea as ever came out. I don't blame her. You're young. I've been young too. **1962** A. CHRISTIE *Mirror Crack'd* xiv. Where women are concerned, there are as good fish in the sea as ever came out of it.

All is FISH that comes to the net

Everything can be used to advantage. *The* is frequently replaced by a possessive pronoun or phrase. Cf. *All is* GRIST *that comes to the mill*.

c **1520** in *Ballads from MSS* (1868–72) I. 95 Alle ys ffysshe that commyth to the nett. **1564** W. BULLEIN *Dialogue against Fever* 70 Takyng vp commoditie,[1] refusyng nothyng: all is fishe that commeth to the nette. **1680** BUNYAN *Mr. Badman* 19 What was his fathers could not escape his fingers, all was fish that came to his net. **1803** C. DIBDIN *Professional Life* II. 137, I be a jolly fisherman, I takes what I can get, Still going on 'my betters' plan, all's fish that comes to net. **1848** DICKENS *Dombey & Son* ix. 'All's fish that comes to your net, I suppose?' 'Certainly,' said Mr. Brogley. 'But sprats an't whales, you know.' **1936** A. CHRISTIE *Murder in Mesopotamia* xix. I don't know that she cares for one more than the other . . . All's fish that comes to her net at present.

[1] advantage, opportunity.

fish: *see also* BIG fish eat little fish; the CAT would eat fish, but would not wet her feet; LITTLE fish are sweet.

fit: *see* if the CAP fits, wear it.

flag: *see* TRADE follows the flag.

flattery: *see* IMITATION is the sincerest form of flattery.

flea: *see* BIG fleas have little fleas upon their backs to bite them; if you LIE down with dogs, you will get up with fleas; NOTHING should be done in haste but gripping a flea.

flesh: *see* what's BRED in the bone will come out in the flesh.

flew: *see* a BIRD never flew on one wing.

flight: *see* BLESSINGS brighten as they take their flight.

flock: *see* BIRDS of a feather flock together.

flower: *see* APRIL showers bring forth May flowers.

fly: *see* (noun) EAGLES don't catch flies;

HONEY catches more flies than vinegar; (verb) TIME flies; *also* FLEW.

flying: *see* it is good to make a BRIDGE of gold to a flying enemy.

folk: *see* there's NOWT so queer as folk; YOUNG folks think old folks to be fools, but old folks know young folks to be fools.

follow: *see* TRADE follows the flag.

He that FOLLOWS freits, freits will follow him
He that looks for portents of the future will find himself dogged by them: a Scottish proverb.

   c **1700** in J. Pinkerton *Scottish Tragic Ballads* (1781) I. 47 Wha luik to freits,[1] my master deir, Freits will ay follow them. **1721** J. KELLY *Scottish Proverbs* 128 He that follows Freets, Freets will follow him. He that notices superstitious Observations (such as spilling of Salt) . . it will fall to him accordingly. **1804** M. PARK in Lockhart *Life of Scott* (1837) II. i. He answered, smiling, 'Freits (omens) follow those who look to them.' . . Scott never saw him again. **1914** *Times Literary Supplement* 9 Apr. 178 The Kings of Scots have always been beset by omens, and . . to him who follows freits, freits follow.
   [1] omens.

folly: *see* where IGNORANCE is bliss, 'tis folly to be wise.

fonder: *see* ABSENCE makes the heart grow fonder.

A FOOL and his money are soon parted
   1573 T. TUSSER *Husbandry* (rev. ed.) ix. A foole and his money be soone at debate: which after with sorow repents him too late. **1587** J. BRIDGES *Defence of Government in Church of England* xv. 1294 A foole and his money is soone parted. **1616** T. DRAXE *Adages* 166 A foole, and his money are soone parted. **1771** SMOLLETT *Humphry Clinker* I. 174 She tossed her nose in distain, saying, she

supposed her brother had taken him into favour . . : that a fool and his money were soon parted. **1816** SCOTT *Antiquary* III. x. 'It's a capital piece; it's a Joe Manton, that cost forty guineas.' 'A fool and his money is soon parted, nephew.' **1945** F. THOMPSON *Lark Rise* xi. Copperplate maxims: 'A fool and his money are soon parted'; 'Waste not, Want not' . . and so on. **1981** C. BERMANT *Patriarch* xx. I can see now I was a fool, perhaps even a greedy fool, and a fool and his money are soon parted.

A FOOL at forty is a fool indeed
   [**1557** R. EDGEWORTH *Sermons* 301 When he [Rehoboam] begonne hys raigne he was one and fortye yeares of age. . . And he that hath not learned some experience or practice and trade of the world by that age will neuer be wise.] **1725** YOUNG *Universal Passion* II. 16 Be wise with speed; A fool at forty is a fool indeed. **1751** N. COTTON *Visions in Verses* 13 He who at fifty is a fool, Is far too stubborn grown for school. **1908** L. MITCHELL *New York Idea* III. 112, I shall come or not [to your wedding] as I see fit. And let me add, my dear brother, that a fool at forty is a fool indeed. **1982** B. EMECHETA *Destination Biafra* i. 'What is it they say about a fool at forty?' 'I don't think you'll be a fool forever, sir.'

There's no FOOL like an old fool
   1546 J. HEYWOOD *Dialogue of Proverbs* II. ii. F4[v] But there is no foole to the olde foole, folke saie. **1721** J. KELLY *Scottish Proverbs* 256 No fool to an old Fool. Spoken when Men of advanc'd Age behave themselves, or talk youthfully, or wantonly. **1732** T. FULLER *Gnomologia* no. 3570 No Fool like the old Fool. **1814** SCOTT *Waverley* III. xv. And troth he might hae ta'en warning, but there's nae fule like an ould fule. **1910** KIPLING *Rewards & Fairies* 257 'There are those who have years without knowledge.' 'Right,' said Puck. 'No fool like an old fool.' **1979** L. BARNEA *Reported Missing* xvii. It is too late for that. As they say, no fool like an old fool.

A FOOL may give a wise man counsel

a **1350** *Ywain & Gawain* (EETS) l. 1477
Bot yit a fole that litel kan,[1] May
wele cownsail another man. **1641**
D. FERGUSSON *Scottish Proverbs* (STS) no.
84 A fool may give a wyse man a
counsell. **1721** J. KELLY *Scottish Proverbs*
25 A Fool may give a wise Man counsel
by a time. An Apology of those who
offer their Advice to them, who may be
supposed to excel them in Parts and
Sense. **1818** SCOTT *Heart of Midlothian*
IV. viii. If a fule may gie a wise man
a counsel, I wad hae him think twice
or he mells[2] wi' Knockdunder. **1942**
E. P. OPPENHEIM *Man who changed Plea*
xvii. Aren't we all fools . . in one or two
things? . . Even a fool, though, can
sometimes give good advice.
  [1] knows.   [2] meddles.

fool: *see also* CHILDREN and fools tell the
truth; FORTUNE favours fools; a man
who is his own LAWYER has a fool for his
client; MORE people know Tom Fool
than Tom Fool knows; SIX hours' sleep
for a man, seven for a woman, and eight
for a fool; YOUNG folks think old folks to
be fools, but old folks know young folks
to be fools; *also* FOOLS.

foolish: *see* PENNY wise and pound foolish.

**FOOLS and bairns should never see half-
done work**
They may mistakenly judge the quality
of the finished article from its awkward-
ness while it is being produced.
  1721 J. KELLY *Scottish Proverbs* 108
Fools should not see half done Work.
Many fine Pieces of Work will look . .
aukward when it is a doing. **1818** W.
SCOTT *Letter* Dec. (1933) V. 265 'Bairns
and fools' . . according to our old canny
proverb should never see half done
work. **1902** A. BENNETT *Anna of Five
Towns* viii. No stage of the manufacture
was incredible by itself, but the result
was incredible . . authenticating the adage that fools and children should never
see anything till it is done. **1913**
A. & J. LANG *Highways & Byways in Border* ix. To the lay eye improvement is yet

barely perceptible. 'Fools and bairns',
however, they tell us, 'should never see
half-done work.'

**FOOLS ask questions that wise men can-
not answer**
  1666 G. TORRIANO *Italian Proverbs* 249
One fool may ask more than seven wise
men can answer. **1738** SWIFT *Polite Conversation* II. 156 'Miss, can you tell
which is the white Goose?' . . 'They say,
a Fool will ask more Questions than the
wisest body can answer.' **1821** SCOTT
Pirate II. v. Bryce Snaelsfoot is a cautious man. . . He knows a fool may
ask more questions than a wise man
cares to answer. **1871** J. S. JONES *Life of
J. S. Batkins* liv. Bean appeared always
to be fond of Amanda. . . I asked him
one day. . . He looked at me, and said,
'Batkins, fools ask questions that wise
men cannot answer.'

**FOOLS build houses and wise men live
in them**
  1670 J. RAY *English Proverbs* 91 Fools
build houses, and wise men buy them.
1721 J. KELLY *Scottish Proverbs* 110 Fools
Big[1] Houses and wise Men buy them. I
knew a Gentleman buy 2000 l. worth
of Land, build a House upon it, and
sell both House and Land to pay
the Expences of his building. **1875**
A. B. CHEALES *Proverbial Folk-Lore* 43
*Fools build houses, and wise men live in
them* is another proverb on this subject;
it is partly true. **1911** W. F. BUTLER
Autobiography xix. The adage says that
fools build houses for other men to live
in. Certainly the men who build the big
house of Empire for England usually get
the attic . . for their own lodgment.
  [1] build.

FOOLS for luck
The construction is apparent from quot.
1834. Cf. FORTUNE *favours fools.*
  [**1631** JONSON *Bartholomew Fair* II. ii.
Bring him a six penny bottle of Ale;
they say, a fooles handsell[1] is lucky.]
1834 *Narrative of Life of David Crockett*
xiii. The old saying—'A fool for luck,
and a poor man for children.' **1854**

J. B. JONES *Life of Country Merchant* xix. They attribute your good fortune to the old hackneyed adage, 'A fool for luck'. **1907** D. H. LAWRENCE *Phoenix II* (1968) 6 'You'll make our fortunes.' 'What!' he exclaimed, 'by making a fool of myself? They say fools for luck. What fools wise folks must be.' **1927** N. MARTIN *Mosaic Earring* xviii. I don't know how you fell on it that you were the spittin' image of this young millionaire, a fool for luck, maybe. [Cf. **1981** T. BARLING *Bikini Red North* xi. All fools are lucky; isn't that the adage?]
   [1] gift.

FOOLS rush in where angels fear to tread
   1711 POPE *Essay on Criticism* l. 625 No Place so Sacred from such Fops is barr'd, Nor is Paul's Church more safe than Paul's Church-yard: Nay, fly to Altars; there they'll talk you dead; For Fools rush in where Angels fear to tread. **1858** G. J. McREE *Iredell's Life & Correspondence* II. 277 Rash presumption illustrates the line, 'Fools rush in where angels fear to tread'. **1922** JOYCE *Ulysses* 649 Prying into his private affairs on the *fools step in where angels* principle. **1943** H. McCLOY *Do not Disturb* ii. The folly of the officious is proverbial: don't rush in where angels fear to tread. **1967** M. DRABBLE *Jerusalem the Golden* iii. The handling of the 'Fools rush in where angels fear to tread' theme, though not wholly elegant, showed ambition.

forbear: *see* BEAR and forbear.

FOREWARNED is forearmed
   [L. *praemonitus, praemunitus*, forewarned, forearmed.] *c* **1425** J. ARDERNE *Treatises of Fistula* (EETS) 22 He that is warned afore is noght bygiled. *c* **1530** J. REDFORD *Wit & Science* l. 1093 Once warnd, half armd folk say. **1587** GREENE *Card of Fancy* IV. 23, I giue thee this Ring of golde, wherin is written . . *Præmonitus, Premunitus* . . inferring this sense, that hee which is forewarned by friendlie counsoule of imminent daungers, is fore-armed against all future mishappe. *a* **1661** T. FULLER *Worthies* (Devon) 272 Let all ships passing thereby be fore-armed because fore-warned thereof. **1885** 'LE JEMLYS' *Shadowed to Europe* xxv. 'Forewarned is forearmed,' he thought, as he complimented himself upon his success in baffling the attempt to ensnare him. **1978** T. SHARPE *Throwback* vi. I have summoned Lockhart to be present next week. . . Forewarned is forearmed as the old saying is.

forget: *see* a BELLOWING COW soon forgets her calf.

forgive: *see* to ERR is human (to forgive divine).

foot: *see* it is not SPRING until you can plant your foot upon twelve daisies; *also* FEET.

foretold: *see* LONG foretold long last, short notice soon past.

fork: *see* FINGERS were made before forks.

FORTUNE favours fools
   [L. *fortuna favet fatuis*, fortune favours fools.] **1546** J. HEYWOOD *Dialogue of Proverbs* II. vi. I1^v They saie as ofte, god sendeth fortune to fooles. **1563** B. GOOGE *Eclogues* E5 But Fortune fauours Fooles as old men saye And lets them lyue And take the wyse awaye. **1738** GAY *Fables* 2nd Ser. II. xii. 'Tis a gross error, held in schools, That Fortune always favours fools. **1960** O. MANNING *Great Fortune* I. vi. Fortune favours fools . . . We were forced to tarry while he slumbered.

FORTUNE favours the brave
   [Attributed to Simonides of Keos (e.g. τυχὰ τολμῶσιν ἀρήγει) from Claudian *Epistles* IV. 9 *fors iuvat audentes, Cei sententia vatis*, fortune favours the brave, according to the poet of Keos; VIRGIL *Aeneid* X. 284 *audentes fortuna iuvat*.] *c* **1385** CHAUCER *Troilus & Criseyde* IV. 600 Thenk ek Fortune, as wel thiselves woost, Helpeth hardy man to his enprise. *c* **1390** GOWER *Confessio Amantis* VII. 4902 And seith, 'Fortune unto the bolde Is favorable forto helpe'.

a **1625** BEAUMONT & FLETCHER *Prophetess* IV. vi. He is the scorn of Fortune: but you'll say, That she forsook him for his want of courage, But never leaves the bold. **1724** A. RAMSAY *Works* (1953) II. 288 Fortune aye favours the active and bauld. **1752** in W. Johnson *Papers* (1939) IX. 86 Make no doubt but Fortune will favour the brave. **1885** TROLLOPE *Dr. Thorne* II. vii. Fortune, who ever favours the brave, specially favoured Frank Gresham. **1970** D. M. DAVIN *Not Here, not Now* II. ii. 'May I see you home?' . . 'Fortune favours the brave, they say.'

fortune: *see also* EVERY man is the architect of his own fortune; OPPORTUNITY never knocks twice at any man's door.

forty: *see* a FOOL at forty is a fool indeed; LIFE begins at forty; SAINT Swithun's day if thou be fair for forty days it will remain.

foul: *see* it's an ILL bird that fouls its own nest.

FOUR eyes see more than two

Two people are more observant than one alone. Cf. TWO *heads are better than one.*

[L. *plus vident oculi, quam oculus,* eyes see more than one eye.] **1591** A. COLYNET *True Hist. Civil Wars France* 37 Two eyes doo see more then one. **1592** G. DELAMOTHE *French Alphabet* II. 45 Foure eyes can see more then two. **1642** T. FULLER *Holy State* IV. v. Matters of inferiour consequence he will communicate to a fast friend, and crave his advice; for two eyes see more than one. **1898** F. M. MULLER *Auld Lang Syne* 80 But who has ever examined any translation from any language, without finding signs of . . carelessness or ignorance? Four eyes see more than two. **1943** J. WEBB *No Match for Murder* v. It might be worth my while to string along. . . Four eyes are better than two.

four: *see also* there goes more to MARRIAGE than four bare legs in a bed.

There's no such thing as a FREE lunch

Originally a colloquial axiom in US economics, though now in general use. The proverb implies that you cannot get something for nothing.

1969 *Newsweek* 29 Dec. 52, I was taught . . the first and only law of economics: 'There is no such thing as a free lunch.' **1971** *New Yorker* 25 Sept. 76 There is no such thing as a free lunch. . . The idea has proved so illuminating for environmental problems that I am borrowing it from its original source, economics. **1979** *New York Times Magazine* 9 Sept. 16 A theory which became popular in the late 1970's . . held that big effects could be produced by small actions. For example, a tax cut of $1 billion would raise the national income by $5 billion, increase the Government revenue by $2 billion, reduce unemployment and restrain inflation. This theorem was discovered written on the back of a napkin in a MacDonald's restaurant. That gave rise to the axiom 'There ain't no such thing as a free lunch, but there is a cheap one.' **1979** L. ST. CLAIR *Obsessions* xi. There's no such thing as a free lunch. So, in return for your help, what do you ask?

free: *see also* the BEST things in life are free; THOUGHT is free.

freit (omen): *see* he that FOLLOWS freits, freits will follow him.

Frenchman: *see* one ENGLISHMAN can beat three Frenchmen.

fresh: *see* don't THROW out your dirty water until you get in fresh.

Friday: *see* Monday's CHILD is fair of face.

A FRIEND in need is a friend indeed

[Cf. ENNIUS *Hecuba* XIII. 210 *amicus certus in re incerta cernitur,* a sure friend is known in unsure times.] *c* **1035** *Durham Proverbs* (1956) 10 Æt thearfe man sceal freonda cunnian [friend shall be known in time of need]. *a* **1400** *Titus & Vespasian* (1905) 98, I shal the save

When tyme cometh thou art in nede;
Than ogh men frenshep to shewe in
dede. *a* **1449** LYDGATE *Minor Poems*
(EETS) II. 755 Ful weele is him that fyn-
dethe a freonde at neede. **1678** J. RAY
English Proverbs (ed. 2) 142 A friend
in need[1] is a friend indeed. **1773**
R. GRAVES *Spiritual Quixote* II. VIII. XX.
(*heading*) A Friend in Need is a Friend
indeed. **1866** C. READE *Griffith Gaunt*
III. xv. You came to my side when I was
in trouble. . . A friend in need is a friend
indeed. **1935** E. BELL *Fish on Steeple* iv.
If they's one thing I do it's never lay
down on my friends. I say a friend in
need is a friend indeed.

[1] friend in need, i.e. a friend who helps
when one is in need.

friend: *see also* the BEST of friends must
part; SAVE us from our friends; SHORT
reckonings make long friends.

frost: *see* so many MISTS in March, so many
frosts in May.

When all FRUIT falls, welcome haws
Used specifically of a person who takes
of necessity an older or otherwise unsuitable lover.

1721 J. KELLY *Scottish Proverbs* 350
When all Fruit fa's welcome ha's. . .
Spoken when we take up with what's
coarse, when the good is spent. **1914**
K. F. PURDON *Folk of Furry Farm* vii.
'Lame of a leg, and grey in the head! . .
That's a fancy man for a girl to take!'
'Marg was none too young herself . .
and when all fruit fails, welcome haws!
She wanted someone.'

fruit: *see also* he that would EAT the fruit
must climb the tree; SEPTEMBER blow
soft, till the fruit's in the loft; STOLEN
fruit is sweet; the TREE is known by its
fruit.

FULL cup, steady hand
Used especially to caution against spoiling a comfortable or otherwise enviable
situation by careless action.

c **1025** *Durham Proverbs* (1956) 15 Swa

fulre fæt swa hit mann sceal fægror
beran [the more full the cup, the more
carefully must one carry it]. *c* **1325**
Proverbs of Hending in *Anglia* (1881) IV.
293 When the coppe is follest, thenne
ber hire feyrest. **1721** J. KELLY *Scottish
Proverbs* 346 When the Cup's full carry it
even. When you have arrived at Power
and Wealth, take a care of Insolence,
Pride, and Oppression. **1732** T. FULLER
Gnomologia no. 122 A full Cup must be
carried steadily. **1889** C. M. YONGE in
Monthly Packet Christmas 46 Poor
things! They were so happy—so open-
hearted. I did long to caution them. 'Full
cup, steady hand.' **1903** G. H. KNIGHT
Master's Questions xxi. I would listen . .
to this question . . whenever . . I am
eagerly reaching out my hands to grasp
what may satisfy an unlikely ambition.
All hands are not steady enough to carry
a full cup.

**It's ill speaking between a FULL man and
a fasting**
A hungry man is never on good terms
with a well-fed man; in quot. 1824, used
as an incitation to eat.

a **1641** D. FERGUSSON *Scottish Proverbs*
(STS) no. 1349 Thair is nothing betuix a
bursten body and a hungered. **1824**
SCOTT *Redgauntlet* I. xi. Ye maun eat and
drink, Steenie . . for we do little else
here, and it's ill speaking between a fou
man and a fasting. **1934** J. BUCHAN *Free
Fishers* ii. It's ill speaking between a full
man and a fasting, but two fasting men
are worse at a crack.

One FUNERAL makes many
Cf. One WEDDING *brings another.*
1894 BLACKMORE *Perlycross* I. vii. It
has been said, and is true too often . .
that one funeral makes many. A strong
east wind . . whistled through the
crowd of mourners.

funeral: *see also* DREAM of a funeral and
you hear of a marriage.

fury: *see* HELL hath no fury like a woman
scorned.

When the FURZE is in bloom, my love's in tune

A corollary of *When the* GORSE *is out of bloom, kissing's out of fashion.*

[Cf. *c* **1225** in *Englische Studien* (1902) XXXI. 5 Whanne bloweth[1] the brom, thanne wogeth[2] the grom; Whanne bloweth the furs, thanne wogeth he wurs.] **1752** *Poor Robin's Almanack* Aug.

B3[v] Dog-days are in he'll say's the reason Why kissing now is out of season: but Joan says furze[3] in bloom still, and she'll be kiss'd if she's her will. **1908** *Spectator* 9 May 740 At almost any season of the year gorse can be found in . . flower. . . When the furze is in bloom, my love's in tune.

[1] flowers.   [2] woos.   [3] gorse.

G

gain: *see* (noun) one man's LOSS is another man's gain; there's no great LOSS without some gain; (verb) what you LOSE on the swings you gain on the roundabouts; NOTHING venture, nothing gain.

game: *see* LOOKERS-ON see most of the game.

gamekeeper: *see* an old POACHER makes the best gamekeeper.

gander: *see* what is SAUCE for the goose is sauce for the gander.

GARBAGE in, garbage out
Garbage is a colloquial term in data processing for 'incorrect input' which will, according to the proverb, inevitably produce faulty output. The acronymic form *GIGO* is also found.
1964 *CIS Glossary of Automated Typesetting. & Related Computer Terms* (Composition Information Services, L.A.) 15 The relationship between input and output is sometimes—when input is incorrect—tersely noted by the expression 'garbage in, garbage out'. **1966** E. J. & J. A. McCARTHY *Integrated Data Processing Systems* v. Many data processing departments put their best operators on verifiers because they wish to avoid the effect of the GIGO principle (Garbage In—Garbage Out). **1978** *Omni* Oct. 30 Computer scientists have a favorite proverb: 'Garbage in, garbage out.' It means that bad input data can be manipulated to produce nearly any output desired, but it will be useless. **1979** *Management Services in Government* Nov. 198 There are two ways to make the best of their [typists'] skills. . . The second reflected the old computer adage, 'garbage in, garbage out'. Typists can get through their work more quickly when they are working from good clear source documents.

garment: *see* SILENCE is a woman's best garment.

gate: *see* a CREAKING door hangs longest; one man may STEAL a horse, while another may not look over a hedge.

gather: *see* a ROLLING stone gathers no moss.

gathered: *see* where the CARCASE is, there shall the eagles be gathered together.

geese: *see* on SAINT Thomas the Divine kill all turkeys, geese, and swine.

generation: *see* from CLOGS to clogs is only three generations; from SHIRTSLEEVES to shirtsleeves in three generations.

It takes three GENERATIONS to make a gentleman
[**1598** J. KEPERS tr. *A. Romei's Courtier's Academy* 187 He may bee called absolutely noble, who shall have lost the memory of his ignobilitie . . during the reuolution of three generations. **1625** F. MARKHAM *Five Decades of Honour* ii. Three perfit descents, do euer so conclude a perfit Gentleman of Blood.] **1823** J. F. COOPER *Pioneers* I. xviii. You will find it no easy matter to make a gentleman of him. The old proverb says, that 'it takes three generations to make a gentleman'. **1915** W. S. MAUGHAM *Of Human Bondage* xxvii. He remembered his uncle's saying that it took three generations to make a gentleman: it was a companion proverb to the silk purse and the sow's ear. **1940** 'M. INNES' *Comedy of Terrors* i. It has always been possible to make a gentleman in three generations; nowadays . . the thing is done in two.

generous: *see* be JUST before you're generous.

GENIUS is an infinite capacity for taking pains
[**1858** CARLYLE *Frederick the Great* I. iv. iii. 'Genius' . . means transcendent

capacity of taking trouble, first of all.] **1870** J. E. HOPKINS *Work amongst Working Men* iv. Gift, like genius, I often think, only means an infinite capacity for taking pains. **1959** M. BRADBURY *Eating People is Wrong* iv. Genius is an infinite capacity for taking pains. But we should still foster it, however much of an embarrassment it may be to us. **1974** T. SHARPE *Porterhouse Blue* xiv. The modern fashion [of research] comes, I suppose, from a literal acceptance of the ridiculous dictum that genius is an infinite capacity for taking pains.

gentleman: *see* when ADAM delved and Eve span, who was then the gentleman?; an ENGLISHMAN'S word is his bond; it takes three GENERATIONS to make a gentleman.

gently: *see* DRIVE gently over the stones; EASY does it; if you gently touch a NETTLE it'll sting you for your pains.

get: *see* you cannot get BLOOD from a stone; if you don't like the HEAT, get out of the kitchen; if you LIE down with dogs, you will get up with fleas; the MORE you get, the more you want; what a NEIGHBOUR gets is not lost; you cannot get a QUART into a pint pot; you don't get SOMETHING for nothing; *also* GOT.

Never look a GIFT horse in the mouth
A horse's age is commonly gauged by the state of its teeth. The proverb warns against questioning the quality or use of a lucky chance or gift.

[*a* **420** JEROME *Commentary on Epistle to Ephesians* Preface, *noli .. ut vulgare proverbium est, equi dentes inspicere donati,* do not, as the common proverb says, look at the teeth of a gift horse.] *a* **1510** J. STANBRIDGE *Vulgaria* (EETS) 27 A gyuen hors may not [be] loked in the tethe. **1546** J. HEYWOOD *Dialogue of Proverbs* I. v. B2ᵛ Where gyfts be gyuen freely, est west north or south, No man ought to loke a geuen hors in the mouth. **1659** N.R. *Proverbs* 80 No man ought to look a guift Horse¹ in the mouth. **1710** S. PALMER *Proverbs* 40

Never look a Gift Horse in the Mouth. **1892** G. & W. GROSSMITH *Diary of a Nobody* xviii. I told him it was a present from a dear friend, and one mustn't look a gift-horse in the mouth. **1971** T. SHARPE *Riotous Assembly* ix. The whole thing was a . . mystery. 'Oh well, never look a gift horse in the mouth,' he thought.

¹ i.e. a horse given as a present.

gift: *see also* FEAR the Greeks bearing gifts.

gill: *see* every HERRING must hang by its own gill.

girl: *see* BOYS will be boys.

GIVE a thing, and take a thing, to wear the Devil's gold ring
A rhyme used by schoolchildren when someone gives something and then asks for it back.

[Cf. PLATO *Philibus* 19E καθάπερ οἱ παῖδες, ὅτι τῶν ὀρθῶς δοθέντων ἀφαίρεσις οὐκ ἔστι, as with children, there is no taking away of what has been rightly given.] **1571** J. BRIDGES *Sermon at Paul's Cross* 29 Shal we make God to say the worde, and eate his worde? to giue a thing, and take a thing, little children say, This is the diuels goldring, not Gods gift. **1611** R. COTGRAVE *Dict. French & English* s.v. Retirer, To giue a thing and take a thing; to weare the diuells gold-ring. **1721** J. KELLY *Scottish Proverbs* 120 Give a Thing, and take a Thing, Is the ill Man's Goud Ring. A Cant among Children, when they demand a Thing again, which they had bestowed. **1894** *Notes & Queries* 8th Ser. VI. 155 Another saying among boys is—Give a thing and take a thing, To wear the devil's gold ring. **1959** I. & P. OPIE *Lore & Language of Schoolchildren* viii. It is a cardinal rule amongst the young that a thing which has been given must not be asked for again. . . [Somerset] Give a thing, take a thing, Dirty man's plaything. . . [Cheshire] Give a thing, take a thing, Never go to God again.

GIVE and take is fair play

1778 F. BURNEY *Evelina* I. xxv. This here may be a French fashion . . but Give and Take is fair in all nations. **1832** MARRYAT *Newton Forster* III. x. Give and take is fair play. All I say is, let it be a fair stand-up fight. **1873** 'TWAIN' & WARNER *Gilded Age* xxxiii. She thought that 'give and take was fair play', and to parry an offensive thrust with a sarcasm was a neat and legitimate thing to do.

GIVE the Devil his due

Also in the phrase *to give the Devil his due*.

1589 LYLY *Pap with Hatchet* III. 407 Giue them their due though they were diuels . . and excuse them for taking anie money at interest. **1596** NASHE *Saffron Walden* III. 36 Giue the diuell his due. **1642** *Prince Rupert's Declaration* 2 The Cavaliers (to give the Divell his due) fought very valiantly. **1751** SMOLLETT *Peregrine Pickle* I. xvii. You always used me in an officer-like manner, that I must own, to give the devil his due. **1936** H. AUSTIN *Murder of Matriarch* xxiii. To give the devil his due . . I don't think that Irvin planned to incriminate anyone else. **1978** R. L. HILL *Evil that Men Do* vi. Giving the devil his due will always jostle the angels.

give: *see also* it is BETTER to give than to receive; give CREDIT where credit is due; give a DOG a bad name and hang him; give a man ROPE enough and he will hang himself; never give a SUCKER an even break.

He GIVES twice who gives quickly

[PUBLILIUS SYRUS *Sentertia* 235 *inopi beneficium bis dat, qui dat celeriter*, he gives twice who gives quickly to the needy. Cf. *c* **1385** CHAUCER *Legend of Good Women* Prologue l. 451 For whoso yeveth a yifte, or dooth a grace, Do it by tyme,[1] his thanks ys wel the more.] **1553** T. WILSON *Art of Rhetoric* 65ᵛ He geueth twise, that geueth sone and chearefully. **1612** T. SHELTON tr. *Cervantes' Don Quixote* I. iv. It is an old proverb, that hee that gives quickly,

gives twice. **1775** BOSWELL *Life of Johnson* I. 443, I did really ask the favour twice; but you have been even with me by granting it so speedily. *Bis dat qui cito dat*. **1907** *Spectator* 22 June 979 The Union Jack Club . . needs £16,000. . . He gives twice who gives quickly. **1980** *Times* 17 Oct. 13 'He gives twice who gives quickly.' . . We have everything to gain by generous action at once.

[1] *by tyme*, in good time, early.

Those who live in GLASS houses shouldn't throw stones

If you live in a house made of glass yourself, do not throw stones at another: i.e. do not criticize or slander another if you are vulnerable to retaliation.

[*c* **1385** CHAUCER *Troilus & Criseyde* ii. 867 Who that hath an hed of verre,[1] Fro cast of stones war hym in the werre!] **1640** G. HERBERT *Outlandish Proverbs* no. 196 Whose house is of glasse, must not throw stones at another. **1754** J. SHEBBEARE *Marriage Act* II. lv. Thee shouldst not throw Stones, who hast a Head of Glass thyself. . . Thee canst have no Title to Honesty who lendest the writings to deceive Neighbour Barter. **1778** T. PAINE in *Pennsylvania Packet* 22 Oct. 1 He who lives in a glass house, says a Spanish proverb, should never begin throwing stones. **1861** TROLLOPE *Framley Parsonage* I. vi. Those who live in glass houses shouldn't throw stones. . . Mr. Robarts's sermon will be too near akin to your lecture to allow of his laughing. **1979** S. KNIGHT *Requiem at Rogano* xv. It was a favourite joke of my father . . that he was once mildly rebuked by the mayor for apparent hypocrisy. 'Alderman Carew,' the old mayor . . said, 'men who live in glass houses—.'

[1] glass.

All that GLITTERS is not gold

The variant form illustrated in quot. 1943 (etc.) is also common.

[L. *non omne quod nitet aurum est*, not all that shines is gold.] *c* **1220** *Hali Meidenhad* (EETS) 11 Nis hit nower neh gold al that ter[1] schineth. *c* **1390**

CHAUCER *Canon's Yeoman's Tale* l. 962 But al thyng which that shineth as the gold Nis nat gold, as that I have herd it told. **1596** SHAKESPEARE *Merchant of Venice* II. vii. 65 All that glisters is not gold, Often have you heard that told. *c* **1628** W. DRUMMOND *Works* (1711) 222 All is not Gold which glittereth. **1773** D. GARRICK in Goldsmith *She stoops to Conquer* Prologue, All is not gold that glitters. Pleasure seems sweet, but proves a glass of bitters.[2] **1847** C. BRONTË *Jane Eyre* II. ix. I wished to put you on your guard. It is an old saying that 'all is not gold that glitters'. **1880** *Dict. English Proverbs* (Asprey Reference Library) 39 All that glitters is not gold. **1933** E. B. BLACK *Ravenelle Riddle* iv. All that glitters is not gold. . . Every bird who calls himself an American doesn't happen to be one. **1943** W. BUCHANAN–TAYLOR *Shake it Again* ii. They would not need to be informed that 'all is not gold that glitters'. **1980** *Times* 19 Jan. 18 A young woman, perhaps grasping the unseemly spectacle of it all, suggested that all that glitters was not gold.

[1] i.e. there.   [2] bitter-tasting medicine.

glove: *see* a CAT in gloves catches no mice.

GO abroad and you'll hear news of home

1678 J. RAY *English Proverbs* (ed. 2) 345 You must goe into the countrey to hear what news at London. **1887** HARDY *Woodlanders* I. iv. Well, what was the latest news at Shottsford. . . As the saying is, 'Go abroad and you'll hear news of home.'

go: *see also* EASY come, easy go; LIGHT come, light go; QUICKLY come, quickly go; he that would go to SEA for pleasure, would go to hell for a pastime; don't go near the WATER until you learn how to swim; the WEAKEST go to the wall; many go out for WOOL and come home shorn; *also* GOES.

You cannot serve GOD and Mammon

Cf. NO *man can serve two masters.*

[MATTHEW vi. 24 Ye cannot serue God and Mammon.[1]] **1531** W. BONDE *Pil-*

grimage of Perfection (rev. ed.) III. vii. No person may serue god eternall and also the Mammonde of iniquite: whiche is golde and syluer and other richesse. **1860** TROLLOPE *Framley Parsonage* II. i. Lady Lufton . . would say of Miss Dunstable that it was impossible to serve both God and Mammon. **1979** J. RATHBONE *Euro-Killers* xvii. No man can serve two masters: for . . he will hold to one, and despise the other. Ye cannot serve God and Mammon.

[1] the Aramaic word for 'riches', taken by medieval writers as the proper name of the devil of covetousness. Now used generally of wealth regarded as an evil influence.

Where GOD builds a church, the Devil will build a chapel

1560 T. BECON *Works* I. 516[v] For commonly, where so ever God buildeth a church, the Deuyll wyl builde a Chappell iuste by. **1701** DEFOE *True-born Englishman* 4 Wherever God erects a House of Prayer, The Devil always builds a Chapel there: And 'twill be found upon Examination, The latter has the largest Congregation. **1903** G. H. KNIGHT *Master's Questions* xiii. Nowhere does the devil build his little chapels more cunningly than close under the shadow of the great temple of Christian liberty. A thing in itself completely right and good, may be, in its effects on others, completely evil.

GOD helps them that help themselves

[AESCHYLUS *Fragments* 395 φιλεῖ δὲ τῷ κάμνοντι συσπεύδειν θεός, God likes to assist the man who toils.] **1545** R. TAVERNER tr. *Erasmus' Adages* (ed. 2) 57 *Dii facientes adiuuant.* The goddes do helpe the doers. **1551** T. WILSON *Rule of Reason* S1[v] Shipmen cal to God for helpe, and God will helpe them, but so not withstandying, if they helpe them selfes. **1736** B. FRANKLIN *Poor Richard's Almanack* (June) God helps them that help themselves. **1892** H. P. LIDDON *Sermons on Some Words of Christ* iii. God does not promise us each and all that . . the ravens shall come to feed us: as the proverb most truly says, He helps them

that help themselves. **1946** R. P. WAR-
REN *All King's Men* (1948) 123 Nobody
ever helped a hick[1] but the hick
himself. . . It is up to you and God, and
God helps those who help themselves.
[1] an ignorant countryman; a silly fellow.

GOD made the country, and man made the town

[VARRO *De Re Rustica* III. i. *nec mirum,
quod divina natura dedit agros, ars humana
aedificavit urbes,* do not wonder that
divine nature gave us the fields, human
art built the cities.] **1667** A. COWLEY in J.
Wells *Poems* 2 My father said . . God the
first Garden made, & the first City,
Cain. **1785** COWPER *Task* I. 40 God
made the country, and man made the
town. **1870** H. TENNYSON *Memoir* 25 Jan.
(1897) II. 96 There is a saying that if God
made the country, and man the town,
the devil made the little country town.
1941 H. MACINNES *Above Suspicion* x.
God made the country, man made the
town. Pity men couldn't learn better.
1977 G. TINDALL *Field & Beneath* i. It has
been said that 'God made the country
and man made the town', but . . the
town is simply disguised countryside.

GOD makes the back to the burden

1822 COBBETT *Weekly Register* 12 Jan.
94 As 'God has made the back to the
burthen,' so the clay and coppice people
make the dress to the stubs and
bushes. **1839** DICKENS *Nicholas Nickleby*
xviii. Heaven suits the back to the
burden. **1939** E. F. BENSON *Trouble for
Lucia* ii. 'Spare yourself a bitty' I've said,
and always she's replied 'Heaven fits
the back to the burden.' **1979** E.
ANTHONY *Grave of Truth* viii. So many
questions and nobody to answer them;
it was a true penance for her. . . God
made the back for the burden. . . An
Irish nun . . had taught them that
saying from her native land.

GOD never sends mouths but He sends meat

1377 LANGLAND *Piers Plowman* B. XIV.
39 For lente neuere was lyf but lyflode[1]
were shapen. **1546** J. HEYWOOD *Dia-*

logue of Proverbs I. iv. B1 God neuer
sendeth mouthe, but he sendeth meat.
1832 J. P. KENNEDY *Swallow Barn* I. xxviii.
God never sends mouths . . but he
sends meat, and any man who has
sense enough to be honest, will never
want wit to know how to live. **1905** A.
MACLAREN *Gospel according to St. Matthew* I. 103 We are meant to be righteous, and shall not in vain desire to be
so. God never sends mouths but He
sends meat to fill them.
[1] livelihood.

GOD sends meat, but the Devil sends cooks

1542 A. BORDE *Dietary of Health* xi. It is
a common prouerbe, God may sende a
man good meate, but the deuyll may
sende an euyll coke to dystrue it. *c* **1607**
T. DELONEY *Thomas of Reading* B3 God
sends meat, and the diuel sends
cookes. **1738** SWIFT *Polite Conversation*
II. 155 This Goose is quite raw: Well,
God sends Meat, but the Devil sends
Cooks. **1822** SCOTT *Nigel* III. iii. That
homely proverb that men taunt my calling with,—'God sends good meat, but
the devil sends cooks.' **1922** JOYCE
Ulysses 169 Milly served me that cutlet
with a sprig of parsley. . . God made
food, the devil the cooks. **1979** *Country
Life* 13 Sept. 807 Another old saying . .
that God sends good meat but the devil
sends the cooks.

GOD's in his heaven; all's right with the world

The standard form is an expression of
satisfaction (see quot. 1841), which has
now largely replaced the consolatory
God is where he was.

1530 J. PALSGRAVE *L'éclaircissement de
la Langue Française* 213 Neuer dispayre
man, god is there as he was. **1612** T.
SHELTON tr. *Cervantes' Don Quixote* I. IV.
iii. God is in heaven. **1678** J. RAY *English Proverbs* (ed. 2) 147 God is where he
was. Spoken to encourage People in any
distress. **1841** BROWNING *Works* (1970)
327 The snail's on the thorn: God's in
his heaven—All's right with the world.
1906 KIPLING *Puck of Pook's Hill* 240

Cheer up, lad. . . God's where He was. **1928** E. WAUGH *Decline & Fall* I. V. When you've been in the soup as often as I have, it gives you a sort of feeling that everything's for the best, really. You know, God's in His heaven; all's right with the world. **1976** L. ALTHER *Kinflicks* x. When you see sunlight playing on the meadows, don't you feel . . that God's in His Heaven and all's right with the world?

GOD tempers the wind to the shorn lamb
God so arranges it that bad luck does not unduly plague the weak or unfortunate. The phrase *to temper the wind (to the shorn lamb)* is also common.

[**1594** H. ESTIENNE *Premices* 47 *ces termes, Dieu mesure le froid à la brebis tondue, sont les propres termes du prouerbe,* these terms, God measures the cold to the shorn sheepe, are the correct terms of the proverb.] **1640** G. HERBERT *Outlandish Proverbs* no. 867 To a close shorne sheep, God gives wind by measure. **1768** STERNE *Sentimental Journey* II. 175 How she had borne it . . she could not tell—but God tempers the wind, said Maria, to the shorn lamb. **1933** V. BRITTAIN *Testament of Youth* I. ii. There is an unduly optimistic proverb which declares that God tempers the wind to the shorn lamb. My subsequent history was hardly to justify such naive faith in the Deity. **1981** *Times* 2 May 4 The wind is tempered to the shorn lamb. Officials . . get full basic salary for the first three months of their 'retirement'.

God: *see also* ALL things are possible with God; EVERY man for himself, and God for us all; MAN proposes, God disposes; MAN's extremity is God's opportunity; the MILLS of God grind slowly, yet they grind exceeding small; the NEARER the church, the farther from God; PROVIDENCE is always on the side of the big battalions; the ROBIN and the wren are God's cock and hen; TAKE the goods the gods provide; put your TRUST in God, and keep your powder dry; the VOICE of the people is the voice of God; a WHISTLING

woman and a crowing hen are neither fit for God nor men; *also* GODS, HEAVEN.

godliness: *see* CLEANLINESS is next to godliness.

Whom the GODS love die young
Cf. The GOOD die young.

[MENANDER *Fragments* 4 ὃν οἱ θεοὶ φιλοῦσιν ἀποθνήσκει νέος, he whom the gods love dies young; PLAUTUS *Bacchides* l. 817 *quem di diligunt, Adolescens moritur,* he whom the gods favour, dies young.] **1546** W. HUGHE *Troubled Man's Medicine* B8ᵛ Most happy be they and best belouid of god, that dye whan they be yong. **1553** T. WILSON *Art of Rhetoric* 40ᵛ Whom god loueth best, those he taketh sonest. **1651** G. HERBERT *Jacula Prudentum* no. 1094 Those that God loves, do not live long. **1821** BYRON *Don Juan* IV. xii. 'Whom the gods love die young,' was said of yore, And many deaths do they escape by this. **1933** V. BRITTAIN *Testament of Youth* II. vii. Oh, Geoffrey, I shall never know anyone quite like you again. . . It's another case of 'whom the Gods love'. **1972** A. PRICE *Colonel Butler's Wolf* xx. 'Whom the gods love die young,' the war taught us that.

The GODS send nuts to those who have no teeth
Said of opportunities or pleasures which come too late to be enjoyed.

[Cf. Fr. *le pain vient à qui les dents faillent,* bread comes to those who lack teeth.] **1929** *American Speech* IV. 463 God gives us nuts to crack when we no longer have teeth. **1967** RIDOUT & WITTING *English Proverbs Explained* 68 The gods send nuts to those who have no teeth. In this life we either have too little of what we do want, or too much of what we don't want or can't use. **1979** 'C. AIRD' *Some die Eloquent* xii. 'Have you any plans for using the money?'. . 'The nuts', said the doctor astringently, 'come when the teeth have gone.'

Whom the GODS would destroy, they first make mad
[EURIPIDES *Fragment* ὃν θεὸς θέλει

ἀπόλεσαι, πρῶτ' ἀπόφρεναι, whom the god wishes to destroy, he first drives out of his mind: L. *quos Deus vult perdere, prius dementat,* those whom God wishes to destroy, he first sends mad.] **1611** JONSON *Catiline* V. 481 A madnesse, Wherewith heauen blinds 'hem, when it would confound 'hem. **1640** G. HERBERT *Outlandish Proverbs* no. 688 When God will punish, hee will first take away the understanding. **1817** BYRON *Letter* 2 Apr. (1976) V. 204 God maddens him whom 'tis his will to lose, And gives the choice of death or phrenzy—Choose! **1875** M. THOMPSON *Hoosier Mosaics* 180 Whom the gods would destroy they first make mad. **1981** *Daily Telegraph* 24 July 4 Already Commonwealth Finance Ministers have elected not to meet on New Zealand's defiled soil. If greater penalties follow, the Commonwealth will confirm that those whom the gods would destroy they first make mad.

He that GOES a-borrowing, goes a-sorrowing

c **1470** in Wright and Halliwell *Reliquiæ Antiquæ* (1841) I. 316 He that fast spendyth must nede borowe; But whan he schal paye ayen, then ys al the sorowe. **1545** R. TAVERNER tr. *Erasmus' Adages* (ed. 2) 46ᵛ He that goeth a borowynge goeth a sorowynge. **1836** MARRYAT *Midshipman Easy* I. viii. You had made your request for the loan . . fully anticipating a refusal, (from the feeling that he who goes a borrowing goes a sorrowing). **1894** J. LUBBOCK *Use of Life* iii. Debt is slavery. 'Who goes a-borrowing goes a sorrowing.' **1925** S. O'CASEY *Juno & Paycock* III. 84 Ah, him that goes a borrowin' goes a sorrowin'! . . An' there isn't hardly a neighbour in the whole street that hasn't lent him money on the strength of what he was going to get, but they're after backing the wrong horse.

goes: *see also* there goes more to MARRIAGE than four bare legs in a bed; PRIDE goes before a fall; what goes UP must come down.

When the GOING gets tough, the tough get going

A favourite family saying of Joseph P. Kennedy, U.S. politician, businessman, and father of the late President.

1962 J. H. CUTLER *'Honey Fitz'* xx. Joe [Kennedy] made his children stay on their toes. . . He would bear down on them and tell them, 'When the going gets tough, the tough get going.' **1970** *New Yorker* 3 Oct. 33 Baron Marcel Bich, the millionaire French pen magnate probably spoke for them all last month when he said, 'When the going gets tough, the tough get going!' ('Quand le chemin devient dur, les durs se cheminent!') **1979** J. CRUMLEY *Last Good Kiss* xvi. 'When the going gets tough, the tough get going?' she asked slyly. 'Make fun if you want to, but that's what character is all about.'

GOLD may be bought too dear

1546 J. HEYWOOD *Dialogue of Proverbs* II. vii. I4 Well (quoth she) a man maie bie golde to dere. **1642** T. FULLER *Holy State* II. xxi. Fearing to find the Proverb true, That Gold may be bought too dear, they returned to their ships. **1889** J. LUBBOCK *Pleasures of Life* (ed. 2) II. ii. It is well worth having . . but it does not requite too great a sacrifice. A wise proverb tells us that gold may be bought too dear.

gold: *see also* it is good to make a BRIDGE of gold to a flying enemy; GIVE a thing, and take a thing, to wear the Devil's gold ring; all that GLITTERS is not gold.

A GOLDEN key can open any door

1580 LYLY *Euphues* II. 71 Who is so ignorant that knoweth not, gold be a key for euery locke, chieflye with his Ladye. **1660** W. SECKER *Nonsuch Professor* II. ix. The gates of the new Jerusalem . . are not got open by golden keys. **1842** TENNYSON *Poems* (1969) 694 Every door is barr'd with gold, and opens but to golden keys. **1945** F. THOMPSON *Lark Rise* xix. Their better-educated neighbours . . did not call on the newly rich family. That was before the days when a golden key could open any door.

golden: *see* SILENCE is golden; SPEECH is silver, but silence is golden.

If you can't be GOOD, be careful

[Cf. **1303** R. BRUNNE *Handlyng Synne* (EETS) l. 8316 The apostle seyth thys autoryte,[1] 'Gyf thou be nat chaste, be thou pryue.[2]' **1528** W. TYNDALE *Obedience of Christian Man* 73 As oure lawears saye, *si non caste tamen caute*, this is, if ye live not chaste, se ye cary[3] clene, and playe the knave secretly.] **1903** A. M. BINSTEAD *Pitcher in Paradise* viii. Always bear in mind what the country mother said to her daughter who was coming up to town to be apprenticed to the Bond Street millinery, 'For heaven's sake be good; but if you can't be good, be careful.' **1907** B. SCOTT *(song-title)* If you can't be good—be careful. **1967** E. WILLIAMS *Beyond Belief* I. iv. Ta ra Alan if you can't be good be careful see you later. **1982** S. GRANT DUFF *Parting of Ways* xvii. Tommy . . gave me a stern warning. . . 'Never meet a German in Prague . . Be good, and if you can't, be very careful.'

[1] an authoritative statement. [2] secret, clandestine. [3] *cary clene*: i.e. 'cary clean'; act or comport (yourself) properly.

A GOOD beginning makes a good ending

c **1300** *South-English Legendary* (EETS) l. 216 This was atte uerste me thingth[1] a god bygynnynge, Ther after was the betere hope to come to god endynge. *c* **1350** *Douce MS* 52 no. 122 Of a gode begynnyng comyth a gode endyng. **1710** S. PALMER *Proverbs* 1 A good Beginning makes a good End. . . 'Tis a great point of Wisdom . . to begin at the right end. **1850** 'M. TENSAS' *Odd Leaves from Life of Louisiana 'Swamp Doctor'* 109, I hope my future lot will be verification of the old adage, that a 'bad beginning makes a good ending', for mine is bad enough. **1934** G. WESTON *His First Million Women* xvi. I was brought up to

believe that 'Of a good beginning cometh a good ending.' . . 'You can't do a good plastering job if your laths aren't right to begin with.'

[1] *me thingth*: it seems to me.

There's many a GOOD cock come out of a tattered bag

The proverb is derived from cock-fighting.

[Cf. **1721** J. KELLY *Scottish Proverbs* 7 An ill Cow may have a good Calf. Bad People may have good Children.] **1883** C. S. BURNE *Shropshire Folklore* xxxvi. There'll come a good cock out of a ragged bag. . . A cockfighting simile, lately used by a farmer, whose buildings were out of repair, but his stock in good condition. **1953** R. SUTCLIFF *Simon* xiv. 'There's many a good cock come out of a tattered bag,' said the dark shape, slowly. There was an instant of . . silence, and then Simon said, 'And a good tune played on an old fiddle.'

The GOOD die young

Cf. *Whom the* GODS *love die young*.

1697 DEFOE *Character of Dr. Annesley* 3 The best of Men cannot suspend their Fate; The Good die early, and the Bad die late. **1814** WORDSWORTH *Excursion* I. 27 The good die first, And they whose hearts are dry as summer dust Burn to the socket. **1852** A. CARY *Clovernook* 39 Sarah . . was dead . . aged nineteen years. . . The old truth was again re-asserted . . in the often repeated verse which followed, that the good die young. **1979** J. LEASOR *Love & Land Beyond* v. The good do die young. That's why people like you and me are left to grow old.

He is a GOOD dog who goes to church

1826 SCOTT *Woodstock* I. i. Bevis . . fell under the proverb which avers, 'He is a good dog which goes to church'; for . . he behaved himself . . decorously. *a* **1895** F. LOCKER-LAMPSON *My Confidences* (1896) 44 Tis said, by men of deep research, He's a good dog who goes to church.

GOOD fences make good neighbours

1640 E. ROGERS *Letter* in *Winthrop Papers* (1944) IV. 282 A good fence help-eth to keepe peace between neigh-bours; but let vs take heed that we make not a high stone wall, to keepe vs from meeting. **1815** H. H. BRACKENRIDGE *Modern Chivalry* (rev. ed.) IV. II. xiii. I was always with him [Jefferson] in his apprehensions of John Bull. . . Good fences restrain fencebreaking beasts, and . . preserve good neigh-bourhoods. **1914** R. FROST *North of Boston* 12 My apple trees will never get across And eat the cones under his pines, I tell him. He only says, 'Good fences make good neighbours.' **1978** T. SHARPE *Throwback* x. 'Hadn't you better go and investigate?' . . Lockhart shook his head. 'Strong fences make good neighbours.'

A GOOD horse cannot be of a bad colour

a **1628** J. CARMICHAELL *Proverbs in Scots* no. 1621 There is gude horse of all hewis. **1653** I. WALTON *Compleat Angler* iv. It is observed by some, that there is no good horse of a bad colour. **1732** T. FULLER *Gnomologia* no. 1713 Good Horses can't be of a bad Colour. **1891** J. L. KIPLING *Beast & Man* viii. 'A good horse is never of a bad colour' . . is wild-ly irreverent from the Oriental point of view. **1912** *Spectator* 28 Dec. 1094 Virgil . . did not hold that 'a good horse can-not be of a bad colour'; he liked bays and grays.

The only GOOD Indian is a dead Indian

Also used deprecatorily of members of various nationalities, etc.

1868 J. M. CAVANAUGH in *Congressional Globe* (U.S.) 28 May 2638, I have never in my life seen a good Indian[1] (and I have seen thousands) except when I have seen a dead Indian. **1886** A. GUR-NEY *Ramble through United States* 29 The Government . . is at length earnestly endeavouring to do tardy justice to the conquered race; but it was distressing to hear again and again from American lips the remark that 'A *good* Indian is a *dead* Indian.' **1895** E. S. ELLIS *People's*

Standard History U.S. IV. lxxxiv. In Jan-uary, 1869, . . Old Toch-a-way . ., a chief of the Comanches, . . [said]: 'Me, Tock-a-way; me good Injun.' . . General [Sheridan] . . set those standing by in a roar by saying: 'The only good Indians I ever saw were dead.' **1915** J. BUCHAN *Salute to Adventurers* v. Never trust an Indian. The only good kind is the dead kind. **1934** G. B. SHAW *On Rocks* Preface 146 'The only good nigger is a dead nigger' say the Americans of the Ku-Klux temperament. **1935** L. I. WILDER *Little House on Prairie* xvii. She did not know why the government made treaties with Indians. The only good In-dian was a dead Indian. **1978** A. PRICE *'44 Vintage* xiii. They've the Indian, Jack—and the only good injun is a dead one. **1980** R. BUTLER *Blood-Red Sun at Noon* II. vi. The only good Jap is a dead Jap.

[1] i.e. North American or 'Red' Indian.

The GOOD is the enemy of the best

Cf. The BEST *is the enemy of the good.*

1912 J. KELMAN *Thoughts on Things Eternal* 108 Every respectable Pharisee proves the truth of the saying that 'the good is the enemy of the best.' . . Christ insists that we shall not be content with a second-best, though it be good. **1939** R. A. HABAS *Morals for Moderns* vii. 'The good', runs the old aphorism, 'is the enemy of the best.' Nowhere is this . . better exemplified than in connection with . . self-deceit.

A GOOD Jack makes a good Jill

1623 W. PAINTER *Palace of Pleasure* C8 A good Iacke alwaies maketh a good Gyll. **1670** J. RAY *English Proverbs* 108 A good Jack makes a good Gill. . . Inferiours imitate the manners of superiours . . wives of their husbands. **1876** I. BANKS *Manchester Man* III. xv. In George Street he was refused admis-sion, Mrs. Ashton justifying her daugh-ter's fight with . . 'A good Jack makes a good Jill.'

GOOD men are scarce

1609 D. TUVILL *Essays Moral & Theo-*

logical 92 Good men are scarce, no age so many brings As Thebes hath gates. **1721** J. KELLY *Scottish Proverbs* 124 Good Folks are scarce, you'll take care of one. Spoken to those who carefully provide against ill Weather, or cowardly shun Dangers. **1836** DICKENS *Sketches by Boz* I. 285 One of the women has agreed to stand a glass round, jocularly observing that 'as good people's wery scarce, what I says is, make the most on 'em.' **1979** 'J. LE CARRÉ' *Smiley's People* xii. Time you had some shut-eye, isn't it? Good men are scarce, I always say.

There's many a GOOD tune played on an old fiddle

a **1902** S. BUTLER *Way of All Flesh* (1903) lxi. Beyond a haricot vein in one of my legs I'm as young as ever I was. Old indeed! There's many a good tune played on an old fiddle. **1953** R. SUT-CLIFF *Simon* xiv. 'There's many a good cock come out of a tattered bag,' said the dark shape, slowly. There was an instant of . . silence, and then Simon said, 'And a good tune played on an old fiddle.' **1979** N. FREELING *Widow* xxx. He looked at her casually. . . 'Not all that bad at that. Many a good tune played on an old fiddle.'

One GOOD turn deserves another

c **1400** in *Bulletin of John Rylands Library* (1930) XIV. 92 O[1] good turne asket another. **1620** J. HALL *Contemplations* V. XIV. 28 One good turne requires another. . . Justly should they haue been set at the vpper end of the table. **1638** T. RANDOLPH *Amyntas* v. vi. One good turne deserves another. **1824** SCOTT *St. Ronan's Well* II. iv. But one good turn deserves another—in that case, you must . . dine with me. **1929** S. T. WARNER *True Heart* II. 151 You've given me the best laugh I've had for months, and one good turn deserves another. **1979** T. SHARPE *Wilt Alternative* xiv. Noblesse oblige? You know, one good turn deserves another and what-not.

[1] one.

GOOD wine needs no bush

A bunch of ivy was formerly the sign of a vintner's shop.

a **1430** LYDGATE *Pilgrimage of Man* (EETS) l. 20415 And at tavernys (with-oute wene[1]) Thys tooknys[2] nor thys bowys grene . . The wyn they mende nat. **1545** R. TAVERNER tr. *Erasmus' Adages* (ed. 2) 42[v] Wyne that is saleable and good nedeth no bushe or garland of yuye[3] to be hanged before. The english prouerbe is thus Good wyne neadeth no signe. **1599** SHAKESPEARE *As You like It* Epilogue 3 If it be true that good wine needs no bush, 'tis true that a good play needs no epilogue. **1711** ADDISON *Spectator* 13 Nov., I was never better pleased than with a plain man's compliment, who upon his friend's telling him that he would like the *Spectator* much better if he understood the motto, replied, that good wine needs no bush. **1845** R. FORD *Handbook for Travellers in Spain* I. ix. Good wine needs neither bush, herald, nor crier. **1979** E. H. GOMBRICH *Sense of Order* i. The old proverb that 'a good wine needs no bush' has its correlate in what advertisers call 'sales resistance' to conspicuous bushes.

[1] doubt.   [2] tokens, signs.   [3] ivy.

good: *see also* ALL good things must come to an end; BAD money drives out good; the BEST is the enemy of the good; BETTER a good cow than a cow of a good kind; BRAG is a good dog, but Holdfast is better; it is good to make a BRIDGE of gold to a flying enemy; a CHANGE is as good as a rest; open CONFESSION is good for the soul; why should the DEVIL have all the best tunes?; DILIGENCE is the mother of good luck; ENOUGH is as good as a feast; EVIL communications corrupt good manners; FAR-FETCHED and dear-bought is good for ladies; there are as good FISH in the sea as ever came out of it; HOPE is a good breakfast but a bad supper; it's an ILL wind that blows nobody any good; JACK is as good as his master; a LIAR ought to have a good memory; LISTENERS never hear any good of themselves; a MISS is as good as a mile; NO news is good news; a NOD's as

good as a wink to a blind horse; there is
NOTHING so good for the inside of a man
as the outside of a horse; see a PIN and
pick it up, all the day you'll have good
luck; any PUBLICITY is good publicity;
the ROAD to hell is paved with good intentions; good SEED makes a good crop;
you can have TOO much of a good thing;
when the WIND is in the east, 'tis neither
good for man nor beast.

goods: see TAKE the goods the gods provide.

goose: see what is SAUCE for the goose is
sauce for the gander; also GEESE.

When the GORSE is out of bloom, kissing's out of fashion
Cf. When the FURZE is in bloom, my love's in tune.

1846 M. A. DENHAM Proverbs relating to
Seasons, &c. 12 When whins[1] are out of
bloom, Kissing's out of fashion. . .
Whins are never out of bloom. **1860** G. J.
WHYTE-MELVILLE Holmby House I. iii.
'When the gorse is out of bloom, young
ladies,' quoth Sir Giles, 'then is kissing
out of fashion!' . . There is no day in the
year when the blossom is off the gorse.
1974 A. DWYER-JOYCE Brass Islands 175
'What's that old jingle about the gorse?'
. . 'When the gorse is out of bloom,
kissing is out of fashion.'
[1] gorse.

What is GOT over the Devil's back is spent under his belly
What is improperly obtained is spent in
foolhardy pleasures or debauchery.

1582 GOSSON Plays Confuted G7[v] That
which is gotten ouer the deuils backe, is
spent vnder his belly. **1607** MIDDLETON
Michaelmas Term IV. i. What's got over
the devil's back (that's by knavery),
must be spent under his belly (that's by
lechery). **1670** J. RAY English Proverbs 80
What is gotten over the Devils back, is
spent under his belly. . . What is got by
oppression or extortion is many times
spent in riot and luxury. **1821** SCOTT
Pirate III. iv. You shall not prevail on me
to go farther in the devil's road with

you; for . . what is got over his back is
spent—you wot how. **1938** J. LINDSAY
1649 cviii. 'What's got over the devil's
back is soon sold under the belly' said
the bully to his girl; 'let down the milk.'

got: see also a PENNY saved is a penny
earned.

grain: see if in FEBRUARY there be no rain,
'tis neither good for hay nor grain.

grandmother: see don't TEACH your grandmother to suck eggs.

While the GRASS grows, the steed starves
Dreams or expectations may be realized
too late.

[Med. L. dum gramen crescit, equus in
moriendo quiescit, while the grass grows,
the horse lies dying.] c **1350** Douce MS
52 no. 20 While the grasse growes, the
goode hors sterues. a **1500** in Wright &
Halliwell Reliquæ Antiquæ (1841) I. 208
While the grasse growes the steede
starves. **1600–1** SHAKESPEARE Hamlet III.
ii. 333 You have the voice of the King
himself for your succession.—Ay, sir,
but 'While the grass grows' – the
proverb is something musty. **1821** J.
GALT Ayrshire Legatees x. Until ye get a
kirk there can be no marriage. But the
auld horse may die waiting for the new
grass. **1911** G. B. SHAW Doctor's Dilemma
III. 56, I shall sell them next year fast
enough, after my one-man-show; but
while the grass grows the steed starves.

The GRASS is always greener on the other side of the fence
[Cf. **1545** R. TAVERNER tr. Erasmus' Ad-
ages (ed. 2) 59[v] Fertilior seges est alieno
semper in aruo. The corne in an other
mans ground semeth euer more fertyll
and plentifull then doth oure owne.]
1959 H. & M. WILLIAMS in J. C. Trewin
Plays of Year XIX. 13 (title) The grass is
greener. **1965** Which? Mar. 91 'The
grass always looks greener on the other
side of the fence,' said another infor-
mant, explaining that while stores who
do practise the system are uneasy about

it . . those outside constantly wonder whether results might not justify it. **1979** *Homes & Gardens* June 171 Everyone else in the world believes that the grass is always greener on the other side of the fence, but I have a wife who knows that her own patch is best.

grease: *see* the SQUEAKING wheel gets the grease.

A GREAT book is a great evil
[CALLIMACHUS *Aetia* III. 72 μέγα βιβλίον μέγα κακόν, a large book is a large evil.] **1628** BURTON *Anatomy of Melancholy* (ed. 3) 7 Oftentimes it falls out . . a great Booke is a great mischiefe. **1711** ADDISON *Spectator* 23 July, We do not expect to meet with any thing in a bulky Volume. . . A great Book is a great Evil. **1909** *British Weekly* 8 Apr. 13 It may be . . said in reference to this unhappy production that a great book is indeed a great evil. **1933** *Oxford English Dictionary* Preface p.vii. If there is any truth in the old Greek maxim that a large book is a great evil, English dictionaries have been steadily growing worse ever since their inception more than three centuries ago.

GREAT minds think alike
1618 D. BELCHIER *Hans Beer-Pot* D1 Though he made that verse, Those words were made before. . . Good wits doe iumpe.[1] **1761** STERNE *Tristram Shandy* III. ix. Great wits jump: for the moment Dr. Slop cast his eyes upon his bag . . the very same thought occurred. **1889** A. JAMES *Journal* 1 Dec. (1964) 61 As great minds jump this proves . . that my Mind *is Great*! **1922** *Punch* 27 Dec. 601 Lord Riddell considers that Mr. H. G. Wells is one of the world's greatest minds. Great minds, as the saying is, think alike.
[1] coincide.

GREAT oaks from little acorns grow
c **1385** CHAUCER *Troilus & Criseyde* II. 1335 As an ook comth of a litel spir,[1] So thorough this lettre . . Encressen gan desir. **1579** GOSSON *School of Abuse* 20ᵛ

But Tall Cedars from little graynes shoote high: great Oakes, from slender rootes spread wide. **1584** J. WITHALS *Dict.* (rev. ed.) D4 Of a nut springes an hasill, and of an Akorn an hie or tall oke. **1732** T. FULLER *Gnomologia* no. 4576 The greatest Oaks have been little Acorns. **1777** D. EVERETT in *Columbian Orator* (1797) 58 Large streams from little fountains flow, Tall oaks from little acorns grow. **1923** *Times* 13 Oct. 7 Here in England, as nowhere else in the world, 'great oaks from little acorns grow'. The oak, as the emblem of British strength, has been symbolic in many ways. **1979** *Oxford Star* 24 May 2 Who are these people to say that 'a foetus is not a human being'? 'Great oaks', they say, 'from little acorns grow.'
[1] shoot.

great: *see also* BIG fleas have little fleas upon their backs to bite them; DEATH is the great leveller; LITTLE strokes fell great oaks; LITTLE thieves are hanged, but great ones escape; there's no great LOSS without some gain; POVERTY is no disgrace, but it is a great inconvenience; THRIFT is a great revenue; TIME is a great healer.

The GREATER the sinner, the greater the saint
1773 R. GRAVES *Spiritual Quixote* II. VII. xi. It was a maxim with Mr. Whitfield, 'The greater the Sinner, the greater the Saint.' **1856** E. HINCHCLIFFE *Barthomley* vi. How well is the old proverb illustrated in this foul seducer. . . 'The greater the sinner, the greater the Saint.'

The GREATER the truth, the greater the libel
c **1787** BURNS *Poems* (1968) I. 349 Dost not know that old Mansfield,[1] who writes like the Bible, Says the more 'tis a truth, sir, the more 'tis a libel? **1828** BULWER-LYTTON *Pelham* I. xxiv. 'You won't catch an old lawyer in such impudence.' 'The greater the truth the greater the libel,' said Warburton, with a sneer. **1882** S. A. BENT *Short Sayings of Great Men* 371 The greater the truth, the

greater the libel. A maxim of the law in vogue . . while Mansfield presided over the King's Bench. . . The maxim is said to have originated in the Star Chamber. **1981** *Times* 9 Apr. 14 Proof of the truth of the words complained of is an absolute defence to a libel action. The adage 'The greater the truth the greater the libel' is a myth.

[1] William Murray, first Earl of Mansfield (1705–93), statesman and judge.

When GREEK meets Greek, then comes the tug of war

1677 N. LEE *Rival Queens* IV. 48 When Greeks joyn'd Greeks, then was the tug of War. **1804** W. IRVING *Journals & Notebooks* (1969) I. 69 Two upright Postillions . . were disputing who was the greatest rogue. . . 'When Greek meets Greek then comes the tug of war.' **1926** A. HUXLEY *Two or Three Graces* 175 When Greek meets Greeks then comes, in this case, an exchange of anecdotes about the deposed sovereigns of eastern Europe—in a word, the tug of bores. **1979** M. A. SCREECH *Rabelais* iii. One is reminded of an adage Erasmus used . . *Magus cum mago*: 'magician meets magician'—Greek, as we say, meets Greek.

Greek: *see also* FEAR the Greeks bearing gifts.

A GREEN Yule makes a fat churchyard

1635 J. SWAN *Speculum Mundi* V. They also say, that a hot Christmas makes a fat Church-yard. **1670** J. RAY *English Proverbs* 42 A green winter makes a fat Church-yard. This Proverb was sufficiently confuted *Anno* 1667, in which the winter was very mild; and yet no mortality . . ensued the Summer or Autumn following. **1721** J. KELLY *Scottish Proverbs* 30 A green yule makes a fat Church-yard. This, and a great many proverbial Observations, upon the Seasons of the Year, are groundless. **1858** G. ELIOT *Amos Barton* in *Scenes of Clerical Life* I. vi. I shouldn't wonder if it takes the old lady off. They say a green Yule makes a fat Churchyard; but so does a white Yule too. **1927** J. BUCHAN

Witch Wood xvii. Every wife in the parish . . quoted dolefully the saw that 'a green Yule makes a fat kirkyard.' **1978** R. WHITLOCK *Calendar of Country Customs* xi. Weather law is also embarrassingly abundant. . . The saying, 'Green Christmas, full Churchyard' is still well known.

greener: *see* the GRASS is always greener on the other side of the fence.

The GREY mare is the better horse
The wife rules, or is more competent than, the husband.

[**1529** MORE *Dialogue of Images* III. V. Here were we fallen in a grete questyon of the law, whyther the gray mare be the better horse . . or whither he haue a wyse face or not that loketh as lyke a foole as an ewe loketh lyke a shepe.] **1546** J. HEYWOOD *Dialogue of Proverbs* II. iv. G4 The grey mare is the better hors. **1836** T. C. HALIBURTON *Clockmaker* 1st Ser. xxi. You can see with half an eye that the 'grey mare is the better horse here'. **1906** J. GALSWORTHY *Man of Property* I. vi. D'you think he knows his own mind? He seems to me a poor thing. I should say the gray mare was the better horse! **1981** V. POWELL *Flora Annie Steel* vii. She did not wish it to seem, to quote an old fashioned expression, that the grey mare was the better horse. . . She strove to avoid prejudicing her husband's position.

grey: *see also* all CATS are grey in the dark.

grieve: *see* what the EYE doesn't see, the heart doesn't grieve over.

grind: *see* the MILL cannot grind with the water that is past; the MILLS of God grind slowly, yet they grind exceeding small.

gripping: *see* NOTHING should be done in haste but gripping a flea.

All is GRIST that comes to the mill
The is frequently replaced by a possessive pronoun or phrase. Cf. *All is* FISH

that comes to the net. The metaphorical expression *grist to one's mill* is also found.

1655 T. FULLER *Church Hist. Britain* III. iii. Forein Casuists bring in a bundle of mortal sins, all grist[1] for their own Mill. **1770** S. FOOTE *Lame Lover* I. 28 Well, let them go on, it brings grist to our mill: for whilst both the sexes stick firm to their honour, we shall never want business. **1896** A. WHYTE *Bible Characters* I. xii. Your stumble, your fall, your misfortune . . all is grist to the mill of the mean-minded man. **1943** A. CHRISTIE *Moving Finger* ix. You're failing to allow for the mentality of a Poison Pen—all is grist that comes to their mill. **1967** RIDOUT & WITTING *English Proverbs Explained* 17 A similar proverb is *all's grist that comes to the mill.* . . Use will be made of everything received. **1979** G. MITCHELL *Mudflats of Dead* iii. All was grist which came to a novelist's mill, and he was still hoping that something, somewhere, would bring him what he still thought of as inspiration.

[1] corn which is to be ground.

ground: *see* BETWEEN two stools one falls to the ground.

grow: *see* while the GRASS grows, the steed starves; GREAT oaks from little acorns grow; ONE for the mouse, one for the crow.

grunt: *see* what can you EXPECT from a pig but a grunt?

guest: *see* FISH and guests stink after three days.

A GUILTY conscience needs no accuser

c **1390** CHAUCER *Canon's Yeoman's Prologue* l. 688 For Catoun[1] seith that he that gilty is Demeth alle thyng be spoke of him. **1597** *Politeuphuia* 10ᵛ A Guilty conscience is a worme that bites and neuer ceaseth. . . A guiltie conscience is neuer without feare. **1721** J. KELLY *Scottish Proverbs* 9 A guilty Conscience self accuses. A Man that has done ill . . shews his Guilt. **1744** *Life & Adventures Matthew Bishop* viii. It is an old saying, a guilty conscience needs no accuser. **1881** D. C. MURRAY *Joseph's Coat* I. viii. 'Where are *you* off to?' asked George with a great effort. . . A guilty conscience needs no accuser.

[1] i.e. Dionysius Cato.

gunner: *see* the COBBLER to his last and the gunner to his linstock.

H

habit: *see* OLD habits die hard.

HALF a loaf is better than no bread
Cf. SOMETHING *is better than nothing.*

1546 J. HEYWOOD *Dialogue of Proverbs* I.
xi. D4ᵛ Throwe no gyft agayne at the
giuers head, For better is halfe a lofe
then no bread. **1636** W. CAMDEN *Remains concerning Britain* (ed. 5) 297 Halfe
a loafe is better than no bread at all.
1681 A. BEHN *Rover* II. II. ii. You know
the Proverb of the half Loaf, Ariadne, a
Husband that will deal thee some Love
is better than one who can give thee
none. **1841** DICKENS *Old Curiosity Shop*
I. xxxiii. 'Mr. Swiveller,' said Quilp,
'being pretty well accustomed to the
agricultural pursuits of sowing wild
oats, Miss Sally, prudently considers
that half a loaf is better than no bread.'
1979 *Guardian* 6 Aug. 10 Half a loaf is
better than no bread at all. The ending of
half a war is immensely better than no
truce at all.

The HALF is better than the whole
A proverb advising economy or restraint.

[HESIOD *Works & Days* xl. πλέον ἥμισυ
παντός, half is more than the whole.]
1550 LATIMER *Sermon before King's Majesty* G3 Ther is a proverbe . . *Dimidium
plus toto*: The halfe somtymes more then
the hole. The meane lyfe is the best lyfe
and the most quyet lyfe of al. **1828** I.
DISRAELI *Curiosities of Literature* 2nd Ser.
I. 419 The half is better than the whole.
1906 A. C. BENSON *From College Window*
v. It is true of conversation as of many
other things, that the half is better than
the whole. People who are fond of talking ought to beware of being lengthy.

**One HALF of the world does not know
how the other half lives**
[**1532** RABELAIS *Pantagruel* II. xxxii. *la
moytié du monde ne sçait comment l'autre
vit*, one half of the world knows not how
the other lives.] **1607** J. HALL *Holy*
Observations xvii. One half of the world
knowes not how the other liues: and
therefore the better sort pitty not the
distressed . . because they knowe
it not. **1640** G. HERBERT *Outlandish
Proverbs* no. 907 Halfe the world
knowes not how the other halfe lives.
1755 B. FRANKLIN *Poor Richard's Almanack* (Preface) It is a common saying,
that One Half of the World does not
know how the other Half lives. **1830**
MARRYAT *King's Own* I. x. It is an old
proverb that 'one half the world do not
know how the other half live'. Add to it,
nor *where* they live. **1945** C. S. LEWIS
That Hideous Strength i. 'I didn't even
know this was Bracton property.' 'There
you are! . . One half of the world
doesn't know how the other half lives.'

HALF the truth is often a whole lie
1758 B. FRANKLIN *Poor Richard's Almanack* (July) Half the Truth is often a great
Lie. **1859** TENNYSON *Poems* (1969) 1107
That a lie which is half a truth is ever the
blackest of lies, That a lie which is all a lie
may be met and fought with outright,
But a lie which is part a truth is a harder
matter to fight. **1875** A. B. CHEALES
Proverbial Folklore 166 Half the truth
is often a whole lie . . is a proverb
which Tennyson has most admirably
versified. **1979** H. HOWARD *Sealed Envelope* xiii. 'You've been lying.' . . 'Half
the truth can be worse than a straight
lie.'

half: *see also* BELIEVE nothing of what you
hear, and only half of what you see; two
BOYS are half a boy, and three boys are
no boy at all; a FAULT confessed is half
redressed; he that has an ILL name is
half hanged; WELL begun is half done.

half-done: *see* FOOLS and bairns should
never see half-done work.

half-way: *see* do not MEET troubles half-
way.

hall: *see* it is MERRY in hall when beards wag all.

Don't HALLOO till you are out of the wood

Do not exult until all danger or difficulty is past.

1770 B. FRANKLIN *Papers* (1973) XVIII. 356 This is Hollowing before you are out of the Wood. **1800** A. ADAMS *Letter* 13 Nov. (1848) 381 It is an old and a just proverb, 'Never halloo[1] until you are out of the woods'. **1866** KINGSLEY *Hereward the Wake* I. iii. Don't holla till you are out of the wood. This is a night for praying rather than boasting. **1908** E. SNEYD KYNNERSLEY *H.M.I.* xxii. The Duke . . wrote Dont halloo till you are out of the wood. **1922** A. MEYNELL in V. Meynell *Memoir* (1929) xix. I whistled before I was out of the wood when I said my cold was better.
[1] (literally) to shout in order to attract attention.

halved: *see* a TROUBLE shared is a trouble halved.

hammer: *see* the CHURCH is an anvil which has worn out many hammers.

One HAND for oneself and one for the ship

A nautical proverb, also used in variant forms in similar contexts: see the explanation in quot. 1902.

1799 *Port Folio* (Philadelphia, 1812) VII. 130 Did I not tell you never to fill both hands at once. Always keep one hand for the owners, and one for yourself. **1822** J. F. COOPER *Pilot* I. vii. The maxim, which says, 'one hand for the owner, and t'other for yourself,' . . has saved many a hearty fellow from a fall that would have balanced the purser's books. **1902** B. LUBBOCK *Round Horn* 58 The old rule on a yard is, 'one hand for yourself and one for the ship,' which means, hold on with one hand and work with the other. **1945** J. BRYAN *Diary* 7 Apr. in *Aircraft Carrier* (1954) 132 Hold tight and brace yourself when you're changing positions. . . One

hand for the plane, and one for yourself. **1968** L. MORTON *Long Wake* i. I did not know then the old adage 'one hand for oneself and one hand for the company.'

The HAND that rocks the cradle rules the world

1865 W. R. WALLACE in J. K. Hoyt *Cyclopædia of Practical Quotations* (1896) 402 A mightier power and stronger Man from his throne has hurled, For the hand that rocks the cradle Is the hand that rules the world. *a* **1916** 'SAKI' *Toys of Peace* (1919) 158 You can't prevent it; it's the nature of the sex. The hand that rocks the cradle rocks the world, in a volcanic sense. **1979** *Guardian* 12 June 9 The hand that rocks the cradle may rule the world but . . the hand itself is controlled by the state.

One HAND washes the other

[EPICHARMUS *Apophthegm* (Kaibel) no. 273 ἁ δὲ χεὶρ τὰν χεῖρα νίζει one hand washes the other: SENECA *Apocolocyntosis* ix. *manus manum lavat*, hand washes hand.] **1573** J. SANFORDE *Garden of Pleasure* 110ᵛ One hand washeth an other, and both wash the face. **1611** R. COTGRAVE *Dict. French & English* s.v. Main, One hand washes the other; applyable to such as giue vpon assurance, or hope, to be giuen vnto; or vnto such as any way serue one anothers turne. **1836** P. HONE *Diary* 12 Mar. (1927) I. 203 Persons in business . . make, as the saying is, 'one hand wash the other'. **1978** H. KEMELMAN *Thursday Rabbi walked Out* v. 'He's touted a lot of business our way.' . . 'You're right,' he said. 'One hand washes the other.'

hand: *see also* a BIRD in the hand is worth two in the bush; COLD hands, warm heart; the DEVIL finds work for idle hands to do; the EYE of a master does more work than both his hands; FULL cup, steady hand; if IFS and ands were pots and pans, there'd be no work for tinkers' hands; MANY hands make light work.

HANDSOME is as handsome does

Handsome denotes chivalrous or genteel behaviour, though it is often popularly taken to refer to good looks. At its second occurrence in the proverb the word is properly an adverb.

c **1580** A. MUNDAY *View of Sundry Examples* in J. P. Collier *John A Kent* (1851) 78 As the ancient adage is, goodly is he that goodly dooth. **1659** N. R. *Proverbs* 49 He is handsome that handsome doth. **1766** GOLDSMITH *Vicar of Wakefield* i. They are as heaven made them, handsome enough if they be good enough; for handsome is that handsome does. **1845** *Spirit of Times* 23 Aug. 297 Handsome is as handsome does. **1873** C. M. YONGE *Pillars of House* II. xvii. 'Don't you think her much better looking than Alda?' 'If handsome is that handsome does.' **1979** A. WILLIAMSON *Funeral March for Siegfried* xxiv. 'But he's such a handsome, *chivalrous*, man.' Handsome is as handsome does, thought York grimly.

HANG a thief when he's young, and he'll no' steal when he's old

1832 A. HENDERSON *Scottish Proverbs* 115 Hang a thief when he's young, and he'll no[1] steal when he's auld. **1896** A. CHEVIOT *Proverbs of Scotland* 126 Hang a thief when he's young, and he'll no steal when he's auld. This was a favourite saying of Lord Justice Clerk Braxfield,[2] who invariably acted upon its teaching. **1979** J. LEASOR *Love & Land Beyond* x. So much killing. . . It reminds me of the Scots proverb, 'Hang a thief when he's young, and he'll no' steal when he's old.'
[1] i.e. not. [2] Robert MacQueen, Lord Braxfield (1722–99), Scottish judge.

hang: *see also* a CREAKING door hangs longest; give a DOG a bad name and hang him; every HERRING must hang by its own gill; give a man ROPE enough and he will hang himself.

One might as well be HANGED for a sheep as a lamb

The proverb alludes to the former penalty for sheep-stealing.

[Cf. **1662** N. ROGERS *Rich Fool* 253 As some desperate Wretches, Who dispairing of life still act the more villainy, giving this desperate Reason of it, As good be hanged for a great deal, as for a little.] **1678** J. RAY *English Proverbs* (ed. 2) 350 As good be hang'd for an old sheep as a young lamb. *Somerset.* **1732** T. FULLER *Gnomologia* no. 683 As good be hang'd for a Sheep as a Lamb. **1836** MARRYAT *Midshipman Easy* II. ii. We may as well be hanged for a sheep as a lamb. . . I vote that we do not go on board. **1841** DICKENS *Barnaby Rudge* liii. Others . . comforted themselves with the homely proverb, that, being hung at all, they might as well be hung for a sheep as a lamb. **1915** D. H. LAWRENCE *Rainbow* vi. One might as well be hung for a sheep as for a lamb. If he had lost this day of his life, he had lost it. **1977** B. PYM *Quartet in Autumn* xv. Letty . . decided that she might as well be hung for a sheep as a lamb and make the most of her meal.

hanged: *see also* if you're BORN to be hanged then you'll never be drowned; CONFESS and be hanged; he that has an ILL name is half hanged; LITTLE thieves are hanged, but great ones escape.

hanging: *see* there are more WAYS of killing a dog than hanging it.

ha'porth: *see* do not spoil the SHIP for a ha'porth of tar.

happen: *see* ACCIDENTS will happen (in the best-regulated families); the UNEXPECTED always happens.

If you would be HAPPY for a week take a wife; if you would be happy for a month kill a pig; but if you would be happy all your life plant a garden

a **1661** T. FULLER *Worthies* Wales 6, I say the Italian-humor, who have a merry Proverb, Let him that would be happy for a Day, go to the Barber; for a Week, marry a Wife; for a Month, buy him a New–horse; for a Year, build him a New–house; for all his Life–time, be an

Honest man. **1809** S. PEGGE *Anonymiana*
II. xix. If you would live well for a week,
kill a hog; if you would live well for a
month, marry; if you would live well all
your life, turn priest. . . Turning priest
. . alludes to the celibacy of the Romish
Clergy, and has a pungent sense, as
much as to say, do not marry at all. **1973**
New Earth Catalog 55 If you would be
happy for a week take a wife; If you
would be happy for a month kill a pig;
But if you would be happy all your life
plant a garden.

Call no man HAPPY till he dies
  [SOPHOCLES *Oedipus Rex* 1530 μηδέν'
ὀλβίζειν, πρὶν ἂν τέρμα τοῦ βίου περάσῃ
μηδὲν ἀλγεινὸν παθών, deem no man
happy, until he passes the end of his life
without suffering grief; OVID *Metamor-
phoses* iii. 135 *dicique beatus Ante obitum
nemo . . debet*, nobody should be called
blessed before his death.] **1545** R.
TAVERNER tr. *Erasmus' Adages* (ed. 2) 53ᵛ
Salon aunsered kynge Cresus, that no
man coulde be named happy, tyl he had
happely and prosperouslye passed the
course of his lyfe. **1565** NORTON & SACK-
VILLE *Gorboduc* III. i. Oh no man happie,
till his ende be seene. **1603** FLORIO tr.
Montaigne's Essays I. xviii. We must ex-
spect of man the latest day, Nor e'er he
die, he's happie, can we say. **1891**
Times 5 Dec. 9 Call no man happy till he
dies is the motto . . suggested by the
career of Dom Pedro [of Brazil]. **1967**
C. S. FORESTER *Hornblower & Crisis* 163
'Call no man happy until he is dead.' . .
He was seventy-two, and yet there was
still time for this dream . . to change to a
nightmare.

happy: *see also* happy is the BRIDE that the
sun shines on; happy is the COUNTRY
which has no history; a DEAF husband
and a blind wife are always a happy
couple.

HARD cases make bad law
Difficult cases cause the clarity of the
law to be obscured by exceptions and
strained interpretations.
  1854 G. HAYES in W. S. Holdsworth

Hist. English Law (1926) IX. 423 A hard
case. But hard cases make bad law.
1945 CHURCHILL in *Hansard* (Commons)
12 June 1478 Well, of course, hard cases
do not make good laws. **1979** *Spectator*
10 Feb. 3 It is said that hard cases make
bad law. So does bad law make hard
cases. Such a case was heard last week
at Oxford Crown Court.

HARD words break no bones
Cf. STICKS *and stones may break my bones,
but words will never hurt me.*
  [Cf. c **1450** *Towneley Play of Noah*
(EETS) l. 380 Thise grete wordis shall
not flay me.] **1697** G. MERITON *Yorkshire
Ale* (ed. 3) 84 Foul words break neay
Banes. **1806** H. H. BRACKENRIDGE *Gazette Publications* 250 Hard words, and
language break nae bane. **1814** G. MOR-
RIS *Letter* 18 Oct. (1889) II. xlix. These . .
are mere words—hard words, if you
please, but they break no bones. **1882**
BLACKMORE *Christowell* III. xvi. 'Scoundrel, after all that I have done—.' 'Hard
words break no bones, my friend.'
1980 G. NELSON *Charity's Child* i. Soft
words! They butter no parsnips. . .
Would you prefer hard ones? . . Hard
words break no bones.

hard: *see also* OLD habits die hard.

harder: *see* the BIGGER they are, the harder
they fall.

hare: *see* FIRST catch your hare; if you RUN
after two hares you will catch neither;
you cannot RUN with the hare and hunt
with the hounds.

HASTE is from the Devil
  1633 J. HOWELL *Familiar Letters* 5 Sept.
(1903) II. 140 As it is a principle in chem-
istry that *Omnis festinatio est a Diabolo*,
All haste comes from Hell, so in . . any
business of State, all rashness and precipitation comes from an ill spirit. **1835**
SOUTHEY *Doctor* III. lxxxiii. If any of my
readers should . . think that I ought to
have proceeded to the marriage without
delay . . I must admonish them in the
words of a Turkish saying, that 'hurry

comes from the Devil, and slow advancing from Allah.' **1929** *Times* 12 Sept. 14 Listening patiently to the views . . [f]or he understood the East; he knew that for an Intelligence officer 'haste is from the devil.'

More HASTE, less speed
The original meaning of *speed* in this proverb is 'quickness in the performance of some action or operation'. Cf. *Make* HASTE *slowly*.

c **1350** *Douce MS* 52 no. 86 The more hast, the worse spede. **1546** J. HEYWOOD *Dialogue of Proverbs* I. ii. A3ᵛ Moste tymes he seeth, the more haste the lesse speede. **1595** *Locrine* (1908) I. ii. My penne is naught; gentlemen, lend me a knife. I thinke the more haste the worst speed. **1705** E. WARD *Hudibras Redivivus* I. i. A mod'rate pace is best indeed. The greater hurry, the worse speed. **1887** BLACKMORE *Springhaven* III. xi. Some days had been spent by the leisurely Dutchman in providing fresh supplies, and the stout bark's favourite maxim seemed to be—'the more haste the less speed.' **1919** S. J. WEYMAN *Great House* xxvii. Tell me the story from the beginning. And take time. More haste, less speed, you know. **1979** P. NIESEWAND *Member of Club* xvi. If they'd taken a bit more time with the terrorist he'd have told them everything. . . More haste, less speed.

HASTE makes waste
Waste properly means the squandering of time, money, etc., though it is also used with reference to material waste.

c **1390** CHAUCER *Tale of Melibee* l. 2244 The prouerbe seith . . in wikked haste is no profit. **1546** J. HEYWOOD *Dialogue of Proverbs* I. ii. A3 Som thyngs . . show after weddyng, that haste maketh waste. **1663** BUTLER *Hudibras* I. iii. *Festina lente*, not too fast; For haste (the Proverb says) makes waste. **1853** R. C. TRENCH *On Lessons in Proverbs* i. Many proverbs, such as Haste makes waste . . have nothing figurative about them. **1940** H. W. THOMPSON *Body, Boots & Brit-*

ches xix. Haste makes waste, Waste makes want, Want makes a poor boy a beggar. **1977** A. McCAFFREY *Mark of Merlin* xi. I . . slammed some logs on the grate. . . Then I had to sweep up the scattered coals. . . 'Haste makes waste,' he chanted from the door.

Make HASTE slowly
Cf. *More* HASTE, *less speed*.

[L. *festina lente*, make haste slowly; after SUETONIUS *Augustus* xxv. 4 *nihil autem minus perfecto duci quam festinationem temeritatemque convenire arbitratur. Crebro itaque illa iactabat:* σπεῦδε βραδέως, he [Augustus] thought that delay and rashness were alike unsuited to a well-trained leader. So he often came out with sayings like 'make haste slowly' [etc.]; cf. *c* **1385** CHAUCER *Troilus & Criseyde* I. 956 He hasteth wel that wisly kan[1] abyde.] **1683** DRYDEN *Poems* (1958) I. 336 Gently make haste. . . A hundred times consider what you've said. **1744** B. FRANKLIN *Poor Richard's Almanack* (Apr.) Make haste slowly. **1938** M. TEAGLE *Murders in Silk* iii. Easy, son. Let's make haste slowly. Does Conner know where the knife came from?

[1] knows how to.

haste: *see also* MARRY in haste and repent at leisure; NOTHING should be done in haste but gripping a flea.

hatched: *see* don't COUNT your chickens before they are hatched.

hate: *see* BETTER a dinner of herbs than a stalled ox where hate is.

What you HAVE, hold
c **1450** *Towneley Play of Killing of Abel* (EETS) l. 142 It is better hold that I haue then go from doore to doore & craue. **1546** J. HEYWOOD *Dialogue of Proverbs* I. x. D1 Hold fast whan ye haue it (quoth she) by my lyfe. **1876** I. BANKS *Manchester Man* I. x. Then . . rang, clear and distinct, Humphrey Chetham's motto— 'Quod tuum, tene!' (What you have, hold!) **1979** *Times* 23 Nov. 5 There had

been a simple 'what we have we hold' approach by the established parties.

You cannot HAVE your cake and eat it

You cannot 'have it both ways': once the cake is eaten, it can no longer be 'had' or retained in one's possession. The positions of *have* and *eat* are often reversed.

1546 J. HEYWOOD *Dialogue of Proverbs* II. ix. L2 I trowe ye raue, Wolde ye bothe eate your cake, and haue your cake? **1611** J. DAVIES *Scourge of Folly* no. 271 A man cannot eat his cake and haue it stil. **1738** SWIFT *Polite Conversation* I. 90 She was handsome in her Time; but she cannot eat her Cake, and have her Cake. **1812** in R. C. Knopf *Document Transcriptions of War of 1812* (1959) VI. 204 We cannot have our cake and eat it too. **1878** TROLLOPE *Is he Popenjoy?* I. viii. You can't eat your cake and have it too. **1959** M. BRADBURY *Eating People is Wrong* vii. You want to have your cake and eat it. Why not, of course? It's an absurd proverb. I always have my cake and eat it. **1980** M. DRABBLE *Middle Ground* 159 Judith cannot eat her cake and have it. Judith liked Hugo for his style . . and she can bloody well suffer for its inconveniences.

have: *see also* the MORE you get, the more you want; NOTHING venture, nothing have; what you SPEND, you have; you can have TOO much of a good thing.

haw: *see* when all FRUIT falls, welcome haws.

HAWKS will not pick out hawks' eyes

1573 J. SANFORDE *Garden of Pleasure* 104 One crowe neuer pulleth out an others eyes. **1817** SCOTT *Rob Roy* III. iii. I wadna . . rest my main dependence on the Hielandmen—hawks winna pike out hawks' een.—They quarrel amang themsells . . but they are sure to join . . against a' civilized folk. **1883** J. PAYN *Thicker than Water* III. xli. Members of his profession . . while warning others of the dangers of the table, seem to pluck from them the flower Safety. (Is it that, since hawks do not peck out hawks' een, they know they can be cured for nothing?) **1915** J. BUCHAN *Salute to Adventurers* vi. I have heard that hawks should not pick out hawks' eyes. What do you propose to gain? **1975** J. O'FAOLAIN *Women in Wall* xiv. The crow doesn't pluck out the crow's eye but poor folk bear the brunt.

hay: *see* if in FEBRUARY there be no rain, 'tis neither good for hay nor grain; MAKE hay while the sun shines; a SWARM in May is worth a load of hay.

head: *see* the FISH always stinks from the head downwards; where MACGREGOR sits is the head of the table; you cannot put an OLD head on young shoulders; a STILL tongue makes a wise head; TWO heads are better than one; YORKSHIRE born and Yorkshire bred, strong in the arm and weak in the head.

heal: *see* PHYSICIAN, heal thyself.

healer: *see* TIME is a great healer.

healthy: *see* EARLY to bed and early to rise, makes a man healthy, wealthy, and wise.

HEAR all, see all, say nowt, tak' all, keep all, gie nowt, and if tha ever does owt for nowt do it for thysen

A proverb now traditionally associated with Yorkshire, with numerous variant forms.

a **1400** *Proverbs of Wisdom* in *Archiv* (1893) XC. 246 Hyre and se, and say nowght. Be ware and wyse, and lye nought . . and haue thy will. **1623** J. WODROEPHE *Spared Hours of Soldier* 276 Heare all, see all, and hold thee still If peace desirest with thy will. **1913** D. H. LAWRENCE *Letter* 1 Feb. (1962) I. 183 It seems queer, that you do it and get no profit. I should think you've forgotten the Yorkshire proverb, 'An' if tha does owt[1] for nowt,[2] do it for thysen.'[3] **1925** *Notes & Queries* 412 The famous Yorkshire motto . . is invariably recited with an air of superior bravado, and will be found upon mugs, post cards, etc. The

authentic version, I believe, is, 'Hear all, see all, say now't, tak' all, keep all, gie[4] now't, and if tha ever does ow't for now't do it for thysen.' **1966** J. BINGHAM *Double Agent* vi. Ducane thought of Sugden saying, 'See all, hear all, say nowt, eat all, sup all, pay nowt, and if tha does owt for nowt, do it for thy'sen.
[1] ought; aught; anything.   [2] nought; nothing.   [3] yourself.   [4] give.

hear: *see also* ASK no questions and hear no lies; BELIEVE nothing of what you hear, and only half of what you see; there's none SO DEAF as those who will not hear; DREAM of a funeral and you hear of a marriage; GO abroad and you'll hear news of home; LISTENERS never hear any good of themselves; SEE no evil, hear no evil, speak no evil.

heard: *see* CHILDREN should be seen and not heard.

heart: *see* ABSENCE makes the heart grow fonder; COLD hands, warm heart; what the EYE doesn't see, the heart doesn't grieve over; FAINT heart never won fair lady; HOME is where the heart is; HOPE deferred makes the heart sick; if it were not for HOPE, the heart would break; PLEASE your eye and plague your heart; it is a POOR heart that never rejoices; put a STOUT heart to a stey brae; the WAY to a man's heart is through his stomach.

If you don't like the HEAT, get out of the kitchen

1952 *Time* 28 Apr. 19 President [Truman] gave a . . down-to-earth reason for his retirement, quoting a favorite expression of his military jester, Major General Harry Vaughan: 'If you don't like the heat, get out of the kitchen.' **1970** *Financial Times* 13 Apr. 25 Property people argue that hoteliers are not facing the facts of economic life, and that if they cannot stand the heat they should get out of the kitchen. **1975** M. BRADBURY *History Man* xiii. He got in the way of justice. . . You know what they say, if you don't like the heat, get out of the kitchen.

HEAVEN protects children, sailors, and drunken men

The proverb is found in various forms.

1861 T. HUGHES *Tom Brown at Oxford* I. xii. Heaven, they say, protects children, sailors, and drunken men; and whatever answers to Heaven in the academical system protects freshmen. **1865** G. MacDONALD *Alec Forbes* III. xi. I canna think hoo he cam' to fa' sae sair; for they say there's a special Providence watches over drunk men and bairns. **1980** S. KING *Firestarter* 57 She didn't even have a bruise—God watches over drunks and small children.

heaven: *see also* CROSSES are ladders that lead to heaven; GOD's in his heaven, all's right with the world; MARRIAGES are made in heaven; *also* GOD.

hedge: *see* one man may STEAL a horse, while another may not look over a hedge.

heir: *see* WALNUTS and pears you plant for your heirs.

HELL hath no fury like a woman scorned

[Cf. *a* 1625 BEAUMONT & FLETCHER *Knight of Malta* I. i. The wages of scorn'd Love is baneful hate.] **1696** C. CIBBER *Love's Last Shift* IV. 71 No Fiend in Hell can match the fury of a disappointed Woman!—Scorned! slighted; dismissed without a parting Pang! **1697** CONGREVE *Mourning Bride* III. 39 Heav'n has no Rage, like Love to Hatred turn'd, Nor Hell a Fury,[1] like a Woman scorn'd. **1886** M. HOLMES *Chamber over Gate* xxvi. You know 'Hell hath no fury,' etc. If your wife should ever wake up to the true state of the case . . I'm afraid she'd be an ugly customer. **1940** G. H. COXE *Glass Triangle* x. If you really want to know who could have wanted to kill him . . start with me. . . You've heard that one about hell having no fury like a woman scorned. **1973** I. MURDOCH *Black Prince* 330 'Hell hath no fury like a woman scorned.' In a way I might have been flattered.
[1] frenzied rage; (personified) in classical

mythology: one of the avenging deities, fearful goddesses from Tartarus who avenged wrong and punished crime.

hell: *see also* the ROAD to hell is paved with good intentions; he that would go to SEA for pleasure, would go to hell for a pastime.

help: *see* EVERY little helps; GOD helps them that help themselves; a MOUSE may help a lion; help you to SALT, help you to sorrow.

hen: *see* the ROBIN and the wren are God's cock and hen; a WHISTLING woman and a crowing hen are neither fit for God nor men.

herb: *see* BETTER a dinner of herbs than a stalled ox where hate is.

heresy: *see* TURKEY, heresy, hops, and beer came into England all in one year.

hero: *see* NO man is a hero to his valet.

Every HERRING must hang by its own gill
Everyone is accountable for his own actions.

   1609 S. HARWARD *MS* (Trinity College, Cambridge) 85 Lett every herring hang by his owne tayle. **1639** J. CLARKE *Parœmiologia Anglo-Latina* 20 Every herring must hang by th'owne gill. **1670** J. RAY *English Proverbs* 102 Every herring must hang by its own gill. . . Every man must give an account for himself. **1890** T. H. HALL CAINE *Bondman* II. ii. Adam, thinking as little of pride, said No, that every herring should hang by its own gills.

He who HESITATES is lost
Early uses of the proverb refer specifically to women.

   1713 ADDISON *Cato* IV. i. When love once pleads admission to our hearts . . The woman that deliberates is lost. **1865** TROLLOPE *Can You forgive Her?* II. x. It has often been said of woman that she who doubts is lost . . never thinking

whether or no there be any truth in the proverb. **1878** J. H. BEADLE *Western Wilds* xxi. In Utah it is emphatically true, that he who hesitates is lost—to Mormonism. **1887** BLACKMORE *Springhaven* xlii. Dolly hesitated, and with the proverbial result. **1920** E. O'NEILL *Beyond Horizon* II. ii. He who hesitates, you know. . . Don't ask me to decide for you. **1980** *Daily Telegraph* 2 Feb. 9 'He who hesitates is lost' . . against Martin Hoffman, one of the fastest analysts and players in the game [of chess].

hid: *see* LOVE and a cough cannot be hid.

Those who HIDE can find
   c **1400** *Seven Sages of Rome* (1845) 68 He may wel fynde that hyde him selven. **1639** J. CLARKE *Parœmiologia Anglo-Latina* 111 They that hide[1] can find. **1842** MARRYAT *Percival Keene* I. iii. 'I could have told you where it was.' 'Yes, yes, those who hide can find.' **1922** JOYCE *Ulysses* 542 (She . . unrolls the potato from the top of her stocking.) Those that hides knows where to find. **1979** 'E. PETERS' *One Corpse too Many* ix. Only those who had hidden here were likely ever to find. The full leafage covered all.
   [1] i.e. hide something: the verb is used absolutely.

The HIGHER the monkey climbs the more he shows his tail
The further an unsuitable person is advanced, the more his inadequacies are apparent. Also found in less polite forms.

   c **1395** WYCLIF *Bible* (1850) Proverbs iii. 35 (*gloss*) The filthe of her foli aperith more, as the filthe of the hynd partis of an ape aperith more, whanne he stieth[1] on high. *c* **1594** BACON *Promus* 102 He doth like the ape that the higher he clymbes the more he shows his ars. **1670** J. RAY *English Proverbs* 57 The higher the Ape goes, the more he shews his tail. . . The higher beggars or base bred persons are advanced, the more they discover the lowness and baseness of their spirits and tempers. **1743** POPE

Dunciad IV. 157 (*note*) The higher you climb, the more you shew your A—. Verified in no instance more than in Dulness aspiring. Emblematized also by an Ape climbing and exposing his posteriors. **1873** TROLLOPE *Phineas Redux* I. xxxiv. He's to be pitchforked up to the Exchequer. . . The higher a monkey climbs—; you know the proverb.
¹ climbs.

hill: *see* BLUE are the hills that are far away.

hindered: *see* MEAT and mass never hindered man.

hindmost: *see* DEVIL take the hindmost; EVERY man for himself, and the Devil take the hindmost.

hire: *see* the LABOURER is worthy of his hire.

HISTORY repeats itself
   1858 G. ELIOT *Janet's Repentance* in *Scenes of Clerical Life* II. x. History, we know, is apt to repeat itself. **1865** H. SEDLEY *Marian Rooke* III. v. i. History, it is said, repeats itself. . . Few but are reminded almost every day . . of something that has gone before. **1957** V. BRITTAIN *Testament of Experience* 11 History tends to defy the familiar aphorism; whether national or personal, it seldom repeats itself. **1971** A. PRICE *Alamut Ambush* xiii. Maybe history repeats itself—but I have to have facts.

history: *see also* happy is the COUNTRY which has no history.

hog: *see* the CAT, the rat, and Lovell the dog, rule all England under the hog.

hold: *see* what you HAVE, hold.

Holdfast: *see* BRAG is a good dog, but Holdfast is better.

HOME is home, as the Devil said when he found himself in the Court of Session
   1832 W. MOTHERWELL in A. Hender-

son *Scottish Proverbs* lxix. Nothing more bitter was ever uttered . . against our Supreme Court of Judicature, than the saying . . Hame is hamely, quo' the Deil, when he fand himself in the Court of Session.¹ **1915** J. BUCHAN *Salute to Adventurers* iv. I saw nothing now to draw me to . . law. . . 'Hame's hame,' runs the proverb, 'as the devil said when he found himself in the Court of Session,' and I had lost any desire for that sinister company.
¹ *Court of Session*: the supreme civil tribunal of Scotland, established in 1532.

HOME is home though it's never so homely
   1546 J. HEYWOOD *Dialogue of Proverbs* I. iv. B1 Home is homely, though it be poore in syght. **1569–70** *Stationers' Register* (1875) I. 192 A ballett intituled home ys homelye be yt neuer so¹ ill. **1670** J. RAY *English Proverbs* 103 Home is home though it be never so homely. **1857** DICKENS *Little Dorrit* II. ix. 'Just as Home is Home though it's never so Homely, why you see,' said Mr. Meagles, adding a new version to the proverb, 'Rome is Rome though it's never so Romely.' **1915** J. WEBSTER *Dear Enemy* 46 Hame is hame, be't ever sae hamely. Don't you marvel at the Scotch?
¹ *never so* (archaic): i.e. ever so

HOME is where the heart is
   1870 J. J. McCLOSKEY in Goldberg & Heffner *Davy Crockett & Other Plays* (1940) 79 'As I am to become an inmate of your home, give me a sort of a panoramic view.' . . 'Well, home, they say, is where the heart is.' **1950** H. M. GAY *Pacific Spectator* IV. 91 'Home is where the heart is,' she said, 'if you'll excuse the bromide.'¹ **1979** K. BONFIGLIOLI *After You with Pistol* xxi. 'Where is "home", please,' I asked. . . 'Home's where the heart is,' he said.
¹ trite remark.

home: *see also* CHARITY begins at home; CURSES, like chickens, come home to roost; EAST, west, home's best; an ENGLISHMAN'S home is his castle; GO abroad and you'll hear news of home;

the LONGEST way round is the shortest way home; there's no PLACE like home; a WOMAN's place is in the home; many go out for WOOL and come home shorn.

HOMER sometimes nods

Nobody can be at his best or most alert all the time.

[HORACE *Ars Poetica* 359 *indignor quandoque bonus dormitat Homerus*, I am indignant when worthy Homer nods.] **1387** J. TREVISA tr. Higden's *Polychronicon* (1874) V. 57 He may take hede that the grete Homerus slepeth somtyme, for in a long work it is laweful to slepe som time. **1677** DRYDEN in *State of Innocence* B1^v Horace acknowledges that honest Homer nods[1] sometimes: he is not equally awake in every line. **1887** T. H. HUXLEY in *Nineteenth Century* Feb. 196 Scientific reason, like Homer, sometimes nods. **1979** D. CLARK *Heberden's Seat* vi. 'We're half asleep, not to have asked where they are before this.' 'Homer nods. . . You can't ask every question.'

[1] becomes drowsy, falls asleep; hence, errs due to momentary lack of attention.

HONESTY is the best policy

1605 E. SANDYS *Europæ Speculum* K3 This over-politick . . order may reach a note higher than our grosse conceipts, who think honestie the best policie. *a* **1763** J. BYROM *Poems* (1773) I. 75 I'll filch no filching;—and I'll tell no lye; Honesty's the best Policy,—say I. **1854** R. WHATELY *Detached Thoughts* II. xviii. 'Honesty is the best policy'; but he who acts on that principle is not an honest man. **1928** J. GALSWORTHY *Swan Song* vi. It had been in their systems just as the proverb 'Honesty is the best policy' was in that of the private banking which then obtained. **1980** J. GASH *Spend Game* ix. When a crisis comes to the crunch I'm full of this alert feeling. I think it's a sort of realization that honesty's the best policy.

HONEY catches more flies than vinegar

Soft or ingratiating words achieve more than sharpness.

1666 G. TORRIANO *Italian Proverbs* 149 Honey gets more flyes to it, than doth viniger. **1744** B. FRANKLIN *Poor Richard's Almanack* (Mar.) Tart Words make no Friends: spoonful of honey will catch more flies than Gallon of Vinegar. **1955** W. C. MACDONALD *Destination Danger* x. I . . know the old saying relative to honey catching more flies than vinegar . . . If this is an act, you might as well save your breath. **1979** M. A. SHARP *Sunflower* xviii. Honey might attract more flies than vinegar, but nothing beat blackmail at securing the absolute allegiance.

There is HONOUR among thieves

[*c* **1622–3** *Soddered Citizen* (1936) l. 305 Theeues haue betweene themselues, a truth, And faith, which they keepe firme, by which They doe subsist. **1703** P. A. MOTTEUX *Don Quixote* II. lx. The old proverb still holds good, Thieves are never rogues among themselves.] **1802** J. BENTHAM *Works* (1843) IV. 225 A sort of honour may be found (according to a proverbial saying) even among thieves. **1823** J. BEE *Dict. Turf* 98 'There is honour among thieves, but none among gamblers,' is very well antithetically spoken, but not true in fact. **1979** R. BERRY *Bishop's Pawn* iv. It wouldn't be the first occasion Pawson had co-operated with the KGB. . . The old saw about honour between thieves would be . . applicable.

The post of HONOUR is the post of danger

a **1533** LD. BERNERS *Huon* (EETS) xx. Where as lyeth grete parelles there lieth grete honour. **1613** T. HEYWOOD *Brazen Age* III. 211 The greater dangers threaten The greater is his honour that breaks through. *a* **1625** J. FLETCHER *Rule Wife* (1640) IV. i. I remembered your old Roman axiom, The more the danger, still the more the honour. [**1711** *Spectator* 1 Dec. 1 We consider Human Life as a State of Probation, and Adversity as the Post of Honour in it.] **1832** A. HENDERSON *Scottish Proverbs* 33 The post of honour is the post of danger. **1905** *British Weekly* 14 Dec. 1 The Chancellorship of the Exchequer . . is pre-eminently the post of danger, and therefore the post of honour in the new Government.

honour: *see also* give CREDIT where credit is due; a PROPHET is not without honour save in his own country.

hop: *see* TURKEY, heresy, hops, and beer came into England all in one year.

HOPE deferred makes the heart sick

c **1395** WYCLIF *Bible* (1850) Proverbs xiii. 13 Hope that is deferrid, tormenteth the soule. *c* **1527** J. RASTELL *Calisto & Melebea* A5ᵛ For long hope to the hart mych troble wyll do. **1557** R. EDGEWORTH *Sermons* 130ᵛ The hope that is deferred, prolonged, and put of, vexeth the minde. **1611** BIBLE *Proverbs* xiii. 12 Hope deferred maketh the heart sicke. **1733** J. TALCOTT in *Collections of Connecticut Hist. Society* (1892) IV. 285 As hope deferred makes the heart sick: so I am in long expectation of your answers. **1889** GISSING *Nether World* II. vii. There was a heaviness at his heart. Perhaps it came only of hope deferred, **1981** *Observer* 26 Apr. 14 If hope deferred makes the heart sick, despair is a poor counsellor also.

HOPE for the best and prepare for the worst

1565 NORTON & SACKVILLE *Gorboduc* I. ii. Good is I graunt of all to hope the best, But not to liue still dreadles of the worst. **1581** W. AVERELL *Charles & Julia* D7 To hope the best, and feare the worst, (loe, such is Loouers gaines). **1706** E. WARD *Third Volume* 337 This Maxim ought to be carest, Provide against the worst, and hope the best. **1813** J. JAY *Correspondence* (1893) IV. 367 To hope for the best and prepare for the worst, is a trite but a good maxim. **1836** E. HOWARD *Rattlin the Reefer* II. xxix. The youngest of us cannot always escape—hoping, trusting, relying on the best, we should be prepared for the worst. **1929** J. BUCHAN *Courts of Morn-*

ing II. vi. 'You have the ultimate safeguard. You need not fear the worst.' 'I hope for the best,' she said. **1981** *Times* 29 Jan. 3 Hoping for the best, preparing for the worst.

HOPE is a good breakfast but a bad supper

1661 W. RAWLEY *Resuscitatio* (ed. 2) 298 But, said the fisher men, we had hope then to make a better gain of it. Saith Mr. [Francis] Bacon well my Maisters, then Ile tell you; hope is a good Breakfast but it is a Bad supper. **1817** H. L. PIOZZI *Autobiography* (1861) II. 188 He was a wise man who said Hope is a good breakfast but a bad dinner. It shall be my supper . . when all's said and done. **1963** *Times* 9 Apr. 1 Hope is a good breakfast, but it is a bad supper.

HOPE springs eternal

1732 POPE *Essay on Man* I. 95 Hope springs eternal in the human breast. Man never Is, but always To be blest. **1865** DICKENS *Our Mutual Friend* II. III. x. Night after night his disappointment is acute, but hope springs eternal in the scholastic breast. **1935** H. SPRING *Rachel Rosing* viii. 'It was understood, wasn't it, that we could not dine together?' 'Oh yes—but you know how it is. Hope springs eternal and so forth.' **1977** A. NEWMAN *Evil Streak* II. 67, I prepared a delicious salad and . . dry martini, his favourite drink. They say hope springs eternal.

If it were not for HOPE, the heart would break

a **1250** *Ancrene Wisse* (1962) 43 Ase me seith, yef hope nere heorte to breke [as one says, if there were not hope, the heart would break]. *c* **1440** *Gesta Romanorum* (EETS) 228 Yf hope wer not, hert schulde breke. **1616** J. WITHALS *Dict.* (rev. ed.) 582 If it were not for hope, the heart would breake. **1748** RICHARDSON *Clarissa* VI. xxix. No harm in hoping, Jack! My uncle says, Were it not for hope, the heart would break. **1911** J. LUBBOCK *Use of Life* (rev. ed.) xv. There is an old proverb that if it were not for Hope the heart would break. Everything may be retrieved except despair.

hope: *see also* while there's LIFE, there's hope; he that LIVES in hope dances to an ill tune.

hopefully: *see* it is BETTER to travel hopefully than to arrive.

You can take a HORSE to the water, but you can't make him drink
The word *the* is frequently omitted from the proverb.
c **1175** *Old English Homilies* (EETS) 1st Ser. 9 Hwa is thet mei thet hors wettrien the him self nule drinken [who can give water to the horse that will not drink of its own accord]? **1546** J. HEYWOOD *Dialogue of Proverbs* I. xi. D3 A man may well bryng a horse to the water, But he can not make hym drynke without he will. **1658** E. PHILLIPS *Mysteries of Love & Eloquence* 160 A man may lead his Horse to water, but he cannot make him drink unless he list. **1857** TROLLOPE *Barchester Towers* III. i. 'Well,' said she . . 'one man can take a horse to water but a thousand can't make him drink.' **1930** W. S. MAUGHAM *Cakes & Ale* v. 'He thinks he'd be mayor himself,' said the people of Blackstable. . . My uncle remarked that you could take a horse to the water but you couldn't make him drink. **1970** J. MITFORD in *Atlantic* (1979) July 50 The dropout rate [for the course] must be close to 90 percent. I guess you can take a horse to the water, but you can't make him drink.

horse: *see also* don't CHANGE horses in midstream; never look a GIFT horse in the mouth; a GOOD horse cannot be of a bad colour; while the GRASS grows, the steed starves; the GREY mare is the better horse; the MAN who is born in a stable is a horse; a NOD's as good as a wink to a blind horse; there is NOTHING so good for the inside of a man as the outside of a horse; if you can't RIDE two horses at once, you shouldn't be in the circus; a SHORT horse is soon curried; it is too late to shut the STABLE-door after the horse

has bolted; one man may STEAL a horse, while another may not look over a hedge; THREE things are not to be trusted; if TWO ride on a horse, one must ride behind; for WANT of a nail the shoe was lost; if WISHES were horses, beggars would ride.

horseback: *see* set a BEGGAR on horseback, and he'll ride to the Devil.

HORSES for courses
Originally an expression in horse-racing: different horses are suited to different race-courses. Now widely used in other contexts.
1891 A. E. T. WATSON *Turf* vii. A familiar phrase on the turf is 'horses for courses'. . . The Brighton Course is very like Epsom, and horses that win at one meeting often win at the other. **1929** *Daily Express* 7 Nov. 18 Followers of the 'horses for courses' theory will be interested in the acceptance of Saracen, Norwest and Sir Joshua. **1963** *Punch* 18 Sept. 430 People enjoy what they are capable of enjoying—horses for courses. **1976** H. WILSON *Governance of Britain* ii. He must concentrate on the doctrine of horses for courses . . in using the specialist knowledge of individual ministers.

hot: *see* a LITTLE pot is soon hot; STRIKE while the iron is hot.

hound: *see* you cannot RUN with the hare and hunt with the hounds.

One HOUR's sleep before midnight is worth two after
1640 G. HERBERT *Outlandish Proverbs* no. 882 One houres sleepe before midnight is worth three after. **1670** J. RAY *English Proverbs* 37 One hours sleep before midnight's worth two hours after. **1829** COBBETT *Advice to Young Men* I. xxxviii. It is said by the country-people that one hour's sleep before midnight is worth more than two are worth after midnight; and this I believe to be a fact.

hour: *see also* the DARKEST hour is just before the dawn; six hours' sleep for a man, seven for a woman, and eight for a fool.

When HOUSE and land are gone and spent, then learning is most excellent
Cf. LEARNING *is better than house and land*.

1752 S. FOOTE *Taste* I. i. It has always been my Maxum . . to give my Children Learning enough; for, as the old Saying is, When house and Land are gone and spent, then Learning is most excellent. **1896** S. BARING-GOULD *Broom-Squire* xxvi. I have . . got Simon to write for me, on the fly-leaf. . . When land is gone, and money is spent, Then learning is most excellent.

house: *see also* BETTER one house spoiled than two; an ENGLISHMAN's home is his castle; FOOLS build houses and wise men live in them; those who live in GLASS houses shouldn't throw stones; LEARNING is better than house and land.

human: *see* to ERR is human (to forgive divine).

hundred: *see* the BUYER has need of a hundred eyes, the seller of but one.

HUNGER drives the wolf out of the wood
[Cf. early 14th-cent. Fr. *la fains enchace le louf dou bois*, hunger chases the wolf from the wood.] **1483** CAXTON *Cato* B6[v] As hunger chaceth the wolfe out of the wode thus sobrete[1] chaseth the deuyl fro the man. **1591** J. FLORIO *Second Fruits* 125 Hunger driues the wolfe out of the wood, if I had not great neede of monie, you should neuer haue them so dog cheape. **1748** SMOLLETT *Gil Blas* (1749) IV. XII. vii. This one . . I own is the child of necessity. Hunger, thou knowest, brings the wolf out of the wood. **1872** BROWNING *Works* (1897) III. 323 Hunger, proverbs say, allures the wolf from the wood.
[1] i.e. sobriety.

HUNGER is the best sauce
[CICERO *De Finibus* II. xxviii. *cibi condimentum esse famem*, hunger is the spice

of food.] **1530** A. BARCLAY *Eclogues* (EETS) II. 743 Make hunger thy sause be thou neuer so nice, For there shalt thou finde none other kind of spice. **1539** R. TAVERNER *Garden of Wisdom* I. B1 He [Socrates] sayd, the beste sawce is hungre. **1555** R. EDEN tr. *P. Martyr's Decades of New World* II. iii. (*margin*) Hunger is the best sauce. **1850** KINGSLEY *Alton Locke* I. ix. If hunger is, as they say, a better sauce than any Ude invents, you should spend . . months shut out from every glimpse of Nature, if you would taste her beauties. **1929** F. M. McNEILL *Scots Kitchen* iii. Mere hunger, which is the best sauce, will not produce cookery, which is the art of sauces. **1939** L. I. WILDER *By Shores of Silver Lake* xxi. 'The gravy is extra good too.' 'Hunger is the best sauce,' Ma replied modestly. **1966** MRS. L. B. JOHNSON *White House Diary* Mar. (1970) 346 We . . arrived hungry as bears at about 3:15. Appetite is the best sauce.

A HUNGRY man is an angry man
c **1641** D. FERGUSSON *Scottish Proverbs* (STS) no. 553 Hungry men ar angry. **1659** J. HOWELL *Proverbs* (English) 13 A hungry man, an angry man. **1738** SWIFT *Polite Conversation* II. 119 'I'm hungry.' . . 'And I'm angry, so let us both go fight.' **1909** *Spectator* 22 May 824 The Acharnians[1] . . made fun of the Athenians. . . 'A hungry man is an angry man' . . and the Athenians were certainly hungry. **1981** B. MARLEY in *Times* 17 Oct. 7 A hungry mob is an angry mob, a pot a cook but the food not enough.
[1] a play by Aristophanes.

hunt: *see* you cannot RUN with the hare and hunt with the hounds.

hurt: *see* what you don't KNOW can't hurt you; STICKS and stones may break my bones but words will never hurt me.

The HUSBAND is always the last to know
Wife is also used as well as *husband*.

1604 MARSTON *What you Will* I. i. A cuckold . . a thing that's hoodwinked with kindness. . . He must be the last must know it. **1659** N.R. *Proverbs* 95 The good man is the last that knows whats amisse at home. **1756** STERNE *Tristram Shandy* VIII. iv. 'It is with love as with cuckoldom'—the suffering party is at least the third, but generally the last who knows anything about the matter. **1936** M. MITCHELL *Gone with Wind* liv. I thought surely the whole town knew by now. Perhaps they all do, except you. You know the old adage: 'The wife is always the last one to find out.' **1959** M. SUMMERTON *Small Wilderness* i. That over-worked truism about the wife being the last to know, wasn't in my case strictly accurate. **1979** C. MACLEOD *Family Vault* iii. 'Do you mean he hasn't heard?' Leila whooped. 'They say the husband's always the last to know,' Harry chimed in.

husband: *see also* a DEAF husband and a blind wife are always a happy couple.

I

ice: see the RICH man has his ice in the summer and the poor man gets his in the winter.

An IDLE brain is the Devil's workshop
Cf. *The* DEVIL *finds work for idle hands to do.*

a **1602** W. PERKINS *Works* (1603) 906 The idle bodie and the idle braine is the shoppe[1] of the deuill. **1732** T. FULLER *Gnomologia* no. 3053 Idle Brains are the Devil's Workhouses. **1855** H. G. BOHN *Hand-Book of Proverbs* 311 An idle brain is the devil's workshop. **1859** S. SMILES *Self-Help* viii. Steady employment . . keeps one out of mischief, for truly an idle brain is the devil's workshop.
   [1] i.e. workshop.

IDLE people have the least leisure
Cf. *The* BUSIEST *men have the most leisure.*

1678 J. RAY *English Proverbs* (ed. 2) 161 Idle folks have the most labour. **1853** R. S. SURTEES *Sponge's Sporting Tour* lvii. 'Got a great deal to do,' retorted Jog, who, like all thoroughly idle men, was always dreadfully busy. **1855** H. G. BOHN *Hand-Book of Proverbs* 414 Idle folks have the least leisure. **1908** *Spectator* 10 Oct. 535 The difference between leisureliness and laziness runs parallel with that between quickness and haste. 'Idle people', says the proverb, 'have the least leisure.'

idle: see also the DEVIL finds work for idle hands to do; it is idle to SWALLOW the cow and choke on the tail.

IDLENESS is the root of all evil
Cf. MONEY *is the root of all evil.*

[*c* **1390** CHAUCER *Second Nun's Prologue* l. 1 The ministre and the norice[1] unto vices, which that men clepe[2] in Englissh ydlenesse.] **1422** J. YONGE in *Secreta Secretorum* (1898) 158 Idylnysse is the . . rote of vicis. **1538** T. BECON *Governance of Virtue* B8[v] Idleness . . is the well-spring and root of all vice.

1707 G. FARQUHAR *Beaux' Strategem* I. i. Idleness is the Root of all Evil; the World's wide enough, let 'em bustle. **1850** DICKENS *David Copperfield* x. 'The boy will be idle there,' said Miss Murdstone, looking into a pickle-jar, 'and idleness is the root of all evil.' **1874** TROLLOPE *Phineas Redux* II. xxxvi. I much prefer downright honest figures. Two and two make four; idleness is the root of all evil . . and the rest of it.
   [1] i.e. nurse.   [2] call.

If IFS and ands were pots and pans, there'd be no work for tinkers' hands
Used as a humorous retort to an over-optimistic conditional expression. *ands*: the conjunction *and* 'if', of which *an* is a weakened form, is employed irregularly here as a noun to denote 'an expression of condition or doubt'.

1850 KINGSLEY *Alton Locke* I. x. 'If a poor man's prayer can bring God's curse down.'. . 'If ifs and ans were pots and pans.' **1886** *Notes & Queries* 7th Ser. I. 71 There is also the old doggerel—If ifs and ands Were pots and pans Where would be the work for Tinkers' hands? **1924** *Times* 30 May 9 If he might vary an old saw he would say, 'If "ifs and ands" could create employment, then there would be little use for the Minister of Labour to tinker at it.' **1981** J. ASHFORD *Loss of Culion* xvi. As my old aunt used to say, 'If ifs and ands were pots and pans, there'd be no work for tinkers' hands.'

Where IGNORANCE is bliss, 'tis folly to be wise
Now frequently abbreviated to *Ignorance is bliss.*

1742 GRAY *Poems* (1966) 10 Thought would destroy their paradise. No more; where ignorance is bliss, 'Tis folly to be wise. **1900** E. J. HARDY *Mr. Thomas Atkins* xxiv. Never did soldiers set out for a war in better spirits than did ours . . against the Boers. They . . afforded a pathetic illustration of the proverb: 'Where ignorance is bliss 'tis folly to be wise.' **1925** S. O'CASEY *Juno & Paycock* II. 49 'You ought to be ashamed o' yourself . . not to know the History o' your country.' . . 'Where ignorance's bliss 'tis folly to be wise.' **1979** A. JUTE *Reverse Negative* 57 He didn't want to know anything. Ignorance is bliss.

IGNORANCE of the law is no excuse for breaking it

[Legal maxim: *ignorantia iuris neminem excusat,* ignorance of the law excuses nobody.] *c* **1412** HOCCLEVE *De Regimene Principum* (EETS) 92 Excuse schal hym naght his ignorance. **1530** C. ST. GERMAN *Dialogues in English* II. xlvi. Ignorance of the law though it be inuincible doth not excuse. **1616** T. DRAXE *Adages* 100 The ignorance of the law excuseth no man. *a* **1654** J. SELDEN *Table-Talk* (1689) 30 Ignorance of the Law excuses no man; not that all Men know the Law, but because 'tis an excuse every man will plead, and no man can tell how to confute him. **1830** N. AMES *Mariner's Sketches* xxviii. Ignorance of the law excuses nobody. . . The gates of mercy are forever shut against them. **1979** *Private Eye* 17 Aug. 6 [He] was fined £5 at Marylebone Court when he learned that ignorance of the law is no excuse for breaking it.

It's an ILL bird that fouls its own nest

A condemnation of a person who vilifies his own family, country, etc.

[Med. Lat. *nidos commaculans inmundus habebitur ales,* the bird is unclean that soils its nest.] *a* **1250** *Owl & Nightingale* (1960) l. 99 Dahet habbe[1] that ilke best that fuleth his owe nest. *c* **1400** N. BOZON *Moral Tales* (1889) 205 Hyt ys a fowle brydde that fylyth hys owne neste. **1591** H. SMITH *Preparative to Marriage* 82 It becommeth not any woman to set light by her husband, nor to publish his infirmities for they say, it is an euill bird that defileth his owne nest. **1670** J. RAY *English Proverbs* 62 It's an ill bird that beraies its own nest. **1817** SCOTT *Rob Roy* II. xiii. Where's the use o' vilifying

ane's country. . . It's an ill bird that files its ain nest. **1926** *Times* 7 Sept. 17 Nothing . . can excuse the bad taste of Samuel Butler's virulent attack upon his defenceless family. . . It's an ill bird that fouls its own nest. **1981** G. MITCHELL *Death-Cap Dancers* xiv. It's a dirty bird that fouls its own nest. Your loyalty does you credit.

[1] *dahet habbe:* a curse on.

He that has an ILL name is half hanged

Cf. *Give a* DOG *a bad name and hang him.*

a **1400** in C. Brown *Religious Lyrics of XIVth Century* (1957) 193 Ho-so hath a wicked name Me semeth for sothe half hongid he is. **1546** J. HEYWOOD *Dialogue of Proverbs* II. vi. I2 He that hath an yll name, is halfe hangd. **1614** T. ADAMS *Devil's Banquet* IV. 156 It is a very ominous and suspitious thing to haue an ill name. The Prouerbe saith, he is halfe hanged. **1897** M. A. S. HUME *Raleigh* xii. Were not . . an ill name half hanged . . he would have been acquitted.

It's ILL waiting for dead men's shoes

The earlier form of the proverb, exemplified in quots. *c* 1549 and 1721, is no longer found. The metaphorical phrase *to wait for dead men's shoes* in also illustrated below.

1530 J. PALSGRAVE *L'éclaircissement de la Langue Française* 306[v] Thou lokest after deed mens shoes. *c* **1549** J. HEYWOOD *Dialogue of Proverbs* I. xi. C5 Who waitth for dead men shoen, shal go long barfote. **1721** J. KELLY *Scottish Proverbs* 148 He goes long bare Foot that wears dead Mens Shoon. Spoken to them who expect to be some Man's Heir, to get his Place, or Wife, if he should dye. **1758** A. MURPHY *Upholsterer* I. ii. You have very good pretensions; but then its waiting for dead Men's Shoes. **1815** SCOTT *Guy Mannering* II. xvi. That's but sma' gear, puir thing; she had a sair time o't with the auld leddy. But it's ill waiting for dead folk's shoon. **1912** E. V. LUCAS *London Lavender* iv. I pointed out that I was executor to no fewer than three persons. . . 'It's ill waiting for dead men's shoes,' Naomi quoted.

ILL weeds grow apace

[Cf. 14th-cent. Fr. *male herbe croist*, bad grass thrives.] *c* **1470** in *Anglia* (1918) XLII. 200 Wyl[d] weed ys sone y-growe. *Creuerat herba satis, que nil habet utilitatis.* **1546** J. HEYWOOD *Dialogue of Proverbs* I. x. C4ᵛ Ill weede growth fast Ales,¹ wherby the corne is lorne.² **1578** J. FLORIO *First Fruits* 31ᵛ An yl weede groweth apace. **1594** SHAKESPEARE *Richard III* II. iv. 13 'Ay,' quoth my uncle Gloucester 'Small herbs have grace: great weeds do grow apace.' . . I would not grow so fast, Because sweet flow'rs are slow and weeds make haste. **1738** SWIFT *Polite Conversation* I. 23 'Don't you think Miss is grown?' . . 'Ay; ill Weeds grow a-pace.' **1905** A. MACLAREN *Gospel according to St. Matthew* II. 208 The roots of the old lay hid, and, in due time, showed again above ground. 'Ill weeds grow apace.'

¹ i.e. Alice. ² lost.

It's an ILL wind that blows nobody any good

A sailing metaphor frequently invoked to explain good luck arising from the source of others' misfortune.

1546 J. HEYWOOD *Dialogue of Proverbs* II. ix. L1 An yll wynde that blowth no man to good, men saie. **1591** SHAKESPEARE *Henry VI Pt. 3* II. v. 55 Ill blows the wind that profits nobody. **1655** T. FULLER *Church Hist. Britain* II. ii. It is an ill wind which bloweth no man Profit. He is cast on the Shoar of Freezland . . where the Inhabitants . . were by his Preaching converted to Christianity. **1660** J. TATHAM *Rump* II. i. 'Tis an ill Wind they say bloughs no body good. **1832** S. WARREN *Diary of Late Physician* I. i. My good fortune (truly it is an ill wind that blows *nobody* any good) was almost too much for me. **1979** J. SCOTT *Angels in your Beer* xxviii. It is an ill wind that blows nobody any good, but then John Quinlan . . was about as close to being a nobody as anyone could get.

ill: *see also* BAD news travels fast; EVIL doers are evil dreaders; it's ill speaking between a FULL man and a fasting; he that LIVES in hope dances to an ill tune; it is ill SITTING at Rome and striving with the Pope; a sow may whistle, though it has an ill mouth for it; never SPEAK ill of the dead; *also* BAD.

IMITATION is the sincerest form of flattery

1820 C. C. COLTON *Lacon* I. 113 Imitation is the sincerest of flattery. **1843** R. S. SURTEES *Handley Cross* I. xv. Imitation is the sincerest of flattery. **1979** R. CASSILIS *Arrow of God* III. xiv. Oh, yess. Imitation sincerest form of flattery. Bootlickers.

impossible: *see* the DIFFICULT is done at once, the impossible takes a little longer.

impression: *see* FIRST impressions are the most lasting.

IN for a penny, in for a pound

1695 E. RAVENSCROFT *Canterbury Guests* v. i. It concerns you to . . prove what you speak. . . In for a Penny, in for a Pound. **1815** SCOTT *Guy Mannering* III. vii. Sampson . . thought to himself, in for a penny in for a pound, and he fairly drank the witch's health in a cupfull of brandy. **1841** DICKENS *Old Curiosity Shop* II. lxvi. Now, gentlemen, I am not a man who does things by halves. Being in for a penny, I am ready as the saying is to be in for a pound. **1979** P. NIESEWAND *Member of Club* viii. 'Do you want to go and have a look, sir?' . . 'Why not? . . In for a penny, in for a pound.'

inclined: *see* as the TWIG is bent, so is the tree inclined.

inconvenience: *see* POVERTY is no disgrace, but it is a great inconvenience.

index: *see* the EYES are the window of the soul.

Indian: *see* the only GOOD Indian is a dead Indian.

infinite: *see* GENIUS is an infinite capacity for taking pains.

inside: *see* there is NOTHING so good for the inside of a man as the outside of a horse.

intention: *see* the ROAD to hell is paved with good intentions.

invention: *see* NECESSITY is the mother of invention.

Ireland: *see* ENGLAND'S difficulty is Ireland's opportunity.

iron: *see* STRIKE while the iron is hot.

J

Every JACK has his Jill

1611 R. COTGRAVE *Dict. French & English* s.v. Demander, Like will to like; a Iacke lookes for a Gill. **1619** in C. W. Bardsley *Curiosities of Puritan Nomenclature* (1880) i. The proverb is, each Jacke shall have his Gill. **1670** J. RAY *English Proverbs* 108 Every Jack must have his Gill. . . It ought to be written *Jyll*. **1855** G. J. WHYTE-MELVILLE *General Bounce* ii. 'Every Jack has his Gill,' if he and she can only find each other out at the propitious moment. **1940** H. W. THOMPSON *Body, Boots & Britches* xix. Every Jack has his Jill; If one won't, another will.

JACK is as good as his master

Jack is variously used as a familiar name for a sailor, a member of the common people, a serving man, and one who does odd jobs.

1706 J. STEVENS *Spanish & English Dict.* s.v. Pedro, Peter is as good as his Master. Like Master, like Man. **1868** READE & BOUCICAULT *Foul Play* II. xx. Is it the general opinion of seamen before the mast? Come, tell us. Jack's as good as his master in these matters. **1905** W. C. RUSSELL *Old Harbour Town* xi. If the crew are to be carried away to an unbeknown place, they all go below to a man, for Jack's as good as his master when it comes to his having to do something which he didn't agree for. **1936** W. HOLTBY *South Riding* I. iv. She was far from thinking Jack as good as his master and explained failure in plebeian upstarts by saying with suave contempt: 'Well, what can you expect? Wasn't bred to power.' **1980** *Times* 15 Nov. 6 A subtly English kind of egalitarianism . . a man is as good as his masters (if not better).

Jack: *see also* a GOOD Jack makes a good Jill; all WORK and no play makes Jack a dull boy.

JAM tomorrow and jam yesterday, but never jam today

1871 'L. CARROLL' *Through Looking-Glass* v. 'The rule is, jam to-morrow and jam yesterday—but never jam to-day.' 'It *must* come sometimes to "jam to-day",' Alice objected. 'No, it can't,' said the Queen. **1951** 'J. WYNDHAM' *Day of Triffids* xii. Just put the Americans into the jam-tomorrow-pie-in-the-sky department awhile. **1979** *Guardian* 9 June 10 The manageress of the launderette calls me darling. . . 'Jam yesterday, jam tomorrow, but never jam today.'

jaw (rush of water): *see* JOUK and let the jaw go by.

jest: *see* many a TRUE word is spoken in jest.

jewel: *see* FAIR play's a jewel.

Jill: *see* a GOOD Jack makes a good Jill; every JACK has his Jill.

job: *see* if a THING's worth doing, it's worth doing well.

join: *see* if you can't BEAT them, join them.

JOUK and let the jaw go by

A Scottish proverb counselling prudent action when trouble threatens. The phrase *to jouk and let the jaw go by* is also found.

1721 J. KELLY *Scottish Proverbs* 189 Juck,[1] and let the jaw[2] go o'er you. That is, prudently yield to a present Torrent. **1817** SCOTT *Rob Roy* II. xii. Gang your ways hame, like a gude bairn—jouk and let the jaw gae by. **1927** J. BUCHAN *Witch Wood* xv. A man must either jouk and let the jaw go bye, as the owercome[3] says, or he must ride the whirlwind.

[1] stoop, duck. [2] a rush of water. [3] a common or hackneyed expression.

JOVE but laughs at lovers' perjury

[[TIBULLUS] *Elegies* III. vi. *periuria ridet amantum Iuppiter,* Jupiter laughs at lovers' perjuries. Cf. *a* **1500** in W. W. Skeat *Chaucerian & Other Pieces* (1897) 311 Your [lovers'] othes laste No lenger than the wordes ben ago! And god, and eke his sayntes, laughe also.] *c* **1550** tr. *A. S. Piccolomini's Lady Lucres* E4[v] Pacorus . . confesseth the faut asketh forgeuenes and . . ryghte well knewe he that Jupyter rather laughethe, then taketh angerlye the periuringe of louers. *c* **1595** SHAKESPEARE *Romeo & Juliet* II. ii. 92 At lovers' perjuries, They say Jove laughs. **1700** DRYDEN *Poems* (1958) IV. 1487 Love endures no Tie, And Jove but laughs at Lovers Perjury! **1922** *Evening Standard* 17 Oct. 5 Perjury in the Divorce Court has been openly permitted to the upper classes for many years, following the maxim . . that 'Jove but laughs at lovers' perjury.' **1973** I. MURDOCH *Black Prince* III. 299 Zeus, they say, mocks lovers' oaths.

No one should be JUDGE in his own cause

[Legal maxim: *nemo debet esse iudex in propria causa,* no one should be judge in his own cause.] *c* **1449** R. PECOCK *Repressor of Blaming of Clergy* (1860) II. 381 Noman oughte be iuge in his owne cause which he hath anentis[1] his neighbour. [**1604** SHAKESPEARE *Measure for Measure* V. i. 166 In this I'll be impartial; be you judge Of your own cause.] **1775** J. WESLEY *Letter* 3 Nov. (1931) VI. 186 No man is a good judge in his own cause. I believe I am tolerably impartial. **1928** *Times* 22 Aug. 9 The principle that no judge could be a judge in his own case was generally accepted. The chairman of a meeting was in a quasi-judicial capacity. **1981** *Daily Telegraph* 16 May 18 The maxim that no one should be judge in his own cause.

[1] against.

JUDGE not, that ye be not judged

[MATTHEW vii. 1 Iudge not, that ye be not iudged.] **1481** CAXTON *Reynard* (1880) xxix. Deme[1] ye noman, and ye shal not be demed. **1509** H. WATSON *Ship of Fools* H1 Judge not but yf that ye wyl be juged. **1925** A. CLUTTON-BROCK *Essays on Life* x. The saying, Judge not, that ye be not judged,' is . . a statement of fact. Nothing makes us dislike a man so much as the knowledge that he is always judging us and all men. **1979** C. DEXTER *Service of all Dead* i. 'Judge not that ye be not judged.' Judge not—at least until the evidence is unequivocal.

[1] judge.

June: *see* a DRIPPING June sets all in tune.

Be JUST before you're generous

1745 E. HAYWOOD *Female Spectator* II. VII. 35 There is, I think, an old saying, that we 'ought to be just before we are generous'. **1780** SHERIDAN *School for Scandal* IV. i. Be just before you are generous. **1834** MARRYAT *Peter Simple* I. xi. I owe every farthing of my money. . . . There's an old proverb—be just before you're generous. **1908** *Spectator* 4 Apr. 529 A likeable man is tempted to be generous before he is just.

justify: *see* the END justifies the means.

K

Why KEEP a dog and bark yourself?

1583 B. MELBANCKE *Philotimus* 119 It is smal reason you should kepe a dog, and barke your selfe. **1670** J. RAY *English Proverbs* 81 What? keep a dog and bark my self. That is, must I keep servants, and do my work my self. **1738** SWIFT *Polite Conversation* I. 17 'Good Miss, stir the Fire.' . . 'Indeed your Ladyship could have stirr'd it much better.'. . 'I won't keep a Dog and bark myself.' **1933** A. CHRISTIE *Thirteen at Dinner* xviii. Why keep a dog and bark yourself? **1974** W. GARNER *Big enough Wreath* ix. If money talks louder than words, no wonder Lowell hardly ever speaks. Why keep a dog and bark?

KEEP a thing seven years and you'll always find a use for it

1623 W. PAINTER *Palace of Pleasure* C5 Things of small value the old proverb say, Wise men seuen yeares will carefully vp lay. **1663** T. KILLIGREW *Parson's Wedding* in *Comedies & Tragedies* (1664) 100 According to the Proverb; Keep a thing seven years, and then if thou hast no use on't throw't away. **1816** SCOTT *Antiquary* II. vi. They say, keep a thing seven year, an' ye'll aye find a use for't. **1945** F. THOMPSON *Lark Rise* xx. 'I don't know that I've any use for it.' 'Use! Use! . . Keep a thing seven years and you'll always find a use for it!'

KEEP no more cats than will catch mice

1673 J. DARE *Counsellor Manners* lxii. If thou hast a regard to Thrift, keep no more Cats than will kill Mice. **1678** J. RAY *English Proverbs* (ed. 2) 350, I will keep no more cats than will catch mice (*i.e.* no more in family then will earn their living). *Somerset.* **1710** S. PALMER *Proverbs* 358 Keep no more Cats than will Catch Mice. Ecquipage and Attendance . . must be agreeable to Character, Dignity and Fortune. **1910** KIPLING *Rewards & Fairies* 73 The King keeps no cats that don't catch mice. She must sail the seas, Master Dawe.

KEEP your shop and your shop will keep you

1605 G. CHAPMAN et al. *Eastward Ho* A2ᵛ I . . garnished my shop . . with good wholsome thriftie sentences; As, 'Touchstone, keepe thy shopp, and thy shoppe will keepe thee.' **1712** ADDISON *Spectator* 14 Oct., Sir William Turner . . would say, Keep your Shop and your Shop will keep you. **1759** GOLDSMITH in *Bee* 17 Nov. 214, I would earnestly recommend this adage to every mechanic in London, 'Keep your shop, and your shop will keep you.' **1905** H. G. WELLS *Kipps* III. iii. A little bell jangled. 'Shop!' said Kipps. 'That's right. Keep a shop and the shop'll keep you.' **1943** S. V. BENÉT *Western Star* I. 20, I keep my shop but my shop doth not keep me. Shall I give such chances [of making a fortune] the go-by and walk the roads? **1976** H. KEMELMAN *Wednesday Rabbi got Wet* vii. 'When I was home, Dad cared a lot more about the store than he did about me,' he said bitterly. She nodded. . . 'That's because a store, if you take care of it, it takes care of you. Your father lives from that store, and your grandfather before him.'

keep: *see also* a man is known by the COMPANY he keeps; EXPERIENCE keeps a dear school; THREE may keep a secret, if two of them are dead; put your TRUST in God, and keep your powder dry.

keeper: *see* FINDERS keepers (losers weepers).

keeping: *see* FINDING's keepings.

key: *see* a GOLDEN key can open any door.

kick: *see* CORPORATIONS have neither bodies to be punished nor souls to be damned.

kill: *see* it is the PACE that kills; it is not WORK that kills; but worry.

killed: *see* CARE killed the cat; CURIOSITY killed the cat.

KILLING no murder
Quot. 1657 is the title of a pamphlet asserting that the assassination of Cromwell, the Protector, would be lawful and laudable.

1657 SEXBY & TITUS (*title*) Killing noe murder. **1800** M. EDGEWORTH *Castle Rackrent* p. xliv. In Ireland, not only cowards, but the brave 'die many times before their death'. There killing is no murder. **1831** E. J. TRELAWNY *Adventures of Younger Son* II. xxix. Arabs, with whom, if precedents and time can make a thing lawful, killing is no murder. **1908** *Times Literary Supplement* 4 June 179 The exception is the share which he took in the conspiracy of Orsini against Napoleon III. . . It was probably a case to which Holyoake would have applied the doctrine of 'killing no murder'. **1961** C. COCKBURN *View from West* vi. The British . . made, in England, propaganda out of the phrase—attributed to the Irish—'killing no murder', they were not foolish enough to take their own propaganda seriously.

killing: *see also* there are more WAYS of killing a cat than choking it with cream; there are more WAYS of killing a dog than choking it with butter; there are more WAYS of killing a dog than hanging it.

kind: *see* BETTER a good cow than a cow of a good kind.

The KING can do no wrong
Altered to *Queen* when appropriate.
[Legal maxim: *rex non potest peccare*, the king can do no wrong; cf. *c* **1538** T. STARKEY *England in Reign of King Henry VIII* (EETS) I. iv. Wyl you make a kyng to have no more powar then one of hys lordys? Hyt ys commynly sayd . . a kyng ys aboue hys lawys.] *a* **1654** J. SELDEN *Table-Talk* (1689) 27 The King can do no wrong, that is no Process[1] can

be granted against him. **1765** W. BLACKSTONE *Commentaries on Laws of England* I. vii. The King can do no wrong. . . The prerogative of the crown extends not to do any injury: it is created for the benefit of the people, and therefore cannot be exerted to their prejudice. **1888** C. M. YONGE *Beechcroft at Rockstone* II. xxii. 'So, Aunt Jane is your Pope.' 'No; she's the King that can do no wrong,' said Gillian, laughing. **1952** 'M. COST' *Hour Awaits* 191 It was very different with Augustus. . . We had always expected that. . . In his case, was it not rather a matter of the king can do no wrong. **1981** *Times* 28 July 14 The Queen [of Holland] has no power but some influence. . . 'The Queen can do no wrong. The ministers are responsible.'
[1] a formal commencement of an action at law.

king: *see also* a CAT may look at a king; in the COUNTRY of the blind, the one-eyed man is king; a PECK of March dust is worth a king's ransom.

kingdom: *see* in the COUNTRY of the blind, the one-eyed man is king.

KINGS have long arms
[Gr. μακραὶ τυράννων χεῖρες, rulers' hands reach a long way; OVID *Heroides* xvii. *an nescis longas regibus esse manus?* know you not that kings have far-reaching hands?] **1539** R. TAVERNER tr. *Erasmus' Adages* A4[v] Kynges haue longe handes. They can brynge in men, they can pluck in thinges, though they be a great weye of. **1578** LYLY *Euphues* I. 221 Knowest thou not Euphues that kinges haue long armes, and rulers large reches? **1752** B. FRANKLIN *Poor Richard's Almanack* (Jan.) Kings haue long Arms, but misfortune longer. **1927** P. B. NOYES *Pallid Giant* iii. 'How will you insure Markham's safety if he takes refuge here?' . . 'Governments, proverbially, have long arms.'

kirtle: *see* NEAR is my kirtle, but nearer is my smock.

KISSING goes by favour

1616 T. DRAXE *Adages* 62 Kissing commeth by fauour. **1621** BURTON *Anatomy of Melancholy* II. iii. Offices are not alwaies given . . for worth. [*note*] Kissing goes by Favour. **1721** J. KELLY *Scottish Proverbs* 225 Kissing goes by Favour. Men shew Regard, or do Service, to People as they affect. **1880** BLACKMORE *Mary Anerley* II. iii. 'I should like . . to give you one kiss, Insie.' . . Before he could give reason in favour of a privilege which goes proverbially by favour, the young maid was gone. **1929** 'L. THAYER' *Dead Man's Shoes* i. Kissing goes by favor all along the line.

kissing: *see also* when the GORSE is out of bloom, kissing's out of fashion.

kitchen: *see* if you don't like the HEAT, get out of the kitchen.

kitten: *see* WANTON kittens make sober cats.

knock: *see* OPPORTUNITY never knocks twice at any man's door.

You should KNOW a man seven years before you stir his fire

1803 C. DIBDIN *Professional Life* I. p. xi, It is a well-meant saying, that you should know a man seven years before you stir his fire; or, in other words, before you venture at too much familiarity. **1902** V. S. LEAN *Collectanea* IV. 204 You may poke a man's fire after you've known him seven years, but not before.

What you don't KNOW can't hurt you

1576 G. PETTIE *Petit Palace* 168 Why should I seeke to take him in it? . . So long as I know it not, it hurteth mee not. **1908** E. WALTER *Easiest Way* III. 66 What a fellow doesn't know doesn't hurt him, and he'll love you just the same. **1979** 'S. WOODS' *This Fatal Writ* 54 'No, this is interesting. . . I didn't know—' 'What you don't know can't hurt you,' said Maitland.

KNOW thyself

[Gr. γνῶθι σαυτόν (or σεαυτόν), motto inscribed on the 6th-cent. BC temple of Apollo at Delphi and quoted by several ancient writers (some attributing it to Solon): see esp. Pausanias x. 24 and Juvenal *Satires* xi.; L. *nosce teipsum*.] **1387** J. TREVISA tr. Higden's *Polychronicon* (1865) I. 241 While the cherle smoot the victor, he schulde ofte seie to hym in this manere: . . Knowe thyself. **1545** R. ASCHAM *Toxophilus* II. 36 Knowe thy selfe: that is to saye, learne to knowe what thou arte able, fitte and apt vnto, and folowe that. **1732** POPE *Essay on Man* II. 1 Know then thyself, presume not God to scan; The proper study of Mankind is Man. **1849** BULWER-LYTTON *Caxtons* III. XVI. x. 'Know thyself,' said the old philosophy. 'Improve thyself,' saith the new. *a* **1930** D. H. LAWRENCE *Last Poems* (1932) 266 At last we escape the barbed wire enclosure of Know Thyself, knowing we can never know.

You never KNOW what you can do till you try

1818 COBBETT *Year's Residence in USA* II. vi. A man knows not what he can do 'till he tries. **1890** M. WILLIAMS *Leaves of Life* I. xiii. On hearing the verdict he . . shouted out: 'I told you so! You never know what you can do till you try.' **1968** D. FRANCIS *Forfeit* xiv. 'Ty, you aren't fit to drive.' 'Never know what you can do till you try.'

know: *see also* BETTER the devil you know than the devil you don't know; one HALF of the world does not know how the other half lives; the HUSBAND is always the last to know; come LIVE with me and you'll know me; MORE people know Tom Fool than Tom Fool knows; NECESSITY knows no law; it is a WISE child that knows its own father; *also* KNOWN.

KNOWLEDGE is power

Similar in form to MONEY *is power*.

[Cf. PROVERBS xxiv. 5 A man of knowledge encreaseth strength; **1597** BACON *De Haeresibus* x. *nam et ipsa scientia potes-*

tas est, for knowledge itself is power.]
1598 in Bacon *Essays* 27^v Knowledge it
selfe is a power whereby he [God]
knoweth. **1806** B. RUSH *Letter* 25 Nov.
(1951) II. 935 The well-known aphorism
that 'knowledge is power'. **1853** BUL-
WER-LYTTON *My Novel* I. II. iii. He . .
said half aloud, — 'Well, knowledge is
power!' **1968** P. BEER *Mrs Beer's House*
xi. 'In a few years' time they'll only be
getting married,' to which Father replied darkly and to his own complete
satisfaction, 'Knowledge is power.'

knowledge: *see also* a LITTLE knowledge is
a dangerous thing.

known: *see* a man is known by the COM-
PANY he keeps; the TREE is known by its
fruit.

L

The LABOURER is worthy of his hire

[LUKE x. 7 The labourer is worthy of his hire.] c 1390 CHAUCER *Summoner's Tale* l. 1973 The hye God, that al this world hath wroght, Seith that the werkman worthy is his hyre. 1580 J. BARET *Alveary* D697 *Digna canis pabulo*. . . A Prouerbe declaring that the laborer is worthie of his hire: it is taken as well of the labour of the mind, as of the bodie. 1824 SCOTT *St. Ronan's Well* I. x. Your service will not be altogether gratuitous, my old friend—the labourer is worthy of his hire. 1935 E. F. BENSON *Lucia's Progress* xi. I shall certainly spend a great deal of it, keeping some for myself—the labourer is worthy of his hire. 1980 *Times* 4 Mar. 7 Forget haggling. . . The labourer is worthy of his hire.

ladder: *see* CROSSES are ladders that lead to heaven.

lady: *see* FAINT heart never won fair lady; FAR-FETCHED and dear-bought is good for ladies; the OPERA isn't over till the fat lady sings.

lamb: *see* GOD tempers the wind to the shorn lamb; one might as well be HANGED for a sheep as a lamb; MARCH comes in like a lion, and goes out like a lamb.

Lancashire: *see* what MANCHESTER says today, the rest of England says tomorrow.

Every LAND has its own law

a 1628 J. CARMICHAELL *Proverbs in Scots* no. 469 Everie land hes the laich. 1721 J. KELLY *Scottish Proverbs* 92 Every land hath its own Laugh, and every Corn its own Caff.[1] Every Country hath its own Laws, Customs, and Usages. 1916 *British Weekly* 2 Nov. 84 'Every land', says the old Scottish proverb, 'has its ain lauch.' And every class has its own mode of thought and expression.

[1] i.e. chaff.

land: *see also* you BUY land you buy stones, you buy meat you buy bones; when HOUSE and land are gone and spent, then learning is most excellent; LEARNING is better than house and land.

lane: *see* it is a LONG lane that has no turning.

large: *see* a GREAT book is a great evil; LITTLE pitchers have large ears.

lark: *see* if the SKY falls we shall catch larks.

The LAST drop makes the cup run over

See also the next proverb.

1655 T. FULLER *Church Hist. Britain* XI. ii. When the Cup is brim full before, the last (though least) superadded drop is charged alone to be the cause of all the running over. 1855 H. G. BOHN *Hand-Book of Proverbs* 509 The last drop makes the cup run over. 1876 J. PAYN *Halves* I. x. An application of her brother-in-law for a five-pound note . . was the last drop that caused Mrs. Raeburn's cup of bitterness to overflow. 1888 C. M. YONGE *Beechcroft at Rockstone* I. i. Valetta burst out crying at this last drop that made the bucket overflow.

It is the LAST straw that breaks the camel's back

The metaphor is also used allusively, especially in the phrase *the last straw*.

1655 J. BRAMHALL *Defence of True Liberty of Human Actions* 54 It is the last feather may be said to break an Horses back. 1793 in *Publications of Colonial Society of Massachusetts* (1954) XXXVI. 298 It is certainly true that the last feather will sink the camel. 1848 DICKENS *Dombey & Son* ii. As the last straw breaks the laden camel's back, this piece of underground information crushed the sinking spirits of Mr. Dombey. 1876 I. BANKS *Manchester Man* III. xv. The last straw breaks the camel's back. 1940 'J. J. CONNINGTON' *Four Defences* xiii. 'This final droplet turns the scale.' . . 'The last straw that breaks the camel's back?' condensed the Counsellor. **1969** P. ROTH *Portnoy's Complaint* 53 Heshie's [plaster] cast was later referred to as 'the straw that broke the camel's back', whatever that meant.

last: *see* (noun) let the COBBLER stick to his last; the COBBLER to his last and the gunner to his linstock.

last: *see also* (adjective) there are no BIRDS in last year's nest; the HUSBAND is always the last to know; the THIRD time pays for all; (adverb) he LAUGHS best who laughs last; he who LAUGHS last, laughs longest.

lasting: *see* FIRST impressions are the most lasting.

late: *see* (adjective) the EARLY man never borrows from the late man; it is NEVER too late to learn; it is NEVER too late to mend; it is too late to shut the STABLE-door after the horse has bolted; (adverb) BETTER late than never.

LAUGH and the world laughs with you; weep and you weep alone

An alteration of the sentiment expressed by HORACE *Ars Poetica* 101 *ut ridentibus arrident, ita flentibus adsunt humani voltus*, men's faces laugh on those who laugh, and correspondingly weep on those who weep; ROMANS xii. 15 Reioyce with them that doe reioice, and weepe with them that weepe.

1883 E. W. WILCOX in *Sun* (New York) 25 Feb. 3 Laugh, and the world laughs with you; Weep, and you weep alone. For the sad old earth must borrow its mirth, But has trouble enough of its own. **1907** 'O. HENRY' *Trimmed Lamp* 211 Laugh, and the world laughs with you; weep, and they give you the laugh. **1912** 'SAKI' *Chronicle of Clovis* 127 The proverb 'Weep and you weep alone,' broke down as badly on application as most of its kind. **1979** *Daily Telegraph* 3 Nov. 13 One of the silliest of all proverbs is 'Laugh and the world laughs with you'—certainly, so far as books are concerned. There is nothing more individual than taste in humour.

Let them LAUGH that win

An older version of the next two proverbs.

1546 J. HEYWOOD *Dialogue of Proverbs* I. v. B2 He laugth that wynth. *a* **1596** G. PEELE *Clyomon & Clamides* F1 But I zay to you my nabor[1] . . wel let them laugh that win. **1777** *Bonner & Middleton's Bristol Journal* 5 July 3 The old Proverb says, let them laugh that wins.—They glory over us, by saying that our Fund is almost exhausted—that is our look out not theirs. **1873** TROLLOPE *Phineas Redux* I. xxxvii. 'You are laughing at me, I know.' 'Let them laugh that win.'

[1] i.e. neighbour.

laugh: *see also* JOVE but laughs at lovers' perjury; LOVE laughs at locksmiths.

He LAUGHS best who laughs last

See also the two adjacent proverbs.

c **1607** *Christmas Prince* (1923) 109 Hee laugheth best that laugheth to the end. **1715** VANBRUGH *Country House* II. v. Does she play her jests upon me too!—but mum, he laughs best that laughs last. **1822** SCOTT *Peveril* IV. iii. Your Grace knows the French proverb, 'He laughs best who laughs last.' **1979** L. BARNEA *Reported Missing* xxvi. 'Whew! Well, that's over and done with!'. . 'He laughs best who laughs last.'

He who LAUGHS last, laughs longest

A modern development of the preceding proverb.

1912 J. MASEFIELD *Widow in Bye Street* IV. 66 In this life he laughs longest who laughs last. **1943** J. LODWICK *Running to Paradise* xxx. He who laughs last laughs longest, and in another four days I was able to look at my mug in the mirror without wincing. **1951** M. DE LA ROCHE *Renny's Daughter* ix. 'We'll see. He who laughs last, laughs. . .' So worked up was Eugene Clapperton that he could not recall the last word of the proverb.

One LAW for the rich and another for the poor

1830 MARRYAT *King's Own* I. xi. Is there nothing smuggled besides gin? Now, if the husbands and fathers of these ladies,—those who have themselves enacted the laws,—wink at their *infringement*, why should not others do so? . . There cannot be one law for the rich and another for the poor. 1913 *Spectator* 8 Nov. 757 The idea prevails abroad that there is one law for the 'rich' Englishman and another for the 'poor' foreigner. 1979 *Guardian* 10 Oct. 10 The strong sense that there was one law for the rich and another for the poor [at the trial].

law: *see also* HARD cases make bad law; IGNORANCE of the law is no excuse for breaking it; every LAND has its own law; NECESSITY knows no law; NEW lords, new laws; POSSESSION is nine points of the law; SELF-preservation is the first law of nature.

A man who is his own LAWYER has a fool for his client

1809 *Port Folio* (Philadelphia) Aug. 132 He who is always his own counsellor will often have a fool for his client. 1850 L. HUNT *Autobiography* II. xi. The proprietor of the *Morning Chronicle* pleaded his own cause, an occasion in which a man is said to have 'a fool for his client'. 1911 *British Weekly* 21 Dec. 386 There is a popular impression, for which there is a good deal to be said, that a man who is his own lawyer has a fool for his client. 1975 D. BAGLEY *Snow Tiger* xiii. You must have heard the saying that the man who argues his own case has a fool for a lawyer.

lay: *see* it is easier to RAISE the Devil than to lay him.

LAY-OVERS for meddlers

An answer to an impertinent or inquisitive child and others. The expression is found chiefly in the north of England, and in the US. *Lay-overs*, also contracted to *layers* or *layors*, are light blows

or smacks given to the meddlesome (but see also quot. 1854).

[1699 B. E. *New Dict. Canting Crew* s.v., *Lare-over*, said when the true Name of the thing must (in decency) be concealed.] 1785 F. GROSE *Classical Dict. Vulgar Tongue* s.v. *Lareovers, Lareovers for medlers*, an answer frequently given to children . . as a rebuke for their impertinent curiosity. 1854 A. E. BAKER *Glossary of Northamptonshire Words & Phrases* I. 389 *Lay-o'ers-for-medlers*, . . a contraction of *lay-overs*, *i.e.* things laid over, covered up, or protected from medlers. 1882 NODAL & MILNER *Glossary of Lancashire Dialect* 179 'What have yo' getten i' that bag?' 'Layers-for-medlers—does ta want to know?' 1936 M. MITCHELL *Gone with Wind* xxxii. When they asked who was going to lend the money she said: 'Layovers catch meddlers,' so archly they all laughed.

lazy: *see* LONG and lazy little and loud.

lead: *see* when the BLIND lead the blind, both shall fall into the ditch; CROSSES are ladders that lead to heaven; all ROADS lead to Rome.

leap: *see* LOOK before you leap.

learn: *see* LIVE and learn; it is NEVER too late to learn; NEVER too old to learn; we must learn to WALK before we can run; don't go near the WATER until you learn how to swim.

LEARNING is better than house and land

Cf. When HOUSE and land are gone and spent, then learning is most excellent.

1773 D. GARRICK in Goldsmith *She stoops to Conquer* A3ᵛ When ign'rance enters, folly is at hand; Learning is better far than house and land. 1800 M. EDGEWORTH *Castle Rackrent* 19, I . . thanked my stars I was not born a gentleman to so much toil and trouble—but Sir Murtagh took me up short with his old proverb, 'learning is better than house or land.' 1859 J. R. PLANCHÉ *Love & Fortune* 8 'Learning is better than house and land.' A fact that I never could understand.

learning: *see also* when HOUSE and land are gone and spent, then learning is most excellent; a LITTLE knowledge is a dangerous thing; there is no ROYAL road to learning.

LEAST said, soonest mended

c **1460** in W. C. Hazlitt *Remains of Early Popular Poetry* (1864) III. 169 Who sayth lytell he is wyse . . And fewe wordes are soone amend. **1555** J. HEYWOOD *Two Hundred Epigrams* no. 169 Lyttle sayde, soone amended. *a* **1641** D. FERGUSSON *Scottish Proverbs* (STS) no. 946 Littl said is soon mended. **1776** T. COGAN *John Buncle, Junior* I. vi. Mum's the word; least said is soonest mended. **1818** SCOTT *Heart of Midlothian* I. vi. A fine preaching has he been at the night . . but maybe least said is sunest mended. **1960** MISS READ *Fresh from Country* xii. A quiet word . . should . . stop any further tale-bearing, and I really think it's a case of 'least said, soonest mended.'

least: *see also* IDLE people have the least leisure.

There is nothing like LEATHER

Also used literally.

1692 R. L'ESTRANGE *Fables of Aesop* cccxlviii. There was a council of mechanics called to advise about the fortifying of a city. . . Up starts a currier;[1] Gentlemen, says he, when y'ave said all that can be said, there's nothing in the world like leather. **1837** F. PALGRAVE *Merchant & Friar* iv. King Log [the birch] was . . forgotten. . . 'Depend upon it, Sir, there is nothing like leather.' **1909** *Votes for Women* 22 Oct. 63 Nothing like leather for Suffragettes' wear.—Miss M. Roberta Mills makes Ties, Bags, Belts, [etc.]. **1935** V. S. PRITCHETT *(title)* Nothing like leather.
[1] one who dressed and coloured leather.

leg: *see* there goes more to MARRIAGE than four bare legs in a bed; everyone STRETCHES his legs according to the length of his coverlet.

leisure: *see* the BUSIEST men have the most leisure; IDLE people have the least leisure; there is LUCK in leisure; MARRY in haste and repent at leisure.

lend: *see* DISTANCE lends enchantment to the view.

LENGTH begets loathing

1742 C. JARVIS *Don Quixote* II. II. ix. The rest I omit, because length begets loathing. *a* **1895** F. LOCKER-LAMPSON *My Confidences* (1896) 43 'Length begets loathing.' I well remember the sultry Sunday evenings when . . we simmered through Mr. Shepherd's long-winded pastorals.

length: *see also* everyone STRETCHES his legs according to the length of his coverlet.

lengthen: *see* as the DAY lengthens, so the cold strengthens.

The LEOPARD does not change his spots

[JEREMIAH xiii. 23 Can the Ethiopian change his skinne? or the leopard his spots?] **1546** J. BALE *First Examination of Anne Askewe* 38 Their olde condycyons wyll they change, whan the blackemoreæne change hys skynne, and the catte of the mountayne[1] her spottes. **1596** SHAKESPEARE *Richard II* I. i. 174 Rage must be withstood. . . Lions make leopards tame.—Yea, but not change his spots. **1869** A. HENDERSON *Latin Proverbs* 317 Pardus maculas non deponit, a leopard does not change his spots. **1979** J. SCOTT *Clutch of Vipers* iv. He always was a dirty old man . . and the leopard doesn't change his spots.
[1] catte of the mountayne, leopard.

less: *see* of two EVILS choose the less; more HASTE, less speed.

LET well alone

Well is normally considered here as a

noun ('what is well'), rather than an adverb.

c **1570** *Scoggin's Jests* (1626) 76 The shomaker thought to make his house greater. . . They pulled downe foure or fiue postes of the house. . . Why said Scoggin, when it was well you could not let it alone. **1822** M. EDGEWORTH *Letter* 12 Jan. (1971) 317 Joanna quoted to me the other day an excellent proverb applied to health: 'Let well alone.' **1829** T. L. PEACOCK *Misfortunes of Elphin* ii. This immortal work . . will stand for centuries. . . It is well: it works well: let well alone. **1910** KIPLING *Rewards & Fairies* 108 Poor Dad wouldn't let well alone. He kept saying, 'Philadelphia, what does all this mean?' **1980** A. JUTE *Reverse Negative* 243, I thought of the vanity of not letting well alone, of not being content with my life.

let: *see also* let the COBBLER stick to his last; let the DEAD bury the dead; let them LAUGH that win; LIVE and let live; let SLEEPING dogs lie; SPARE at the spigot, and let out at the bung-hole.

leveller: *see* DEATH is the great leveller.

A LIAR ought to have a good memory

[QUINTILIAN *Institutionis Oratoriae* IV. ii. *mendacem memorem esse oportet*, a liar ought to have a good memory.] *a* **1542** WYATT in *Poetical Works* (1858) p.xxxvii, They say, 'He that lie well must have a good remembrance, that he agree in all points with himself, lest he be spied.' *c* **1690** R. SOUTH *Twelve Sermons* (1722) IV. 167 Indeed, a very rational Saying, That a lyar ought to have a good Memory. **1721** J. KELLY *Scottish Proverbs* 50 A Lyar should have a good Memory. Lest he tell the same Lye different ways. **1945** F. THOMPSON *Lark Rise* xiii. 'A liar ought to have a good memory,' they would say.

libel: *see* the GREATER the truth, the greater the libel.

lick: *see* if you can't BEAT them, join them.

If you LIE down with dogs, you will get up with fleas

An assertion that human failings, such as dishonesty and foolishness, are contagious.

[L. *qui cum canibus concumbunt cum pulicibus surgent*, he who lies with dogs will rise with fleas.] **1573** J. SANFORDE *Garden of Pleasure* 103ᵛ *Chi va dormir con i cani, si leua con i pulici.* He that goeth to bedde wyth Dogges, aryseth with fleas. **1640** G. HERBERT *Outlandish Proverbs* no. 343 Hee that lies with the dogs, riseth with fleas. **1721** J. KELLY *Scottish Proverbs* 129 He that sleeps with Dogs, must rise with Fleas. If you keep Company with base and unworthy Fellows, you will get some Ill by them. **1791** 'P. PINDAR' *Rights of Kings* 32 To this great truth, a Universe agrees, 'He who lies down with dogs, will rise with fleas'. **1842** C. J. LEVER *Jack Hinton* xxii. If you lie down with dogs, you'll get up with fleas, and that's the fruits of travelling with a fool. **1931** C. W. SYKES tr. *Kober's Circus Nights & Circus Days* ix. It's like this; if you go to bed with your dogs, you're bound to get up with fleas. If you once work for a small salary, the news gets round. **1968** A. WHITNEY *Every Man has his Price* iii. I've always believed the saying 'When you lie down with dogs, you get up with fleas'.

lie: *see* (noun) ASK no questions and hear no lies; HALF the truth is often a whole lie.

lie: *see also* (verb) as you MAKE your bed, so you must lie upon it; as a TREE falls, so shall it lie; TRUTH lies at the bottom of a well.

LIFE begins at forty

1932 W. B. PITKIN *Life begins at Forty* i. Life begins at forty. This is the revolutionary outcome of our New Era. . . Today it is half a truth. Tomorrow it will be an axiom. **1945** *Zionist Review* 14 Dec. 6 Among Palestine pioneers, life does not 'begin at forty'. **1952** 'M. COST' *Hour Awaits* 142 Life begins at forty. . . I know you're only in your thirties, but it leaves a nice margin.

LIFE isn't all beer and skittles

1855 T. C. HALIBURTON *Nature & Human Nature* I. ii. 'This life ain't all beer and skittles.' Many a time . . when I am disappointed sadly I say that saw over. **1857** T. HUGHES *Tom Brown's Schooldays* I. ii. Life isn't all beer and skittles. **1931** A. CHRISTIE *Sittaford Mystery* xxvi. 'It's an experience, isn't it?' 'Teach him life can't be all beer and skittles,' said Robert Gardner maliciously.

While there's LIFE there's hope

[CICERO *Ad Atticum* IX. x. *dum anima est, spes esse dicitur,* as the saying is, while there is life there is hope.] **1539** R. TAVERNER tr. *Erasmus' Adages* 36[v] The sycke person whyle he hath lyfe, hath hope. **1670** J. RAY *English Proverbs* 113 While there's life, there's hope, he cry'd; Then why such haste? so groan'd and dy'd. **1868** READE & BOUCICAULT *Foul Play* I. xi. They lost, for a few moments, all idea of escaping. But . . 'while there's life there's hope.' **1979** L. MEYNELL *Hooky & Villainous Chauffeur* viii. 'I don't want to go on to ninety-five.' 'Not now you don't; but you wait till you're ninety-four. While there's life there's hope, as they say.'

life: *see also* ART is long and life is short; the BEST things in life are free; if you would be HAPPY for a week take a wife; my SON is my son till he gets him a wife, but my daughter's my daughter all the days of her life; VARIETY is the spice of life.

LIGHT come, light go

Cf. EASY *come, easy go.*

c **1390** CHAUCER *Pardoner's Tale* l. 781 And lightly as it comth, so wol we spende. *a* **1475** J. FORTESCUE *Works* (1869) I. 489 For thyng that lightly cometh, lightly goeth. **1546** J. HEYWOOD *Dialogue of Proverbs* II. ix. L1 Lyght come lyght go. **1712** J. ARBUTHNOT *John Bull still in his Senses* iv. A thriftless Wretch, spending the Goods and Gear that his Fore-Fathers won with the Sweet of their Brows; light come, light go. **1861** C. READE *Cloister & Hearth* II. x. Our honest customers are the thieves. . . With them and with their purses 'tis lightly come, and lightly go. **1937** G. HEYER *They found Him Dead* iv. He was a bad husband to her. Light come light go.

light: *see also* (adjective) MANY hands make light work.

LIGHTNING never strikes the same place twice

1857 P. H. MYERS *Prisoner of Border* xii. They did not hit me at all. . . Lightning never strikes twice in the same place, nor cannon balls either, I presume. **1979** M. YORK *Death in Account* x. His bank had been raided the moment his back had been turned. 'Well, lightning never strikes the same place twice. . . I expect we'll be safe enough now.'

LIKE breeds like

1557 R. EDGEWORTH *Sermons* 178[v] Wyth a frowarde[1] synner, a man shall be naughtye[2] . . for lyke maketh like. *c* **1577** *Misogonus* 2[v] The like bredes the like (eche man sayd). **1842** TENNYSON *Poems* (1969) 703 Like men, like manners: Like breeds like, they say. **1931** 'D. FROME' *Strange Death of Martin Green* xiv. Murder is an awfully bad thing for anybody to get away with, even once. Like breeds like.

[1] refractory, evilly-disposed.   [2] wicked.

LIKE will to like

[HOMER *Odyssey* XVII. 218 ὡς αἰεὶ τὸν ὁμοῖον ἄγει θεὸς ὡς τὸν ὁμοῖον, the god always brings like to like; CICERO *De Senectute* III. vii. *pares autem vetere proverbio cum paribus facillime congregantur,* according to the old proverb equals most easily mix together.] *a* **1400** *Legends of Saints* (STS) I. 226 In proverbe I haf hard say That lyk to lyk drawis ay. *c* **1450** *Proverbs of Good Counsel* in *Book of Precedence* (EETS) 70 This proverbe dothe specify, 'Lyke wyll to lyke in eche company'. **1648** HERRICK *Hesperides* 378 Like will to like, each Creature loves his kinde. **1822** SCOTT *Peveril* II. ii. How could I help it? like will to like—the boy would come—the girl would see him. **1855** T. C. HALIBURTON *Nature & Human Nature* I. xi. Jessie had a repugnance to the union. . . 'Jessie . . nature, instead of forbiddin' it approves of it; for like takes to like.' **1922** S. J. WEYMAN *Ovington's Bank* xxxi. He's learned this at your d—d counter, sir! That's where it is. It's like to like. **1981** R. BARNARD *Mother's Boys* xiv. Mrs. Hodsden's connection with his house will be quite plain to you when you meet my husband. Like clings to like, they say. . . And those two certainly cling.

like *see also* (adjective) like FATHER, like son; like MASTER, like man; like MOTHER, like daughter; like PEOPLE, like priest.

like: *see also* (verb) if you don't like the HEAT, get out of the kitchen.

linen: *see* never CHOOSE your women or your linen by candlelight; one does not WASH one's dirty linen in public.

lining: *see* every CLOUD has a silver lining.

link: *see* a CHAIN is no stronger than its weakest link.

linstock (a forked staff to hold a lighted match): *see* the COBBLER to his last and the gunner to his linstock.

lion: *see* a LIVE dog is better than a dead lion; MARCH comes in like a lion, and goes out like a lamb; a MOUSE may help a lion.

lip: *see* there's MANY a slip 'twixt cup and lip.

LISTENERS never hear any good of themselves
    1647 *Mercurius Elencticus* 26 Jan.—2 Feb. 76 The old Proverb is, Hearkners never heare good of them selves. **1678** J. RAY *English Proverbs* (ed. 2) 75 Listners ne'er hear good of themselves. **1839** DICKENS *Nicholas Nickleby* xlii. 'If it is

fated that listeners are never to hear any good of themselves,' said Mrs. Browdie, 'I can't help it, and I am very sorry for it.' **1907** E. NESBIT *Enchanted Castle* v. He . . opened the door suddenly, and there . . was Eliza. . . 'You know what listeners never hear,' said Jimmy severely. **1977** A. NEWMAN *Evil Streak* IV. 178 They say listeners never hear any good of themselves but there is no excuse for . . ingratitude.

There is no LITTLE enemy
    [*c* **1390** CHAUCER *Tale of Melibee* l. 2512 Ne be nat necligent to kepe thy persone, nat oonly fro thy gretteste enemys, but fro thy leeste enemy. Senek seith: 'A man that is well avysed, he dredeth his leste enemy.'] **1659** J. HOWELL *Proverbs* 8 There's no enemy little, viz. we must not undervalue any foe. **1733** B. FRANKLIN *Poor Richard's Almanack* (Sept.) There is no little enemy. **1887** J. LUBBOCK *Pleasures of Life* I. v. To be friendly with every one is another matter; we must remember that there is no little enemy.

LITTLE fish are sweet
    1830 R. FORBY *Vocabulary of East Anglia* 434 'Little fish are sweet.'—It means small gifts are always acceptable. **1914** K. F. PURDON *Folk of Furry Farm* vii. 'They'll sell at a loss,' he went on, with a sigh, 'but sure, little fish is sweet! and the rent has to be made up.' **1981** J. BINGHAM *Brock* 92 Wealthy proprietor of the *Melford Echo* and three or four small newspapers in the country. ('Little fish are sweet, old boy.')

A LITTLE knowledge is a dangerous thing
The original *learning* is also used instead of *knowledge*.
    1711 POPE *Essay on Criticism* l. 215 A little Learning is a dang'rous Thing; Drink deep, or taste not the Pierian[1] Spring. **1829** P. EGAN *Boxiana* 2nd Ser. II. 4 The sensible idea, that 'A little learning is a dangerous thing!' **1881** T. H. HUXLEY *Science & Culture* iv. If a little knowledge is dangerous, where is

the man who has so much as to be out of danger? **1974** T. SHARPE *Porterhouse Blue* xviii. His had been an intellectual decision founded on his conviction that if a little knowledge was a dangerous thing, a lot was lethal.

[1] of Pieria, a region in northern Greece traditionally regarded as the home of the Muses.

LITTLE pitchers have large ears
Children overhear much that is not meant for them. A pitcher's *ears* are its handles. Cf. WALLS *have ears.*

1546 J. HEYWOOD *Dialogue of Proverbs* II. v. G4^v Auoyd your children, small pitchers haue wide eares. **1594** SHAKESPEARE *Richard III* II. iv. 37 Good madam, be not angry with the child.— Pitchers have ears. **1699** B.E. *New Dict. Canting Crew* s.v. Pitcher-bawd, Little Pitchers have large ears. **1840** R. H. BARHAM *Ingoldsby Legends* 1st Ser. 226 A truth Insisted on much in my earlier years, To wit, 'Little pitchers have very long ears!' **1972** A. PRICE *Colonel Butler's Wolf* i. He watched her shoo her sisters safely away. . . He had been lamentably careless in forgetting that little pitchers had large ears.

A LITTLE pot is soon hot
A small person is easily roused to anger or passion.

1546 J. HEYWOOD *Dialogue of Proverbs* I. xi. D2 It is wood[1] at a woorde, little pot soone whot. **1594–8** SHAKESPEARE *Taming of Shrew* IV. i. 6 Now were not I a little pot and soon hot, my very lips might freeze to my teeth. **1670** J. RAY *English Proverbs* 115 A little pot's soon hot. . . Little persons are commonly cholerick. **1884** C. READE *Perilous Secret* II. xv. Cheeky little beggar, But . . 'a little pot is soon hot.'

[1] mad, furious.

LITTLE strokes fell great oaks
[Cf. ERASMUS *Adages* I. viii. *multis ictibus deiicitur quercus,* the oak is felled by many blows.] *c* **1400** *Romaunt of Rose* l. 3688 For no man at the firste strok Ne may nat felle down an ok. **1539** R. TAVERNER tr. *Erasmus' Adages* 26^v Wyth

many strokes is an oke ouerthrowen. Nothyng is so stronge but that lyttell and lyttell maye be brought downe. **1591** SHAKESPEARE *Henry VI, Pt. 3* II. i. 54 And many strokes, though with a little axe, Hews down and fells the hardest-timber'd oak. By many hands your father was subdu'd. **1757** B. FRANKLIN *Poor Richard Improved: 1758* (Mar.) Stick to it steadily and you will see great Effects; for . . Little Strokes fell great Oaks. **1869** C. H. SPURGEON *John Ploughman's Talk* xxii. 'By little strokes Men fell great oaks.' By a spadeful at a time the navvies digged . . the embankment. **1981** *Family Circle* Feb. 57 From the cradle to the grave we are reminded that . . great oaks are only felled by a repetition of little strokes.

LITTLE thieves are hanged, but great ones escape
1639 J. CLARKE *Parœmiologia Anglo-Latina* 172 Little theeves are hang'd, but great ones escape. **1979** *Daily Telegraph* 22 Nov. 18 In view of the Blunt affair, I am reminded of the proverb, 'Little thieves are hanged but great ones escape.'

LITTLE things please little minds
[OVID *Ars Amatoria* 159 *parva leves capiunt animos,* small things enthral light minds.] **1576** G. PETTIE *Petit Palace* 139 A litle thyng pleaseth a foole. **1584** LYLY *Sappho & Phao* II. iv. Litle things catch light mindes. **1845** DISRAELI *Sybil* II. ii. Little things affect little minds. Lord Marney . . was kept at the station which aggravated his spleen. **1880** C. H. SPURGEON *John Ploughman's Pictures* 81 Precious little is enough to make a man famous in certain companies . . for . . little things please little minds. **1963** D. LESSING *Man & Two Women* 74 Small things amuse small minds. **1973** *Galt Toy Catalogue* 35 As the saying goes— Little things please little minds.

little: *see also* (adjective) BIG fish eat little fish; BIG fleas have little fleas upon their backs to bite them; BIRDS in their little nests agree; little BIRDS that can sing and won't sing must be made to sing; EVERY little helps; GREAT oaks from little acorns grow; LONG and lazy little and loud; MANY a little makes a mickle; MUCH cry and little wool; (adverb) LOVE me little, love me long.

LIVE and learn

[**1575** G. GASCOIGNE *Glass of Government* II. 88 We live to learne, for so Sainct Paule doth teach.] *c* **1620** in *Roxburghe Ballads* (1871) I. 60 A man may liue and learne. **1771** SMOLLETT *Humphry Clinker* III. 168 'Tis a true saying, live and learn—O woman, what chuckling and changing have I seen! **1894** J. LUBBOCK *Use of Life* vi. No doubt we go on learning as long as we live: 'Live and learn,' says the old proverb. **1978** F. WELDON *Praxis* xx. 'Well,' she said vaguely, 'you live and learn.'

LIVE and let live

1622 G. DE MALYNES *Ancient Law-Merchant* I. xlv. According to the Dutche prouerbe . . Leuen ende laeten leuen, To liue and to let others liue. **1641** D. FERGUSSON *Scottish Proverbs* (STS) no. 582 Live and let live. **1678** J. RAY *English Proverbs* (ed. 2) 170 Live and let live, *i.e.* Do as you would be done by. Let such pennyworths as your Tenants may live under you. **1762** SMOLLETT *Sir Launcelot Greaves* II. xvi. He deals very little in physic stuff, . . whereby he can't expect the pothecary to be his friend. You knows, master, one must live and let live, as the saying is. **1843** R. S. SURTEES *Handley Cross* II. vii. Live and let live, as the criminal said to the hangman. **1979** C. BRAND *Rose in Darkness* iv. Not that Sari cared two hoots how other people conducted their private lives. Live and let live.

A LIVE dog is better than a dead lion

[ECCLESIASTES ix. 4 To him that is ioyned to all the liuing, there is hope: for a liuing dogge is better then a dead Lion.] *c* **1390** in *Minor Poems of Vernon MS* (EETS) 534 Better is a quik[1] and an hol hounde Then a ded lyon . . And better is pouert with godnes Then

richesse with wikkedness. **1566** J. BARTHLET *Pedigree of Heretics* 2ᵛ A lyuing Dogge, is better than a dead Lion. **1798** 'P. PINDAR' *Tales of Hoy* 41 It was a devil of a trick . . but, 'A living Dog is better than a dead Lion,' as the saying is. **1864** TROLLOPE *Can You forgive Her?* II. vii. He had so often told the widow that care killed the cat, and that a live dog was better than a dead lion. **1928** D. H. LAWRENCE *Woman who rode Away* 132 When the lion is shot, the dog gets the spoil. So he had come in for Katherine, Alan's lioness. A live dog is better than a dead lion. **1939** L. I. WILDER *By Shores of Silver Lake* xi. I'm thankful the paymaster was sensible. Better a live dog than a dead lion.

[1] living.

They that LIVE longest, see most

[**1605–6** SHAKESPEARE *King Lear* v. iii. 325 We that are young Shall never see so much nor live so long.] **1620** T. SHELTON tr. *Cervantes' Don Quixote* II. lii. My Mother was vsed to say, That 'twas needfull to liue long, to see much. **1837** T. HOOK *Jack Brag* III. ii. Them as lives longest sees the most. **1961** N. LOFTS *House at Old Vine* VI. vi. Them that live longest see most. You remember that, young man, if ever you're down on your luck.

Come LIVE with me and you'll know me

1925 S. O'CASEY II. 49, I only seen him twiced; if you want to know me, come an' live with me. **1960** C. S. LEWIS *Four Loves* iii. You must really give no kind of preference to yourself; at a party it is enough to conceal the preference. Hence the old proverb 'come live with me and you'll know me'.

live: *see also* EAT to live, not live to eat; he who FIGHTS and runs away, may live to fight another day; those who live in GLASS houses shouldn't throw stones; one HALF of the world does not know how the other half lives; MAN cannot live by bread alone; a REED before the wind lives on, while mighty oaks do fall; THREATENED men live long.

lived: *see* BRAVE men lived before Agamemnon.

He who LIVES by the sword dies by the sword

[MATTHEW xxvi. 52 All they that take the sword shall perish with the sword; **1601** A. MUNDAY et al. *Death of Robert, Earl of Huntington* L1 Alas for woe: but this is iust heauens doome On those that liue by bloode: in bloode they die.] **1652** R. WILLIAMS *Complete Writings* (1963) IV. 352 All that take the Sword . . shall perish by it. **1804** G. MORRIS *Diary & Letters* (1889) II. xlv. To quote the text, 'Those who live by the sword shall perish by the sword.' **1916** J. BUCHAN *Greenmantle* vi. I did not seek the war . . It was forced on me. . . He that takes the sword will perish by the sword. **1978** 'M. CRAIG' *Were He Stranger* xiii. Mark me, Sydney, he who lives by the sword dies by the sword.

He that LIVES in hope dances to an ill tune

1591 J. FLORIO *Second Fruits* 149 This argument of yours is lame and halting, but doo not you knowe that. He that dooth liue in hope, dooth dance in narrowe scope. **1640** G. HERBERT *Outlandish Proverbs* no. 1006 Hee that lives in hope danceth without musick. **1732** T. FULLER *Gnomologia* no. 2224 He that liveth in Hope, danceth without a Fiddle. **1977** J. AIKEN *Five Minute Marriage* ii. 'He that lives in hope danceth to an ill tune,' remarked Mrs. Andrews, who was full of proverbs.

He LIVES long who lives well

1553 T. WILSON *Art of Rhetoric* 45ᵛ They lyued long enough, that have liued well enough. **1619** W. DRUMMOND *Midnight's Trance* (1951) 29 Who liueth well, liueth long. **1642** T. FULLER *Holy State* I. vi. If he chance to die young, yet he lives long that lives well. **1861** H. BONAR in *Hymns of Faith & Hope* 2nd Ser. 129 He liveth long who liveth well! All other life is short and vain.

load: *see* a SWARM in May is worth a load of hay.

loaf: *see* HALF a loaf is better than no bread; a SLICE off a cut loaf isn't missed.

loathing: *see* LENGTH begets loathing.

lock: *see* it is too late to shut the STABLE-door after the horse has bolted.

locksmith: *see* LOVE laughs at locksmiths.

loft: *see* SEPTEMBER blow soft, till the fruit's in the loft.

London: *see* what MANCHESTER says today, the rest of England says tomorrow.

LONG and lazy, little and loud; fat and fulsome, pretty and proud

c **1576** T. WHYTEHORNE *Autobiography* (1961) 23 Hy women be layzy & low be lowd, fair be sluttish, and fowll be proud. **1591** J. FLORIO *Second Fruits* 189 If long, she is lazy, if little, she is lowde. **1648** HERRICK *Hesperides* 166 Long and lazie. That was the Proverb. Let my mistress be Lasie to others, but be long to me. **1659** J. HOWELL *Proverbs* (English) 10 Long and lazy, little and loud, Fatt and fulsome, prety and proud; in point of women. **1872** BLACKMORE *Maid of Sker* I. xiii. You are long enough, and lazy enough; put your hand to the bridle.

LONG foretold, long last; short notice, soon past

1866 A. STEINMETZ *Manual of Weathercasts* xiv. Old saws[1] about the barometer. Long foretold, long last; short notice, soon past. **1889** J. K. JEROME *Three Men in Boat* v. The barometer is . . misleading. . . Boots . . read out a poem which was printed over the top of the oracle, about 'Long foretold, long last; Short notice, soon past.' The fine weather never came that summer.
[1] traditional maxims or proverbs.

It is a LONG lane that has no turning
Commonly used as an assertion that an

unfavourable situation will eventually change for the better.

1633 *Stationers' Register* (1877) IV. 273 (*ballad*) Long runns that neere turnes. **1670** J. RAY *English Proverbs* 117 It's a long run that never turns. **1732** T. FULLER *Gnomologia* no. 2863 It is a long Lane that never turns. **1748** RICHARDSON *Clarissa* IV. xxxii. It is a long lane that has no turning—Do not despise me for my proverbs. **1849** BULWER-LYTTON *Caxtons* III. XVII. i. I wonder we did not run away. But . . 'It is a long lane that has no turning.' **1945** F. P. KEYES *River Road* VIII. xxxvii. 'You're through in politics, Gervais. You might just as well face it.' . . 'It's a long lane that has no turning.'

long: *see also* (adjective) ART is long and life is short; KINGS have long arms; NEVER is a long time; OLD sins cast long shadows; SHORT reckonings make long friends; a STERN chase is a long chase; he who SUPS with the Devil should have a long spoon; (adverb) he LIVES long who lives well; LOVE me little, love me long; THREATENED men live long.

The LONGEST way round is the shortest way home

[**1580** LYLY *Euphues* II. 96 Thou goest about[1] (but yet the neerest way) to hang me vp for holydayes.] **1635** F. QUARLES *Emblems* IV. ii. The road to resolution lies by doubt: The next way home's the farthest way about. **1776** G. COLMAN *Spleen* II. 24 The longest way about is the shortest way home. **1846** J. K. PAULDING *Letter* 9 May (1962) vii. The Potatoes arrived . . *via* New York . . in pursuance of the Old Proverb, that 'the longest way round is the shortest way home.' **1942** K. ABBEY *And let Coffin Pass* xviii. 'The longest way round is the shortest way home.' . . 'We'll make the best time by skirting the pines.'

[1] *go about*: used equivocatingly to mean 'endeavour' and 'go around or roundabout'. The context is of a person metaphorically described as a hat which can be taken up and put down at will.

longest: *see also* (adjective) BARNABY bright, Barnaby bright, the longest day and the shortest night; (adverb) a CREAKING door hangs longest; they that LIVE longest, see most; he who LAUGHS last, laughs longest.

LOOK before you leap

c **1350** *Douce MS* 52 no. 150 First loke and aftirward lepe. **1528** W. TYNDALE *Obedience of Christian Man* 130 We say . . Loke yer thou lepe, whose literall sence is, doo nothinge sodenly or without avisement. **1567** W. PAINTER *Palace of Pleasure* II. xxiv. He that looketh not before he leapeth, may chaunce to stumble before he sleapeth. **1621** BURTON *Anatomy of Melancholy* II. iii. Looke before you leape. **1836** MARRYAT *Midshipman Easy* I. vi. Look before you leap is an old proverb. . . Jack . . had pitched into a small apiary, and had upset two hives of bees. **1941** C. MACKENZIE *Red Tapeworm* i. Do you remember the rousing slogan which the Prime Minister gave the voters . . on the eve of the last General Election? . . Look Before You Leap. **1979** D. MAY *Revenger's Comedy* ix. Changing horses, love? I should look before you leap.

look: *see also* a CAT may look at a king; the DEVIL looks after his own; never look a GIFT horse in the mouth; a MAN is as old as he feels, and a woman as old as she looks; take care of the PENCE and the pounds will take care of themselves; those who PLAY at bowls must look out for rubbers; one man may STEAL a horse, while another may not look over a hedge.

LOOKERS-ON see most of the game

1529 J. PALSGRAVE in *Acolastus* (EETS) p. xxxviii, It fareth between thee and me as it doth between a player at the chess and a looker on, for he that looketh on seeth many draughts that the player considereth nothing at all. **1597** BACON *Essays* 'Of Followers' 7[v] To take aduise of friends is euer honorable: For lookers on many times see more then gamesters. **1666** G. TORRIANO *Ita-*

lian Proverbs 111 As the English say, The stander by sees more than he who plays. **1850** F. E. SMEDLEY *Frank Fairlegh* vii. Remembering the old adage, that 'lookers-on see most of the game,' I determined . . to accompany him. **1937** A. CHRISTIE *Murder in Mews* 153 If it is true that the looker-on knows most of the game, Mr. Satterthwaite knew a good deal.

lord: *see* EVERYBODY loves a lord; NEW lords, new laws.

What you LOSE on the swings you gain on the roundabouts
A fairground metaphor used in a variety of forms.

1912 P. CHALMERS *Green Days & Blue Days* 19 What's lost upon the roundabouts we pulls up on the swings. **1927** *Times* 24 Mar. 15 By screwing more money out of tax-payers he diminishes their savings, and the market for trustee securities loses on the swings what it gains on the roundabouts. **1978** G. MOORE *Farewell Recital* 129 There are compensations: what you lose on the swings you gain on the roundabouts. And let's face it, a cup of tea or a cup of coffee are all very well but they are not so much fun as polygamy.

You cannot LOSE what you never had
The sentiment is expressed in a number of ways: quot. 1974 represents a local equivalent.

a **1593** MARLOWE *Hero & Leander* I. 276 Of that which hath no being do not boast, Things that are not at all are never lost. **1676** I. WALTON *Compleat Angler* (ed. 5) I. v. 'He has broke all; there's half a line and a good hook lost.' 'I¹ and a good Trout too.' 'Nay, the Trout is not lost, for . . no man can lose what he never had.' **1788** WESLEY *Works* (1872) VII. 41 He only *seemeth* to have this. . . No man can lose what he never had. **1935** *Oxford Dict. English Proverbs* 601 You cannot lose what you never had. **1974** 'J. HERRIOT' *Vet in Harness* viii. 'Only them as has them can lose them,' she said firmly, her head tilted as al-

ways. I had heard that said many times and they were brave Yorkshire words.
¹ i.e. aye.

lose: *see also* a BLEATING sheep loses a bite; the SUN loses nothing by shining into a puddle; a TALE never loses in the telling; you WIN a few, you lose a few.

loser: *see* FINDERS keepers (losers weepers).

One man's LOSS is another man's gain
c **1527** T. BERTHELET tr. *Erasmus' Sayings of Wise Men* D1ᵛ Lyghtly whan one wynneth, an other loseth. **1733** J. BARBER in *Correspondence of Swift* (1965) IV. 189 Your loss will be our gain, as the proverb says. **1821** SCOTT *Pirate* I. vi. Doubtless one man's loss is another man's gain. **1918** D. H. LAWRENCE *Letter* 21 Feb. (1962) I. 544, I am glad to have the money from your hand. But . . one man's gain is another man's loss. **1979** R. LITTELL *Debriefing* vi. Well, their loss is my gain!

There's no great LOSS without some gain
a **1641** D. FERGUSSON *Scottish Proverbs* (STS) no. 1408 Thair was never a grit loss without som small vantag. **1868** W. CLIFT *Tim Bunker Papers* 134 However, 'there is no great loss but what there is some small gain,' and Jake Frink claims that he has got his money's worth in experience. **1937** L. I. WILDER *On Banks of Plain Creek* xxv. The hens . . were eating grasshoppers. . . 'Well, we won't have to buy feed for the hens. . . There's no great loss without some gain.'

lost: *see* 'tis BETTER to have loved and lost, than never to have loved at all; he who HESITATES is lost; what a NEIGHBOUR gets is not lost; there is NOTHING lost by civility; for WANT of a nail the shoe was lost.

lottery: *see* MARRIAGE is a lottery.

loud: *see* LONG and lazy little and loud.

louder: *see* ACTIONS speak louder than words.

LOVE and a cough cannot be hid
[L. *amor tussisque non celantur,* love and a cough cannot be concealed.] *a* **1325** *Cursor Mundi* (EETS) l. 4276 Luken luue at the end wil kith [concealed love will show itself in the end]. **1573** J. SANFORDE *Garden of Pleasure* 98ᵛ Foure things cannot be kept close, Loue, the cough, fyre, and sorrowe. **1611** R. COTGRAVE *Dict. French & English* s.v. Amour, We say, Loue, and the Cough cannot be hidden. **1640** G. HERBERT *Outlandish Proverbs* no. 49 Love and a Cough cannot be hid. **1863** G. ELIOT *Romola* I. vi. If there are two things not to be hidden—love and a cough—I say there is a third, and that is ignorance. **1943** 'P. WENTWORTH' *Chinese Shawl* xxvi. He is in love with Laura Fane. . . Love and a cold cannot be hid.

One cannot LOVE and be wise
[PUBLILIUS SYRUS *Sententiae* xxii. *amare et sapere vix deo conceditur,* to love and to be wise is scarcely allowed to God.] *c* **1527** T. BERTHELET tr. *Erasmus' Sayings of Wise Men* B1ᵛ To have a sadde¹ mynde & loue is nat in one person. **1539** R. TAVERNER tr. *Erasmus' Adages* II. A5 To be in loue & to be wyse is scase graunted to god. **1612** BACON *Essays* 'Of Love' xii. It is impossible to loue and bee wise. **1631** R. BRATHWAIT *English Gentlewoman* 32 The Louer is euer blinded . . with affection . . whence came that vsuall saying One cannot loue and be wise. **1872** G. ELIOT *Middlemarch* II. III. xxvii. If a man could not love and be wise, surely he could flirt and be wise at the same time?
¹ serious.

LOVE begets love
[L. *amor gignit amorem,* love produces love.] **1648** HERRICK *Hesperides* 297 Love love begets, then never be Unsoft to him who's smooth to thee. **1812** E. NARES *I'll consider of It* iii. 'Love' says the proverb, 'produces love.' **1909** A. MACLAREN *Epistle to Ephesians* 275 Love begets love,

and . . if a man loves God, then that glowing beam will glow whether it is turned to earth or turned to heaven.

LOVE is blind
[Cf. PLATO *Laws* 731e τυφλοῦται γὰρ περὶ τὸ φιλούμενον ὁ φιλῶν, he who loves is blind about the one he loves; PLAUTUS *Miles Gloriosus* l. 1260 *Naso poliam haecquidem plus videt quam oculis. Caeca amore est,* 'Indeed he sees better with that nose of this than with his eyes.' 'It is blinded by love.'] *c* **1390** CHAUCER *Merchant's Tale* l. 1598 For love is blynd alday, and may nat see. **1591** SHAKESPEARE *Two Gentlemen of Verona* II. i. 61 If you love her you cannot see her.—Why?—Because Love is blind. **1855** G. BRIMLEY in *Cambridge Essays* iii. There is profound beauty and truth in the allegory that represents love as a blind child. **1978** A. MALING *Lucky Devil* xii. 'How did you ever come to marry an idiot like Irving?' . . 'Love is blind.'

LOVE laughs at locksmiths
Cf. LOVE *will find a way.*
[**1592–3** SHAKESPEARE *Venus & Adonis* l. 576 Were beauty under twenty locks kept fast, Yet love breaks through and picks them all at last.] **1803** C. COLMAN *(title)* Love laughs at locksmiths: an operatic farce. **1901** F. R. STURGIS *Sexual Debility in Man* ix. Love is said to laugh at locksmiths, and incidentally at parental authority, and this young man was no exception. **1922** 'D. YATES' *Jonah & Co* iv. And now push off and lock the vehicle. I know Love laughs at locksmiths, but the average motor-thief's sense of humour is less susceptible.

LOVE makes the world go round
[Cf. Fr. *c'est l'amour, l'amour, l'amour, Qui fait le monde A la ronde* (Dumerson & Ségur *Chansons Nationales & Populaires de France,* 1851, II. 180) it is love, love, love, that makes the world go round.] **1865** 'L. CARROLL' *Alice's Adventures in Wonderland* ix. '"Oh, 'tis love, 'tis love that makes the world go round!"' 'Somebody said,' Alice whispered, 'that

it's done by everybody minding their own business.' **1902** 'O. HENRY' in *Brandur Magazine* 27 Sept. 4 It's said that love makes the world go round. The announcement lacks verification. It's the wind from the dinner horn that does it. **1980** V. CANNING *Fall from Grace* ix. A nice young man . . but gullible and romantic. . . Love makes the world go round.

LOVE me little, love me long

a **1500** in *Archiv* (1900) CVI. 274 Love me lytyll and longe. **1546** J. HEYWOOD *Dialogue of Proverbs* II. ii. G1 Olde wise folke saie, loue me lyttle loue me long. **1629** T. ADAMS *Works* 813 Men cannot brooke poore friends. This inconstant Charitie is hateful as our English phrase premonisheth; Loue me Little, and Loue me Long. **1721** J. KELLY *Scottish Proverbs* 229 Love me little, love me long. A Dissuasive from shewing too much, and too sudden Kindness. **1907** *Times Literary Supplement* 8 Mar. 77 Mrs. Bellew is a lady who cannot love either little or long. She . . tires very quickly of the men who are irresistibly drawn to her.

LOVE me, love my dog

[ST. BERNARD *Sermon: In Festo Sancti Michaelis* iii. *qui me amat, amat et canem meum,* who loves me, also loves my dog.] *a* **1500** in *Archiv* (1893) XC. 81 He that lovyeth me lovyeth my hound. **1546** J. HEYWOOD *Dialogue of Proverbs* II. ix. K4[v] Ye haue bene so veraie[1] a hog, To my frends. What man, loue me, loue my dog. **1692** R. L'ESTRANGE *Fables of Aesop* cvi. Love Me, Love My Dog. . . For there are certain Decencies of Respect due to the Servant for the Master's sake. **1826** LAMB *Elia's Last Essays* (1833) 262 *That you must love me, and love my dog.* . . We could never yet form a friendship . . without the intervention of some third anomaly . . the understood *dog* in the proverb. **1975** D. BAGLEY *Snow Tiger* xvi. I'll have to bring Victor. . . Love me—love my dog.

[1] veritable, real (in emphatic use).

LOVE will find a way

a **1607** T. DELONEY *Gentle Craft* (1648) I. xv. Thus love you see can finde a way, To make both Men and Maids obey. **1661** 'T.B.' (*title*) Love will finde out the way. **1765** in T. Percy *Reliques* III. III. 236 Over the mountains, And over the waves; . . Love will find out the way. **1975** *Listener* 16 Oct. 504 The red-plush curtain fell on a reprise of 'Love will find a way'.

love: *see* (noun) the COURSE of true love never did run smooth; all's FAIR in love and war; when the FURZE is in bloom, my love's in tune; LUCKY at cards, unlucky in love; MONEY is the root of all evil; it is best to be OFF with the old love before you are on with the new; PITY is akin to love; when POVERTY comes in at the door, love flies out of the window; PRAISE the child, and you make love to the mother; the QUARREL of lovers is the renewal of love; (verb) EVERYBODY loves a lord; whom the GODS love die young.

loved: *see* 'tis BETTER to have loved and lost, than never to have loved at all.

Lovell: *see* the CAT, the rat, and Lovell the dog, rule all England under the hog.

lover: *see* JOVE but laughs at lovers' perjury; the QUARREL of lovers is the renewal of love.

There is LUCK in leisure

It is often advisable to wait before acting.

1683 G. MERITON *Yorkshire Dialogue* 9 There's luck in Leizur. **1859** 'SKITT' *Fisher's River* vii. Thinks I, 'There's luck in leisure,' as I've hearn folks say. . . So I jist waited a spell. **1936** J. ESTEVEN *While Murder Waits* xxii. 'You . . won't decide now?' . . 'There's luck in leisure, Victoria.'

There is LUCK in odd numbers

Cf. THIRD time lucky.

[VIRGIL *Eclogues* l. 875 *numero deus impare gaudet,* God delights in an uneven number.] **1598** SHAKESPEARE *Merry Wives of Windsor* v.i. 3 This is the third time; I hope good luck lies in odd numbers. **1837** S. LOVER *Rory O'More* I. (*title-page*) 'There's luck in odd numbers,' says Rory O'More. **1883** J. PAYN *Thicker than Water* I. i. She was . . by no means averse to a third experiment in matrimony. . . 'There was luck in odd numbers.'

luck: *see also* the DEVIL's children have the Devil's luck; DILIGENCE is the mother of good luck; FOOLS for luck; see a PIN and pick it up, all the day you'll have good luck.

LUCKY at cards, unlucky in love

[**1738** SWIFT *Polite Conversation* III. 213 Well, Miss, you'll have a sad Husband, you have such good Luck at Cards.]

1866 T. W. ROBERTSON *Society* II. ii. 'I'm always lucky at cards!' . . 'Yes, I know an old proverb about that. . . Lucky at play, unlucky in—.' *a* **1871** ——*Play* (1889) III. ii. Unlucky in love, lucky at cards. **1941** P. CHEYNEY *Trap for Bellamy* iv. 'Lucky at cards, unlucky in love.' . . I'm going to find out if the proverb's true. . . What are they playing tonight? **1981** *Oxford Mail* 29 Aug. 5 Arthur and Hilda Cover have defied the old proverb by being lucky at cards and lucky in love.

lucky: *see also* it is BETTER to be born lucky than rich; THIRD time lucky.

lunch: *see* there's no such thing as a FREE lunch.

M

Where MACGREGOR sits is the head of the table

Other names are used as well as Mac-Gregor.

[Cf. **1580** LYLY *Euphues* II. 39 When . . Agesilaus sonne was set at the lower end of the table, & one cast it in his teeth as a shame, he answered: this is the vpper end where I sit. **1732** T. FULLER *Gnomologia* no. 4362 That is the upper End, where the chief Person sits.] **1837** R. W. EMERSON *American Scholar* 19 Wherever Macdonald[1] sits, there is the head of the table. Linnaeus makes botany the most alluring of studies and wins it from the farmer and the herbwoman. **1903** K. D. WIGGIN *Rebecca of Sunnybrook Farm* viii. If wherever the MacGregor sat was the head of the table, so . . wherever Rebecca stood was the centre of the stage. **1918** A. G. GARDINER *Leaves in Wind* 197 There are . . people who carry the centre of the stage with them. . . 'Where O'Flaherty sits is the head of the table.' **1940** J. W. BELLAH *Bones of Napoleon* 69 Like Macdonald—where Lord Innes sat was the head of the table. **1980** *Times* 12 May 15 (Letter from His Honour Judge MacGregor) Sir, Where MacGregor[2] sits *is* the head of the table.

[1] i.e. the head of the Macdonald clan.
[2] The proverb is sometimes attributed to Robert MacGregor of Campbell ('Rob Roy': 1671–1734), highland freebooter.

mad: *see* whom the GODS would destroy, they first make mad.

made: *see* GOD made the country, and man made the town; PROMISES, like piecrust, are made to be broken; *also* MAKE.

Mahomet: *see* if the MOUNTAIN will not come to Mahomet, Mahomet must go to the mountain.

MAKE hay while the sun shines

1546 J. HEYWOOD *Dialogue of Proverbs* I. iii. A4 Whan the sunne shynth make hey. **1583** B. MELBANCKE *Philotimus* 24 Yt is well therefore to make hay while the sunne shines. **1835** J. CARLYLE *Letters & Memorials* (1883) I. 21 'It is good to make hay while the sun shines,' which means, in the present case . . to catch hold of a friend while she is in the humour. **1924** E. BAGNOLD *Serena Blandish* vi. The countess's enthusiasm was cooling. Martin . . said warningly, 'You must make hay, my child, while the sun shines.' **1960** MISS READ *Fresh from Country* i. 'Make hay while the sun shines,' quoted Anna aloud. It seemed as encouraging a proverb as any under the circumstances.

As you MAKE your bed, so you must lie upon it

c **1590** G. HARVEY *Marginalia* (1913) 88 Lett them . . go to there bed, as themselues shall make it. **1640** G. HERBERT *Outlandish Proverbs* no. 340 He that makes his bed ill, lies there. **1721** J. KELLY *Scottish Proverbs* 16 As you make your bed, so you lye down. According to your Conditions you have your Bargain. **1832** S. WARREN *Diary of Late Physician* II. vi. As soon as his relatives . . heard . . they told him . . that as he had made his bed, so he must lie upon it. **1978** S. RADLEY *Death & Maiden* iv. He . . had been brought up in a village which maintained a relentless communal belief in the duty of lying on the bed you had made, however uncomfortable.

make: *see also* make HASTE slowly; if you don't make MISTAKES you don't make anything; you cannot make an OMELETTE without breaking eggs; you can't make a SILK purse out of a sow's ear.

male: *see* the FEMALE of the species is more deadly than the male.

Mammon: *see* you cannot serve GOD and Mammon.

MAN cannot live by bread alone

[DEUTERONOMY viii. 3 Man doth not liue by bread alone, but by euery word that proceedeth out of the mouth of the Lord doth man liue; MATTHEW iv. 4 Man shall not liue by bread alone.] **1875** R. W. EMERSON in *North American Review* May–June 418 Man does not live by bread alone, but by faith, by admiration, by sympathy. **1927** J. BUCHAN *Witch Wood* iii. Man canna live by bread alone, but he assuredly canna live without it. **1973** *Galt Toy Catalogue* 35 As the saying goes—Man cannot live by bread alone.

Whatever MAN has done, man may do

[Cf. **1723** S. CRANSTON in G. S. Kimball *Correspondence of Colonial Governors of Rhode Island* (1902) I. 9 But as the Proverb is what hath been may be againe.] **1863** C. READE *Hard Cash* II. xiv. 'Dark Deeds are written in an unknown tongue called 'Lawyerish' . . ; pick it out if you can.' 'Whatever man has done man may do,' said Dr. Sampson stoutly. **1958** M. KENNEDY *Outlaws on Parnassus* x. There is no law against the didactic novel. . . *What you can do you may do* obtains in this field of art as it does in all the others.

A MAN is as old as he feels, and a woman as old as she looks

Both parts of the proverb are sometimes used on their own (see also quot. 1930).

1871 V. LUSH *Thames Journal* 27 Aug. (1975) 114 She is always making me out so much older than I am and that's not fair, for a man is only as old as he *feels* and a woman is only as old as she *looks*. **1891** W. MORRIS *News from Nowhere* iii. 'How old am I, do you think?' 'Well,' quoth I, 'I have always been told that a woman is as old as she looks.' **1907** *Illustrated London News* 25 May 794 The adage that a man is as old as he feels, and a woman as old as she looks, may be said to contain much inherent truth. **1930** A. HUXLEY *Brief Candles* 262 He's

always saying he's so old, which is all nonsense, because you're only as old as you feel. **1981** J. SPARROW *Words on Air* 67 Two ways of looking at old age. . . 'Be your age!' is the first, 'A man is as old as he feels' is the second.

MAN proposes, God disposes

[*c* **1420** T. À. KEMPIS *De Imitatione Christi* I. xix. *homo proponit, sed Deus disponit*, man proposes but God disposes.] *c* **1440** LYDGATE *Fall of Princes* (EETS) I. 3291 A man off malice may a thyng purpose . . But God a-boue can graciousli dispose[1] Ageyn such malice to make resistence. *c* **1450** tr. *T. à Kempis' De Imitatione Christi* (EETS) I. xix. For man purposith & god disposith. **1640** G. HERBERT *Outlandish Proverbs* no. 1 Man Proposeth, God disposeth. **1853** R. C. TRENCH *On Lessons in Proverbs* (ed. 2) iii. A proverb . . Man proposes, God disposes . . that every nation in Europe possesses. **1958** L. DURRELL *Mountolive* IV. 88 In diplomacy one can only propose, never dispose. That is up to God, don't you think?

[1] determine or control the course of affairs or events.

MAN's extremity is God's opportunity

1629 T. ADAMS *Works* 619 Heere is now a deliuery fit for God, a cure for the Almightie hand to vndertake. Mans extremity is Gods opportunitie. **1706** LD. BELHAVEN in Defoe *Hist. Union* (1709) V. 34 Man's Extremity is God's opportunity. . . Some unforeseen Providence will fall out, that may cast the Ballance. **1916** E. A. BURROUGHS *Valley of Decision* viii. In the first winter of the war . . we were all much encouraged by tales of a new thirst for religion among the majority of the men. . . 'Man's extremity, God's opportunity.' **1949** D. SMITH *I capture Castle* xiii. 'You should look in on the church if ever you're mentally run down.' . . 'You mean "Man's extremity is God's opportunity"?' **1980** *Times* 4 Dec. 17 Those extremities which have, until now, been often God's opportunity.

The MAN who is born in a stable is a horse
The sense is reversed in the last quotation.

1833 M. SCOTT *Tom Cringle's Log* I. iv. 'I am an Englishman and no traitor, nor will I die the death of one.' . . 'Truly . . a man does sometimes become a horse by being born in a stable.' **1906** *Times Literary Supplement* 27 Apr. 147 Except on the principle that the man who is born in a stable is a horse, [he] was not an Irishman at all. **1980** J. O'FAOLAIN *No Country for Young Men* ii. Father Casey . . has a theory that the Irish back in Ireland have less claim to Irishness than men like himself. Something to do with . . being born in a stable not necessarily making you a horse.

man: *see also* BETTER be an old man's darling, than a young man's slave; a BLIND man's wife needs no paint; the CHILD is father of the man; in the COUNTRY of the blind, the one-eyed man is king; DO right and fear no man; a DROWNING man will clutch at a straw; the EARLY man never borrows from the late man; every ELM has its man; EVERY man for himself; EVERY man for himself, and God for us all; EVERY man for himself, and the Devil take the hindmost; EVERY man has his price; EVERY man is the architect of his own fortune; EVERY man to his taste; EVERY man to his trade; it's ill speaking between a FULL man and a fasting; a HUNGRY man is an angry man; you should KNOW a man seven years before you stir his fire; a man who is his own LAWYER has a fool for his client; one man's LOSS is another man's gain; MANNERS maketh man; like MASTER, like man; one man's MEAT is another man's poison; MONEY makes a man; a MONEYLESS man goes fast through the market; NINE tailors make a man; NO man can serve two masters; NO man is a hero to his own valet; NO moon, no man; there is NOTHING so good for the inside of a man as the outside of a horse; the RICH man has his ice in the summer and the poor man gets his in the winter; give a man ROPE enough and he will hang himself; SIX hours' sleep for a man, seven for a woman, and eight for a fool; one man may STEAL a horse, while another may not look over a hedge; the STYLE is the man; TIME and tide wait for no man; for WANT of a nail the shoe was lost; the WAY to a man's heart is through his stomach; a WILFUL man must have his way; when the WIND is in the east, 'tis neither good for man nor beast; *also* MEN.

What MANCHESTER says today, the rest of England says tomorrow
The proverb occurs in a variety of forms.

1898 KIPLING *Day's Work* 51 What the horses o' Kansas think to-day, the horses of America will think to-morrow; an' I tell *you* that when the horses of America rise in their might, the day o' the Oppressor is ended. **1902** V. S. LEAN *Collectanea* I. 116 What Lancashire thinks to-day all England will think tomorrow. This was in the days of the Anti–Corn–Law League.[1] Since then the initiative in political movements proceeds from Birmingham. **1944** C. MILBURN *Journal* 24 Aug. in *Diaries* (1979) xiii. Manchester rang its bells yesterday—a day before St. Paul's . . thus justifying its words, so often used: 'What Manchester says today, the rest of England says tomorrow!' **1980** *Listener* 6 Mar. 300 What Manchester does today— . . is the old boast that 'What Manchester does today London thinks tomorrow.'

[1] The Corn Law, restricting the importation of foreign corn, was abolished in 1846. Manchester was considered the home of free trade.

Manchester: *see also* YORKSHIRE born and Yorkshire bred, strong in the arm and weak in the head.

MANNERS maketh man
c **1350** *Douce MS* 52 no. 77 Maner makys man. *c* **1450** in *Archiv* (1931) CLIX. 88 Maners & clothyng makes man. **1509** A. BARCLAY *Ship of Fools* 118 An old prouerbe . . Sayth that good lyfe and maners makyth man. *a* **1661** T. FULLER

Worthies (Hants.) 3 Manners makes a man, Quoth William Wickham.[1] This generally was his Motto, inscribed frequently on the places of his Founding. **1721** J. KELLY *Scottish Proverbs* 246 Meat feeds, Cloth cleeds, but Manners makes the Man. . . Good Meat, and fine Cloaths, without good Breeding, are but poor Recommendations. **1824** BYRON *Don Juan* xv. 18 The difference is, that in days of old Men made the manners; manners now make men. **1966** I. MURDOCH *Time of Angels* xii. A single lamp revealed . . the very large glass-front bookcase superscribed *Manners Makyth Man*.

[1] William of Wykeham (1324–1404), bishop of Winchester and chancellor of England; he was the founder of Winchester College and New College, Oxford.

manners: *see* EVIL communications corrupt good manners; OTHER times, other manners.

MANY a little makes a mickle

The proper version of the next proverb. *Pickle*[1] is also found instead of *little*.

a **1250** *Ancrene Wisse* (1962) 32 Thys ofte as me seith, of lutel muchel waxeth. **1545** R. TAVERNER tr. *Erasmus' Adages* (ed. 2) G5 We commonly say in englyshe: Many a lyttle maketh a great. **1614** W. CAMDEN *Remains concerning Britain* (ed. 2) 310 Many a little makes a micle.[2] **1822** CARLYLE in J. A. Froude *Life* (1884) I. xii. 'Many a little makes a mickle.' It will be a long . . and weary job, but I must plod along. **1905** *Westminster Gazette* 29 Apr. 3 'There is the Tithe Relief. . . But that is a small item.' 'Yes, but many a pickle maks a muckle.'[2'] **1979** C. COLVIN *Maria Edgeworth in France & Switzerland* 196 Many a pickle (or little) makes a mickle.

[1] a small quantity or amount (Scottish).

[2] a great quantity or amount (now only Scottish).

MANY a mickle makes a muckle

A popular corruption of the preceding entry. This alternative form is in fact nonsensical, as *muckle* is merely a

variant of the dialectal *mickle* 'a large quantity or amount'.

1793 G. WASHINGTON *Writings* (1939) XXXII. 423 A Scotch addage, than which nothing in nature is more true . . 'many mickles make a muckle.' **1940** *Huntly Express* 19 Jan. 3 He said at the close of his address 'As the Scots say, and they should know, mony a mickle mak's a muckle.' . . As the Scots know, he had quoted the proverb wrongly. **1968** K. GILES *Death among Stars* v. Many a mickle makes a muckle, as Sir P. says. And Sir P. would be generous.

There's MANY a slip 'twixt cup and lip

[ERASMUS *Adages* I. v. 1 *multa cadunt inter calicem supremaque labra*, many things fall between the cup and the surface of the lips.] **1539** R. TAVERNER tr. *Erasmus' Adages* 15 Many thynges fall betwene the cuppe and the mouth. . . Betwene the cuppe and the lyppes maye come many casualties. **1783** in *Collections of Massachusetts Hist. Society* (1877) 5th Ser. II. 216 Have a care, and remember the old proverb of 'many a slip,' &c. **1840** R. H. BARHAM *Ingoldsby Legends* 1st Ser. 280 Doubtless the adage, 'There's many a slip 'Twixt the cup and the lip,' hath reference to medicine. **1934** F. W. CROFTS *12.30 from Croydon* xii. He had seen nothing in writing on the subject, and he could not help being worried by dark thoughts of slips between cups and lips. **1979** E. KYLE *Summer Scandal* xiii. 'I thought you were here for life.' . . 'There's many a slip between cup and lip.'

MANY hands make light work

[HESIOD *Works & Days* 380 πλείων μὲν πλεόνων μελέτη, more hands mean more work; ERASMUS *Adages* II. iii. 95 *multae manus onus levius reddunt*, many hands make a burden lighter.] *c* **1330** *Sir Beves* (EETS) l. 3352 Ascopard be strong & sterk,[1] Mani hondes maketh light werk! **1678** BUTLER *Hudibras* III. ii. Most Hands dispatch apace, And make light work, (the proverb says). **1721** J. KELLY *Scottish Proverbs* 244 Many Hands make light Work. Because it is

but little to every one. **1923** *Observer* 11 Feb. 9 What is the use of saying that 'Many hands make light work' when the same copy-book tells you that 'Too many cooks spoil the broth'? **1978** 'L. EGAN' *Dream Apart* ix. Well, Lorenzo . . , you back again? Got the family with you this time, hah? Many hands make light work.
¹ physically powerful.

many: *see also* there's many a GOOD tune played on an old fiddle; many a TRUE word is spoken in jest; many go out for WOOL and come home shorn.

MARCH comes in like a lion, and goes out like a lamb
The weather is traditionally wild at the beginning of March, but fair by the end.
a **1625** J. FLETCHER *Wife for Month* (1717) II. i. 'I would chuse March, for I would come in like a Lion.' . . 'But you'd go out like a Lamb when you went to hanging.' **1670** J. RAY *English Proverbs* 41 March hack ham comes in like a lion, goes out like a lamb. **1849** C. BRONTË *Shirley* II. iv. Charming and fascinating he resolved to be. Like March, having come in like a lion, he purposed to go out like a lamb. **1906** E. HOLDEN *Country Diary of Edwardian Lady* (1977) 25 March has come in like a lamb with a warm wind . . from the South-west. **1980** *Eastern Evening News* 3 Apr. 13 That old proverb of 'March coming in like a lion and going out like a lamb' proved to be quite the reverse.

March: *see also* (noun) APRIL showers bring forth May flowers; on the FIRST of March the crows begin to search; so many MISTS in March, so many frosts in May; a PECK of March dust is worth a king's ransom.

march: *see* (verb) an ARMY marches on its stomach.

mare: *see* the GREY mare is the better horse; MONEY makes the mare to go; NOTHING so bold as a blind mare.

market: *see* BUY in the cheapest market and sell in the dearest; a MONEYLESS man goes fast through the market.

MARRIAGE is a lottery
1642 T. FULLER *Holy State* III. xxii. Marriage shall prove no lottery to thee, when the hand of providence chuseth for thee, who, if drawing a blank, can turn into a prize by sanctifying a bad wife unto thee. **1875** SMILES *Thrift* xii. 'Marriage is a lottery.' It may be so, if we abjure the teachings of prudence. **1939** F. SULLIVAN *Sullivan at Bay* 14 What is marriage? . . Marriage is a lottery.

There goes more to MARRIAGE than four bare legs in a bed
c **1549** J. HEYWOOD *Dialogue of Proverbs* I. viii. B1 In house to kepe household, whan folks wyll wed, Mo thyngs belong, than foure bare legs in a bed. **1623** W. CAMDEN *Remains concerning Britain* (ed. 3) 273 Longs more to marriage then foure bare legges in a bed. **1738** SWIFT *Polite Conversation* I. 84 Consider, Mr. Neverout, Four bare Legs in a Bed; and you are a younger Brother. **1958** in M. L. Wolf *Dict. Painting* p. vii, As the old proverb has it, 'there goes more to marriage than four bare legs in a bed.'

marriage: *see also* DREAM of a funeral and you hear of a marriage.

MARRIAGES are made in heaven
1567 W. PAINTER *Palace of Pleasure* xxiii. True it is, that marriages be don in Heaven and performed in earth. **1580** LYLY *Euphues* II. 223 Mariages are made in heauen, though consumated in yearth. **1738** SWIFT *Polite Conversation* I. 78 They say, Marriages are made in Heaven; but I doubt, when she was married, she had no Friend there. **1932** S. GIBBONS *Cold Comfort Farm* i. I prefer the idea of arrangement to that other statement, that marriages are made in Heaven. **1980** 'S. WOODS' *Weep for Her* 187 She's a sentimental sort who believes marriages are made in heaven.

MARRY in haste and repent at leisure

  1568 E. TILNEY *Duties in Marriage* B4 Some haue loued in post hast, that afterwards haue repented them at leysure. **1615** J. DAY *Festivals* x. Marrying in hast, and Repenting by leasure. **1734** B. FRANKLIN *Poor Richard's Almanack* (May) Grief often treads upon the heels of pleasure, Marry'd in haste, we oft repent at leisure. **1872** W. STIRLING-MAXWELL *Works* (1891) VI. xvii. 'Marry in haste and repent at leisure' is a proverb that may be borne in mind with advantage in the choice of a party as well as of a wife. **1979** A. PRICE *Tomorrow's Ghost* ix. It stands to reason you know better what to look for the second time round—'Marry in haste and repent at leisure.'

MARRY in May, rue for aye

  Some earlier related proverbs are also illustrated below. Cf. MAY *chickens come cheeping*.

  [OVID *Fasti* v. 489 *si te proverbia tangunt, mense malum Maio nubere volgus ait*, if proverbs influence you, the common people say it is bad luck to marry in May.] **1675** *Poor Robin's Almanack* May, The Proverb saies . . Of all the Moneths 'tis worst to Wed in May. **1821** J. GALT *Annals of Parish* vi. We were married on the 29th day of April . . on account of the dread that we had of being married in May, for it is said, 'Of the marriages in May, The bairns die of a decay.' **1879** W. HENDERSON *Notes on Folk-Lore of Northern Counties* (rev. ed.) i. The ancient proverb still lives on the lips of the people of Scotland and the Borders—Marry in May, Rue for aye. **1913** E. M. WRIGHT *Rustic Speech* xiii. May . . is an evil month for marriage. . . Marry in May, you'll rue it for aye, is a Devonshire saying. **1981** *Observer Magazine* 28 June 27 On weddings and engagements we are told that May is an unlucky month for getting married, 'Marry in May, rue for aye'.

martin: *see* the ROBIN and the wren are God's cock and hen.

martyr: *see* the BLOOD of the martyrs is the seed of the Church.

mass: *see* MEAT and mass never hindered man.

Like MASTER, like man

  [PETRONIUS *Satyricon* lviii. *qualis dominus, talis et servus*, as is the master, so is the servant.] **1530** J. PALSGRAVE *L'éclaircissement de la Langue Française* 120ᵛ Suche maystre suche man.[1] **1538** ELYOT *Dict.* s.v. Similes, A lewde[2] servaunt with an yll master. . . Lyke master lyke man. **1620** T. SHELTON tr. *Cervantes' Don Quixote* II. x. The Prouerbe be true that sayes, like master, like man', and I may add, 'like lady, like maid'. Lady Hercules was fine, but her maid was still finer. **1979** M. G. EBERHART *Bayou Road* iv. 'Like master, like man,' Marcy's father had said bitterly . . of the disappearance of an entire set of Dresden plates.

  [1] servant.   [2] ill-mannered, foolish.

master: *see also* the EYE of a master does more work than both his hands; JACK is as good as his master; NO man can serve two masters.

MAY chickens come cheeping

  The proverb literally means that the weakness of chickens born in May is apparent from their continual feeble cries. Cf. MARRY *in May, rue for aye*.

  1868 A. HISLOP *Proverbs of Scotland* 223 May birds are aye cheeping. This refers to the popular superstition against marrying in . . May, the children of which marriages are said to 'die of decay'. **1895** S. O. ADDY *Household Tales* II. viii. Children born in the month of May require great care in bringing up, for 'May chickens come cheeping.'

May: *see also* (noun) APRIL showers bring forth May flowers; ne'er CAST a clout till May be out; MARRY in May, rue for aye; so many MISTS in March, so many frosts in May; a SWARM in May is worth a load of hay.

may: *see* (verb) he that WILL not when he may, when he will he shall have nay.

means: *see* the END justifies the means; he who WILLS the end, wills the means.

There is MEASURE in all things
Cf. MODERATION *in all things.*
[HORACE *Satires* I. i. *est modus in rebus*, there is measure in things.] c **1385** CHAUCER *Troilus & Criseyde* II. 715 In every thyng, I woot, there lith mesure.[1] **1598–9** SHAKESPEARE *Much Ado about Nothing* II. i. 59 If the prince be too important, tell him there is measure in every thing. **1616** T. DRAXE *Adages* 131 There is a measure in all things. **1910** KIPLING *Rewards & Fairies* 84 There's no clean hands in the trade. But steal in measure. . . There is measure in all things made.
[1] moderation.

MEAT and mass never hindered man
a **1628** J. CARMICHAELL *Proverbs in Scots* no. 134 A mease[1] of meat hinderit never man. **1639** J. CLARKE *Parœmiologia Anglo-Latina* 273 Meat and mattens hinder no mans journey. **1641** D. FERGUSSON *Scottish Proverbs* (STS) no. 644 Meat and masse never hindred no man. **1817** SCOTT *Rob Roy* III. ii. 'What the devil are ye in sic a hurry for?' said Garschattachin; 'meat and mass never hindered wark.' **1893** R. L. STEVENSON *Catriona* I. xix. Meat and mass never hindered man. The mass I cannot afford you, for we are all good Protestants. But the meat I press on your attention.
[1] mess, serving.

One man's MEAT is another man's poison
[LUCRETIUS *De Rerum Natura* iv. 637 *quod ali cibus est aliis fuat acre venenum*, what is food to one person may be bitter poison to others.] c **1576** T. WHYTHORNE *Autobiography* (1961) 203 On bodies meat iz an otherz poizon. **1604** *Plato's Cap* B4 That ould moth-eaten Prouerbe . . One mans meate, is another mans poyson. a **1721** M. PRIOR *Dialogues of Dead* (1907) 246 May I not nauseate the food which you Covet; and is it not even a Proverb, that what is meat to one Man is Poyson to another. **1883** TROLLOPE *Autobiography* x. It is more true of novels than perhaps of anything else, that one man's food is another man's poison. **1979** J. R. L. ANDERSON *Festival* viii. I see you . . don't use sugar or milk – well, one man's meat, as the old saying goes.

meat: *see also* you BUY land you buy stones, you buy meat you buy bones; GOD never sends mouths but He sends meat; GOD sends meat, but the Devil sends cooks; the NEARER the bone, the sweeter the meat.

meddler: see LAY-OVERS for meddlers.

Do not MEET troubles half-way
Cf. *Never* TROUBLE *trouble till trouble troubles you.*
[Cf. SENECA *Ad Lucilium* XIII. x. *quid iuvat dolori suo occurrere?* what help is it to run out to meet your troubles?; **1598–9** SHAKESPEARE *Much Ado about Nothing* I. i. 82 Are you come to meet your trouble? The fashion of the world is to avoid cost, and you encounter it.] **1896** J. C. HUTCHESON *Crown & Anchor* xvi. I can't see the use of anticipating the worst and trying to meet troubles half-way. **1940** M. SADLEIR *Fanny by Gaslight* III. ii. What happens when she goes? . . Do not meet troubles half way. . . When need arises we will see what can be done. **1980** G. THOMPSON *Murder Mystery* xx. Don't go meeting trouble half-way. There might just be something we can do.

meet: *see* EXTREMES meet; when GREEK meets Greek then comes the tug of war.

memory: *see* a LIAR ought to have a good memory.

So many MEN, so many opinions
[TERENCE *Phormio* II. iv. *quot homines tot sententiae*, so many men, so many opinions.] c **1390** CHAUCER *Squire's Tale* l. 203 As many heddes, as manye wittes

ther been. **1483** *Vulgaria abs Terencio* Q3ᵛ Many men many opinyons. Euery man has his guyse. **1692** R. L'ESTRANGE *Fables of Aesop* ccclviii. So many Men, so many Minds; and this Diversity of Thought must necessarily be attended with Folly, Vanity, and Error. **1754** RICHARDSON *Grandison* VI. xx. Doctors differ. So many persons, so many minds. **1924** 'A. CARP' *Augustus Carp, Esq.* xii. They were all those things, and they would remember the old saying, so many men, so many opinions.

men: *see also* the BEST-laid schemes of mice and men gang aft agley; the BEST of men are but men at best; BRAVE men lived before Agamemnon; the BUSIEST men have the most leisure; DEAD men don't bite; DEAD men tell no tales; GOOD men are scarce; one VOLUNTEER is worth two pressed men; THREATENED men live long; YOUNG men may die, but old men must die.

mend: *see* it is NEVER too late to mend; when THINGS are at the worst they begin to mend.

mended: *see* LEAST said, soonest mended.

mending: *see* a WOMAN and a ship ever want mending.

merrier: *see* the MORE the merrier.

It is MERRY in hall when beards wag all
    c **1300** *King Alisaunder* (EETS) l. 1164 Swithe[1] mury hit is in halle, When the burdes wawen alle! **1546** J. HEYWOOD *Dialogue of Proverbs* II. vii. I3ᵛ It is mery in halle, When berds wag all. **1598** SHAKESPEARE *Henry IV, Pt.* 2 v. iii. 35 Be merry, be merry, my wife has all. . . 'Tis merry in hall when beards wag all. **1738** SWIFT *Polite Conversation* II. 170 Come; they say, 'Tis merry in Hall, when Beards wag all. **1857** TROLLOPE *Barchester Towers* III. iv. ''Twas merry in the hall when the beards wagged all;' and the clerical beards wagged merrily . . that day. **1970** G. E. EVANS (*title*) Where beards wag all.
    [1] SO.

merry: *see also* a CHERRY year a merry year.

Merryman: *see* the best DOCTORS are Dr. Diet, Dr. Quiet, and Dr. Merryman.

mice: *see* the BEST-laid schemes of mice and men gang aft agley; a CAT in gloves catches no mice; when the CAT's away, the mice will play; KEEP no more cats than will catch mice.

mickle: *see* MANY a little makes a mickle; MANY a mickle makes a muckle.

midge: *see* the MOTHER of mischief is no bigger than a midge's wing.

midnight: *see* one HOUR's sleep before midnight is worth two after.

mid-stream: *see* don't CHANGE horses in mid-stream.

MIGHT is right
    [PLATO *Republic* I. 338c φημὶ γὰρ ἐγὼ εἶναι τὸ δίκαιον οὐκ ἄλλο τι ἢ τὸ τοῦ κρείττονος ξυμφέρον, for I [Thrasymachus] say that justice is nothing else than the interest of the stronger; LUCAN *Pharsalia* I. 175 *mensuraque iuris vis erat*, might was the measure of right.] *a* **1327** in T. Wright *Political Songs* (1839) 254 For miht is right, the lond is laweless. **1546** J. HEYWOOD *Dialogue of Proverbs* II. v. H2ᵛ We se many tymes myght ouercomth ryght. **1790** J. TRUSLER *Proverbs Exemplified* 78 The law is so expensive. . . Might too often overcomes right. **1892** J. NICHOL *Carlyle* iv. [In] *Chartism* . . he clearly enunciates 'Might is right'—one of the few strings on which . . he played through life. **1979** *Guardian* 17 May 24 By adult examples, pupils are being taught such evil doctrines as 'Might is right'.

mightier: *see* the PEN is mightier than the sword.

mighty: *see* a REED before the wind lives on, while mighty oaks do fall.

mile: *see* a MISS is as good as a mile.

milk: see why buy a cow when milk is so cheap?; it is no use crying over spilt milk.

The MILL cannot grind with the water that is past

1616 T. DRAXE *Adages* 151 The water that is past, cannot make the mill goe. **1640** G. HERBERT *Outlandish Proverbs* no. 153 The mill cannot grind with the water that's past. **1856** R. C. TRENCH *Poems* 197 Oh seize the instant[1] time; you never will With waters once passed by impel the mill. **1980** G. RICHARDS *Red Kill* xiv. It did no good to think back. The mill cannot grind with the water that is past, as the old people in the mountains used to say.
[1] present.

mill: see also all is grist that comes to the mill.

The MILLS of God grind slowly, yet they grind exceeding small

[Quoted in SEXTUS EMPIRICUS *Against Professors* I. 287 ὀψὲ θεῶν ἀλέουσι μύλοι, ἀλέουσι δὲ λεπτά, the mills of the gods are late to grind, but they grind small.] **1640** G. HERBERT *Outlandish Proverbs* no. 747 Gods Mill grinds slow, but sure. **1870** LONGFELLOW *Poems* (1960) 331 Though the mills of God grind slowly, yet they grind exceeding small; Though with patience he stands waiting, with exactness grinds he all. **1942** 'F. BEEDING' *Twelve Disguises* i. That's my business. . . The mills of God grind slowly, but they grind exceeding small. **1979** G. SWARTOUT *Skeletons* 81 The law should have been allowed to take its course. The mills of the gods grind slowly, it is true, but in time, in time—.

mind: see the EYES are the window of the soul; GREAT minds think alike; LITTLE things please little minds; OUT of sight, out of mind; TRAVEL broadens the mind.

The age of MIRACLES is past

1599 SHAKESPEARE *Henry V* I. i. 67

It must be so; for miracles are ceas'd; And therefore we must needs admit the means How things are perfected. **1602**——*All's Well that ends Well* II. iii. 1 They say miracles are past; and we have our philosophical persons to make modern and familiar things supernatural and causeless. **1840** CARLYLE *On Heroes & Hero Worship* iv. The Age of Miracles past? The Age of Miracles is for ever here! **1974** A. PRICE *Other Paths to Glory* II. viii. Of course he found nothing. . . The age of miracles is long past.

mischief: see the MOTHER of mischief is no bigger than a midge's wing.

MISERY loves company

Now predominantly current in the United States.

[a **1349** R. ROLLE *Meditations on Passion* in C. Horstmann *Yorkshire Writers* (1895) I. 101 It is solace to haue companie in peyne.] **1578** LYLY *Euphues* I. 238 In miserie Euphues it is a great comfort to haue a companion. **1620** T. SHELTON tr. *Cervantes' Don Quixote* II. xiii. If that which is commonly spoken be true, that to haue companions in misery is a lightner of it, you may comfort me. **1775** T. GILBERT *Letter* 4 May in W. B. Clark et al. *Naval Documents of American Revolution* (1964) I. 279 All my Letters are intersepted by those Rebels who want Every one to be kept in Dark like themselves. (Misery Loves Company). **1851** H. D. THOREAU *Journal* 1 Sept. (1949) II. 440 If misery loves company, misery has company enough. **1979** M. McCARTHY *Cannibals & Missionaries* v. Misery loved company, and their own band of liberals could stand some diversification.

MISFORTUNES never come singly

c **1300** *King Alisaunder* (EETS) l. 1276 Men telleth in olde mone[1] The qued[2] comuth nowher alone. **1509** A. BARCLAY *Ship of Fools* 236 Wyse men sayth, and oft it fallyth so . . That one myshap fortuneth neuer alone. **1622** J. MABBE tr. *Aleman's Guzman d'Alfarache* I. iii. Misfortunes seldome come alone. **1711** ADDISON *Spectator* 8 Mar., The Lady . .

said to her Husband with a Sigh, My Dear, Misfortunes never come single. **1791** T. BURR *Letter* 27 July in M. L. Davis *Memoirs of Aaron Burr* (1836) I. 301 We certainly see the old proverb very often verified. 'That misfortunes never come singly,' that poor little woman is a proof. **1894** BLACKMORE *Perlycross* II. vii. As misfortunes never come single, the sacred day robbed him of another fine resource. **1931** 'L. CHARTERIS' *Wanted for Murder* v. Blessings, like misfortunes, never come singly. There was even a packet of Havana cigarettes . . behind the bath salts. **1981** G. MITCHELL *Death-Cap Dancers* v. 'The car . . skidded and hit a tree.' 'Misfortunes never come singly.'

[1] lament.   [2] harm.

A MISS is as good as a mile

The syntax of the proverb has been distorted by abridgement: the original structure is apparent from quot. 1614.

1614 W. CAMDEN *Remains concerning Britain* (ed. 2) 303 An ynche in a misse is as good as an ell.[1] **1655** T. FULLER *Hist. Cambridge* 37 An hairs breadth fixed by a divine-finger, shall prove as effectuall a separation from danger as a miles distance. **1788** *American Museum* Apr. 382 A miss is as good as a mile. **1825** SCOTT *Journal* 3 Dec. (1939) 28 He was very near being a poet—but a miss is as good as a mile, and he always fell short of the mark. **1978** T. SHARPE *Throwback* vii. If you aimed at a grouse it was hit or miss and a miss was as good as a mile.

[1] a measure of length normally equivalent to forty-five inches.

You never MISS the water till the well runs dry

a **1628** J. CARMICHAELL *Proverbs in Scots* no. 1140 Manie wats[1] not quhairof[2] the wel sauris[3] quhill[4] it fall drie. **1659** J. HOWELL *Proverbs* (British) 24 Of the Well we see no want, till either dry, or Water skant. **1721** J. KELLY *Scottish Proverbs* 351 We'll never know the worth of Water 'till the well go dry. **1874** H. LINN *You never miss Water* 5 Do not let your chances, like sunbeams pass you by; For

you never miss the water till the well runs dry. **1971** V. PETERS *Competition for Alan* x. It's a case of not missing the water until the well runs dry.

[1] know.   [2] whereof.   [3] i.e. savours; tastes.   [4] until.

missed: *see* a SLICE off a cut loaf isn't missed.

If you don't make MISTAKES you don't make anything

1896 CONRAD *Outcast of Islands* III. ii. It's only those who do nothing that make no mistakes, I suppose. **1925** *Times* 9 Nov. 17 The comforting assurance that 'a man who never makes mistakes never makes anything.' **1980** M. DRABBLE *Middle Ground* 86 If you don't make mistakes you don't make anything, she said, a motto which Hugo seemed to remember having seen pinned over the desk.

So many MISTS in March, so many frosts in May

1612 A. HOPTON *Concordancy of Years* xxx. Some say, so many mistes in March, so many hoare frosts after Easter. **1678** J. RAY *English Proverbs* (ed. 2) 344 So many frosts in March so many in May. **1978** R. WHITLOCK *Calendar of Country Customs* iii. Many old country beliefs are not content with generalities but strive to be more precise. A well-known proverb is: So many mists in March, So many frosts in May.

mixen: *see* BETTER wed over the mixen than over the moor.

MODERATION in all things

Cf. *There is* MEASURE *in all things.*

[HESIOD *Works & Days* 694 μέτρα φυλάσσεσθαι: καιρὸς δ'ἐπὶ πᾶσιν ἄριστος, observe due measure; moderation is best in all things; PLAUTUS *Poenulus* 238 *modus omnibus rebus . . optimus est habitu,* moderation in all things is the best policy.] **1849** MELVILLE *Mardi* II. lxxvii. I am for being temperate in these things. . . All things in moderation are good; whence, wine in moderation is good. **1879** W. H. G. KINGSTON tr. *Swiss Family Robinson* ii. 'Oh, father, sugar canes. . . Do let us take a lot home to mother.'. . 'Gently there. . . Moderation in all things.' **1980** S. T. HAYMON *Death & Pregnant Virgin* ii. Norfolk . . [is] on the same scale I am. No Niagaras, no hills higher than hills . . ought to be. Moderation in all things.

Monday: *see* Monday's CHILD is fair of face.

MONEY has no smell

L. *non olet*, it [money] does not smell: founded upon a remark of Titus, Vespasian's son, who had criticized a tax on public lavatories. When Vespasian held a coin from the first payment to his son's nose and asked him whether the smell was offensive, Titus said no. Vespasian replied 'And yet it comes from urine' (Suetonius *Vespasian* xxiii.).

1922 A. BENNETT *Mr. Prohack* iii. The associations of the wealth scarcely affected him. He understood in the flesh the deep wisdom of that old proverb . . that money has no smell. **1940** R. CHANDLER *Farewell, my Lovely* xxxiv. He punched the cash-register and dropped the bill into the drawer. They say money don't stink. I sometimes wonder.

MONEY isn't everything

1927 E. O'NEILL *Marco Millions* iii. Money isn't everything, not always. *a* **1947** F. THOMPSON *Still glides Stream* (1948) ii. He said quite angrily that money was not everything, there was the satisfaction of knowing you'd turned out a good job. **1975** J. I. M. STEWART *Young Pattullo* xv. If one owns property one can always have a little money follow one around. But we all know that money isn't everything.

MONEY is power

Cf. KNOWLEDGE *is power.*

1741 N. AMES *Almanack* 4 Laws bear Name, but Money has the Power. **1789** F. AMES *Letter* 16 May in *Works* (1854) I. 39 Money is power, a permanent revenue is permanent power, and the cre-

dit which it would give was a safeguard to the government. **1818** M. EDGEWORTH *Letter* 13 Oct. (1971) 115 Now he had money 'and money is power'. **1930** MEANS & THACKER *Strange Death of President Harding* iv. One can do nothing— be nothing, without money, not even in the White House. Money is power. **1980** J. O'FAOLAIN *No Country for Young Men* i. The lads would have to have . . money if they were to get guns. . . Money was power.

MONEY is the root of all evil

Cf. IDLENESS *is the root of all evil.*

[I TIMOTHY vi. 10 The loue of money is the root of all euill.] *c* **1000** AELFRIC *Homilies* (1843) I. 256 Seo gytsung is ealra yfelra thinga wyrtruma [covetousness is the root of all evil things]. *c* **1449** R. PECOCK *Repressor of Blaming of Clergy* (1860) II. 555 Loue to money . . is worthi to be forborn . . as Poul seith, it is 'the roote of al yuel'. **1616** J. WITHALS *Dict.* (rev. ed.) 546 Riches are the root of all euill. **1777** in L. H. Butterfield et al. *Adams Family Correspondence* (1963) II. 345 Many have been loth to believe . . That Money is the Root of all Evil. **1858** TROLLOPE *Dr. Thorne* I. xii. 'But, doctor, you'll take the money.'. . 'Quite impossible . .' said the doctor,. . valiantly rejecting the root of all evil. **1978** W. L. DEANDREA *Killed in Ratings* ii. Magazines have got these funny little sayings. . . Here's one. 'Money is the root of all evil . . but that's one evil I'm rooting for.'

MONEY makes a man

[L. *divitiae virum faciunt,* wealth makes the man.] *a* **1500** in R. L. Greene *Early English Carols* (1935) 263 Yt ys allwayes sene nowadayes That money makythe the man. *a* **1661** T. FULLER *Worthies* (Hants.) 3 We commonly say . . In the Change,[1] Money makes a man, which puts him in a solvable condition. **1828** BULWER-LYTTON *Pelham* I. xxxiv. The continent only does for us English people to see. . . Here, you know, 'money makes the man.' **1920** D. H. LAWRENCE *Letter* 7 May (1962) I. 629

Money maketh a man; even if he was a monkey to start with.

[1] Exchange.

MONEY makes money

1572 T. WILSON *Discourse upon Usury* 54ᵛ Mony getteth money. *a* **1654** J. SELDEN *Table-Talk* (1689) 57 'Tis a vain thing to say, Money begets not Money; for that no doubt it does. **1776** A. SMITH *Wealth of Nations* I. I. ix. Money, says the proverb, makes money. When you have got a little, it is often easy to get more. **1865** DICKENS *Our Mutual Friend* III. v. We have got to recollect that money makes money, as well as makes everything else. **1935** A. CHRISTIE *Miss Marple's Final Cases* (1979) 60 Everything she did turned out well. Money made money.

MONEY makes the mare to go

a **1500** in R. L. Greene *Early English Carols* (1935) 262 In the heyweyes[1] ther joly[2] palfreys Yt [money] makyght to . . praunce. **1573** J. SANFORDE *Garden of Pleasure* 105ᵛ Money makes the horsse to goe. **1670** J. RAY *English Proverbs* 122 It's money makes the mare to go. **1857** KINGSLEY *Two Years Ago* p. xvi. I'm making the mare go here . . without the money too, sometimes. I'm steward now. **1930** L. MEYNELL *Mystery at Newton Ferry* xiii. 'Tis money makes the mare go. . . They're all after it, every one of them. **1978** *Countryman* Spring 193 Weardale farmer's advice to daughter about to reject a proposal of marriage from a wealthy tradesman: 'Never cock your snoop at money, my lass, 'cos it's money that makes the mare to go'.

[1] i.e. highways. [2] fresh, spirited.

MONEY talks

1666 G. TORRIANO *Italian Proverbs* 179 Man prates, but gold speaks. **1681** A. BEHN *Rover* II. III. i. Money speaks in a Language all Nations understand. **1903** *Saturday Evening Post* 5 Sept. 12 When money talks it often merely remarks 'Good-by'. **1915** P. G. WODEHOUSE *Something Fresh* iii. The whole story took on a different complexion for Joan. Money talks.[1] **1979** P. WAY *Sunrise* xiii. Money talks. So does Her Majesty's Government. Even today.

[1] has influence.

money: *see also* BAD money drives out good; a FOOL and his money are soon parted; where there's MUCK there's brass; you PAYS your money and you takes your choice; TIME is money.

A MONEYLESS man goes fast through the market

The proverb is explained in quot. 1721. The last example is a variation of the original proverb, asserting that a person rushes to wherever what he lacks may be found.

1721 J. KELLY *Scottish Proverbs* 10 A silverless Man goes fast through the Market. Because he does not stay to cheapen[1] or buy. **1732** T. FULLER *Gnomologia* no. 330 A Moneyless Man goes fast thro' the Market. **1977** J. AIKEN *Five-Minute Marriage* iv. Found your way here at last, then, miss, have you? A moneyless mare trots fast to the market.

[1] haggle or bargain for something.

monk: *see* the COWL does not make the monk.

monkey: *see* the HIGHER the monkey climbs, the more he shows his tail; if you PAY peanuts, you get monkeys; SOFTLY, softly, catchee monkey.

moon: *see* NO moon, no man.

moor: *see* BETTER wed over the mixen than over the moor.

MORE people know Tom Fool than Tom Fool knows

Tom Fool is a name traditionally given to a simpleton, or to one who acts the part of a fool, as in a drama or morris dance.

1656 S. HOLLAND *Wit & Fancy* II. i. In all Comedies more know the Clown, then the Clown knows. **1723** DEFOE *Colonel Jack* (ed. 2) 347 It was no satisfaction

to me that I knew not their faces, for
they might know mine . . according to
the old English proverb, 'that more
knows Tom Fool, than Tom Fool
knows'. **1922** *Times* 15 Nov. 10 One of
the candidates . . sent me his election
address. . . More people, thought I,
know Tom Fool than Tom Fool knows.
1980 L. MEYNELL *Hooky & Prancing Horse*
iv. Hooky asked . . 'How's the great
pulsating world of journalism?' Mac
was . . surprised; but he consoled himself with the thought that more people
know Tom Fool than Tom Fool knows.

The MORE the merrier
c **1380** *Pearl* (1953) l. 850 The mo[1] the
myryer, so God me blesse. **1546** J.
HEYWOOD *Dialogue of Proverbs* II. vii. I3
The mo the merier, we all daie here[2] and
se. Ye but the fewer the better fare (saied
he). **1614** T. ADAMS *Devil's Banquet* IV.
196 The company is . . all the Pat-
riarchs, Prophets, Saints. . . The more
the mirrier, yea, and the better cheare
to. **1855** KINGSLEY *Westward Ho!* III. iv.
The old proverb comes true—'the more
the merrier: but the fewer the better
fare.' **1976** L. ALTHER *Kinflicks* xiii.
'Take my word for it. Have another
baby.' . . 'The more, the merrier!'
[1] more.   [2] i.e. hear.

The MORE you get, the more you want
Cf. MUCH *would have more.*
[HORACE *Epistles* II. ii. *quanto plura
parasti, tanto plura cupis,* you want as
much again as you have already got.]
c **1340** R. ROLLE *Psalter* (1884) 97 The
mare that a man has the brennandere[1]
he askis. *a* **1450** *Castle of Perseverance* l.
3268 in *Macro Plays* (EETS) The more he
hadde, the more he cravyd, Whyl the lyf
lefte hym with-Inne. **1578** J. FLORIO
First Fruits 32 The more a man hath, the
more he desireth. **1798** W. MANNING
Key of Liberty (1922) 9 In short he is never
easy, but the more he has the more he
wants. **1940** G. H. COXE *Glass Triangle* x.
I was averaging eighty to a hundred
[dollars] a week. Well, you know how it
is. The more you get the more you want.
[1] more burningly.

more: *see also* more HASTE, less speed;
MUCH would have more; there are more
WAYS of killing a cat than choking it with
cream; there are more WAYS of killing a
dog than choking it with butter; there
are more WAYS of killing a dog than
hanging it.

MORNING dreams come true
[HORACE *Satires* I. x. *post mediam noc-
tem visus, cum somnia vera,* he appeared
to me after midnight, when dreams are
true.] **1540** J. PALSGRAVE *Acolastus*
II. i. After mydnyght men saye, that
dreames be true. **1616** JONSON *Love Restored* VIII. 385 All the morning dreames
are true. **1813** W. B. RHODES *Bombastes
Furioso* III. 7 This morn . . I dreamt (and
morning dreams come true, they say).
1909 A. MACLAREN *Romans* 87 Our high-
est anticipations and desires are not
unsubstantial visions, but morning
dreams, which are proverbially sure to
be fulfilled.

morning: *see also* RED sky at night
shepherd's delight, red sky in the morning shepherd's warning.

moss: *see* a ROLLING stone gathers no
moss.

Like MOTHER, like daughter
Cf. *Like* FATHER, *like son.*
[EZEKIEL xvi. 44 Euery one . . shall vse
this prouerbe against thee, saying, As is
the mother, so is her daughter.] *a* **1325**
Cursor Mundi (EETS) l. 18857 O suilk[1] a
moder, wel slik[1] a child. **1474** CAXTON
Game of Chess II. ii. For suche moder
suche doughter comunely. **1644** R. WIL-
LIAMS *Bloody Tenent of Persecution* xcix.
Is not this as the Prophet speaks, Like
mother, like daughter? **1861** C. READE
Cloister & Hearth II. xvii. 'Mother, you
were so hot against her.'. . 'Ay. . . Like
mother like daughter: cowardice it is our
bane.' **1960** J. PHILIPS *Whisper Town* II.
ii. She had a reputation . . of being
'fast'. 'Like mother, like daughter,'
people said.
[1] such.

The MOTHER of mischief is no bigger than a midge's wing

a **1628** J. CARMICHAELL *Proverbs in Scots* no. 1468 The mother of mischeif, is na mair nor[1] a midgewing. **1796** M. EDGEWORTH *Parent's Assistant* (ed. 2) 149 'The mother of mischief', says an old proverb, 'is no bigger than a midge's wing.' **1858** D. M. MULOCK *Woman's Thoughts about Women* viii. Fatal and vile as her [Gossip's] progeny may be, 'the mother of mischief', says the proverb, 'is no bigger than a midge's wing.'
[1] than.

mother: *see also* DILIGENCE is the mother of good luck; NECESSITY is the mother of invention; PRAISE the child, and you make love to the mother.

If the MOUNTAIN will not come to Mahomet, Mahomet must go to the mountain

1625 BACON *Essays* 'Of Boldness' xii. Mahomet cald the Hill to come to him . . And when the Hill stood still, he was neuer a whit abashed, but said; If the Hill will not come to Mahomet, Mahomet wil go to the hil. **1732** T. FULLER *Gnomologia* no. 2707 If the Mountain will not come to Mahomet, Mahomet must go to the Mountain. **1975** D. BAGLEY *Snow Tiger* xvii. You couldn't go to see him, so the mountain had to go to Mahomet. It was . . important to him.

mountain: *see also* FAITH will move mountains.

A MOUSE may help a lion

The proverb alludes to Aesop's fable of the lion and the rat, which is told by Caxton in *Fables* (1484) 40.

1563 *Mirror for Magistrates* (1938) 274 The mouse may sometyme help the Lyon in nede. . . O prynces seke no foes. **1732** T. FULLER *Gnomologia* no. 264 A Lyon may come to be beholding to a Mouse. **1842** MARRYAT *Percival Keene* I. xvii. A mouse may help a lion, as the fable says. **1935** J. BUCHAN *House of Four Winds* xi. I only offer to show my gratitude by doing what I can. . . A mouse may help a lion.

mouse: *see also* ONE for the mouse, one for the crow; *also* MICE.

mouth: *see* never look a GIFT horse in the mouth; GOD never sends mouths but He sends meat; a sow may whistle, though it has an ill mouth for it.

Out of the MOUTHS of babes—

Young children may speak disconcertingly wisely at times. The proverb is used in a variety of abbreviated and allusive forms, often without a knowledge of the complete Biblical quotations.

[PSALMS viii. 2 Out of the mouth of babes and sucklings hath thou ordained strength; MATTHEW xxi. 16 Iesus saith vnto them [the Pharisees], Yea, haue yee neuer read, Out of the mouth of babes and sucklings thou hast perfected praise.] **1899** KIPLING *Stalky & Co.* 11 In the present state of education I shouldn't have thought any three boys would be well enough grounded. . . But out of the mouths—. **1906** —— *Puck of Pook's Hill* 285 Out of the mouths of babes do we learn. **1979** 'C. AIRD' *Some die Eloquent* xviii. It was something Crosby said. . . 'About the source of the money.' 'Out of the mouths,' conceded Leeyes.

move: *see* FAITH will move mountains.

MUCH cry and little wool

a **1475** J. FORTESCUE *On Governance of England* (1885) x. His hyghnes shall haue theroff, but as hadd the man that sherid is[1] hogge, much crye and litil woll. **1659** J. HOWELL *Proverbs* (English) 13 A Great cry and little wooll, quoth the Devil when he sheard the hogg. **1711** ADDISON *Spectator* 18 Dec., Those . . make the most noise, who have least to sell . . to whom I cannot but apply that old Proverb of *Much cry, but little wool.* **1922** *Punch* 29 Nov. 520 Ministers have taken good care that the adage,

'Much cry and little wool,' shall not apply to them.

[1] *sherid is*, sheared his.

MUCH would have more

Cf. *The* MORE *you get, the more you want.*

c **1350** *Douce MS* 52 no. 65 Mykull[1] wulle more. *a* **1400** *Wars of Alexander* (EETS) l. 4397 Mekill wald have mare as many man spellis.[2] **1597** T. MORLEY *Plain Introduction to Music* II. 70 The Common Prouerb is in me verified, that much would have more. **1732** T. FULLER *Gnomologia* no. 3487 Much would have more; but often meets with less. **1897** J. MCCARTHY *Hist. Own Times* V. 131 Expedition after expedition has been sent out to extend the Egyptian frontier. . . 'Much will have more,' the old proverb says; but in this case . . much is compelled for the sake of . . security to try to have more.

[1] much.   [2] tells.

much: *see also* you can have TOO much of a good thing.

Where there's MUCK there's brass

1678 J. RAY *English Proverbs* (ed. 2) 179 Muck and money go together. **1855** H. G. BOHN *Hand-Book of Proverbs* 564 Where there is muck there is money. **1943** J. W. DAY *Farming Adventure* xii. 'Where there's muck there's money' is as true now as then. But farms today lack the mud. **1967** *Punch* 13 Sept. 396 'Where there's muck there's brass[1]' synopsised for many a North-country businessman the value of dirt in the profit-making process. **1980** *Cosmopolitan* Dec. 4 Our report [on blue-collar jobs for women] . . should bury the myth that women don't like getting dirt on their hands. . . Where there's muck there's brass.

[1] money (slang and dialectal).

muckle: *see* MANY a mickle makes a muckle.

mud: *see* throw DIRT enough, and some will stick.

MURDER will out

Cf. TRUTH *will out.*

c **1325** *Cursor Mundi* (EETS) l. 1084 For-thi[1] men sais into this tyde,[2] Is no man that murthir may hide. *c* **1390** CHAUCER *Nun's Priest's Tale* l. 4242 Mordre wol out that se we day by day. **1596** SHAKESPEARE *Merchant of Venice* II. ii. 73 Truth will come to light; murder cannot be hid long. **1860** W. COLLINS *Woman in White* II. 64 Crimes cause their own detection, do they? And murder will out (another moral epigram), will it? **1978** F. NEUMANN *Seclusion Room* ix. 'Murder will out,' Berman announced, smiling fatuously.

[1] therefore.   [2] time.

murder: *see also* KILLING no murder.

Murphy's law: *see* if ANYTHING can go wrong, it will; the BREAD never falls but on its buttered side.

What MUST be, must be

Cf. Ital. *che sarà sarà,* what will be, will be: (this English form is also used).

c **1390** CHAUCER *Knight's Tale* l. 1466 Whan a thyng is shapen, it shal be. **1519** W. HORMAN *Vulgaria* 20[v] That the whiche muste be wyll be. **1546** J. HEYWOOD *Dialogue of Proverbs* II. i. F3 That shalbe, shalbe. **1616** BEAUMONT & FLETCHER *Scornful Lady* III. i. I must kiss you. . . What must be, must be. **1841** S. WARREN *Ten Thousand a Year* I. i. It's really very inconvenient . . for any of my young men to be absent . . but—I suppose—what must be must be. **1850** DICKENS *David Copperfield* lvii. 'My love,' observed Mr. Micawber, '. . I am always willing to defer to your good sense. What will be—will be.' **1938** N. STREATFEILD *Circus is Coming* ii. Peter is sensible. . . He knows what must be must be. **1981** J. BINGHAM *Brock* 70 Oh well, what must be, must be.

N

nail: *see* ONE nail drives out another; for WANT of a nail the shoe was lost.

name: *see* give a DOG a bad name and hang him; he that has an ILL name is half hanged; NO names, no pack-drill.

nation: *see* happy is the COUNTRY which has no history; the ENGLISH are a nation of shopkeepers.

NATURE abhors a vacuum
[L. *natura abhorret vacuum*, Nature abhors a vacuum.] **1551** CRANMER *Answer to Gardiner* 299 Naturall reason abhorreth vacuum. **1642** T. FULLER *Holy State* v. ii. Queen Joan . . (hating widowhood as much as Nature doth vacuum) maried James King of Majorca. **1686** R. BOYLE *Free Inquiry* VII. 292 The Axiom of the Schools, that Nature abhors a Vacuum. **1771** S. JOHNSON *Letter* 20 June (1952) I. 249 Whatever philosophy may determine of material nature, it is certainly true of intellectual nature, that it abhors a vacuum: our minds cannot be empty. **1975** A. PRICE *Our Man in Camelot* 10 The vacuity of Major Davies's personal file . . was damning. Because, like nature, the CIA abhorred a vacuum.

nature: *see also* you can DRIVE out Nature with a pitchfork, but she keeps on coming back; SELF-preservation is the first law of nature.

nay: *see* he that WILL not when he may, when he will he shall have nay.

NEAR is my kirtle, but nearer is my smock
A justification for looking after one's own closest interests. See also the next proverb.
[PLAUTUS *tunica propior palliost*, my tunic is closer than my cloak.] **1461** *Paston Letters* (1976) II. 228 Nere is my kyrtyl,[1] but nerre is my smok.[2] **1545** R. TAVERNER tr. *Erasmus' Adages* (ed. 2) B7ᵛ The Englysshe prouerbe sayethe thus: nere is my cote, but nerer is my shyrt. **1622** J. HOWELL *Familiar Letters* 1 May (1903) I. 126 That king . . having too many irons in the fire at his own home . . answered them that his shirt was nearer to him than his doublet. **1861** C. READE *Cloister & Hearth* IV. xxix. You must not think all of him and none of yourself. Near is your kirtle, but nearer is your smock.
[1] a woman's gown or skirt.   [2] a woman's undergarment.

NEAR is my shirt, but nearer is my skin
See also the preceding proverb.
c **1570** in J. P. Collier *Old Ballads* (1840) 99 Neerer is my skin then shirte. **1631** J. HENSHAW *Spare Hours of Meditations* 63 His charitie beginnes at home, and there it ends; neere is his coat, but neerer is his skinne. **1712** J. ARBUTHNOT *Lewis Baboon* v. My Shirt (quoth he) is near me, but my Skin is nearer: Whilst I take care of the Welfare of other Folks, no body can blame me, to apply a little Balsam to my own Sores. **1890** T. H. HALL CAINE *Bondman* II. x. 'We can't trust you.' . . 'Not your own brother?' said Jacob. ' "Near is my shirt, but nearer is my skin," as the saying is.'

The NEARER the bone, the sweeter the meat
a **1398** J. TREVISA tr. *Bartholomew's On Properties of Things* (1975) XIX. 1 The nerer the boon the swetter is the flesshe. *a* **1661** T. FULLER *Worthies* (Wales) 2 As the sweetest flesh is said to be nearest the bones, so most delicious vallies are interposed betwixt these Mountains. **1778** in B. Franklin *Writings* (1906) VIII. 258 We all agree the nearer the bone the sweeter the meat. **1945** F. THOMPSON *Lark Rise* i. 'The nearer the bone the sweeter the meat,' they used to say, and they were getting very near the bone. . . Their children . . would have to depend wholly upon

whatever was carved for them from the communal joint. **1979** TREVANIAN *Shiborni* I. 38 A little skinny . . for my taste, but, like my ol' daddy used to say: the closer the bone, the sweeter the meat.

The NEARER the church, the farther from God

1303 R. BRUNNE *Handlyng Synne* (EETS) l. 9242 Tharfor men seye, an weyl ys trowed,[1] 'the nere the cherche, the fyrther fro God.' **1620** T. SHELTON tr. *Cervantes' Don Quixote* II. xlvii. Eat nothing of all this meat . . for this dinner was presented by Nunnes, and it is an olde saying, The neerer the Church, the farther from God. **1879** J. E. HOPKINS *Work amongst Working Men* i. I fear it was a practical comment on the truth of the uncomfortable proverb, 'The nearer the church, the farther from God,' that so bad a district should adjoin one of the great head-quarters of the church.

[1] believed.

NECESSITY is the mother of invention

[**1519** W. HORMAN *Vulgaria* 52 Nede taught hym wytte. *Necessitas ingenium dedit.*] **1545** R. ASCHAM *Toxophilus* II. 18^v Necessitie, the inuentor of all goodnesse (as all authours in a maner, doo saye) . . inuented a shaft heed. **1608** G. CHAPMAN *Tragedy of Byron* IV. i. The great Mother, Of all productions (graue Necessity). **1658** R. FRANCK *Northern Memoirs* (1694) 44 Art imitates Nature, and Necessity is the Mother of Invention. **1726** SWIFT *Gulliver's Travels* IV. x. I soaled my Shoes with wood, which I cut from a Tree. . . No man could more verify the Truth . . That, Necessity is the Mother of Invention. **1861** C. READE *Cloister & Hearth* II. vi. 'But, dame, I found language too poor to paint him. I was fain to invent. You know Necessity is the mother of—.' 'Ay! ay, that is old enough, o' conscience.' **1974** A. PRICE *Other Paths to Glory* I. vi. Necessity has been once more the mother of invention. I have invented Captain Lefevre.

NECESSITY knows no law

[L. *Necessitas non habet legem,* Necessity has no law.] **1377** LANGLAND *Piers Plowman* B. XX. 10 Nede ne hath no lawe, ne neure shal falle in dette. *c* **1530** TYNDALE *Answer to More* B1 Two things are without law, God and necessity. *a* **1555** N. RIDLEY *Lamentation of Miserable Estate of Church* (1556) D4 The latter reason . . includeth a necessitie which after the common sayinge hathe no law. **1680** DRYDEN *Kind Keeper* III. ii. Necessity has no Law; I must be patient. **1776** F. RHINELANDER *Letter* 23 Feb. in H. C. Van Schaack *Life of Peter Van Schaack* (1842) 54 Troops . . quarter themselves in any houses they find shut up. Necessity knows no law. **1864** MRS. H. WOOD *Trevlyn Hold* II. xiv. Necessity has no law, and he was obliged to rise. **1939** 'D. YATES' *Gale Warning* vi. ' "Don't speak to the man at the wheel" is a very good rule.' 'So', said I, 'is "Necessity knows no law."'

need: *see* (noun) a FRIEND in need is a friend indeed; (verb) GOOD wine needs no bush; a GUILTY conscience needs no accuser.

NEEDS must when the Devil drives

c **1450** LYDGATE *Assembly of Gods* (EETS) l. 21 Hit ys oft seyde by hem that yet lyues He must nedys go that the deuell dryues. **1602** SHAKESPEARE *All's Well that ends Well* I. iii. 29 He must needs go that the devil drives. **1835** SOUTHEY *Doctor* III. lxxxiii. Needs must go when the Devil drives. **1843** R. S. SURTEES *Handley Cross* III. xi. Needs must[1] when the devil drives! . . But I'd rather do any thing than injure that poor blue-eyed beauty. **1978** T. SHARPE *Throwback* iii. I don't want to marry the damned woman either, but needs must when the devil drives.

[1] *needs must:* used elliptically for 'one needs must (i.e. must of necessity) go'.

What a NEIGHBOUR gets is not lost

1567 L. WAGER *Mary Magdalene* D4^v There is nothyng lost that is done for such a friende. **1721** J. KELLY *Scottish*

Proverbs 198 It is no tint,[1] a Friend gets. **1891** J. L. KIPLING *Beast & Man* viii. The public at large have reaped much of the crop sown by Government for its own army, but, as the Scottish saying has it, 'What a neighbour gets is not lost.'

[1] *no tint*, not lost.

neighbour: *see also* GOOD fences make good neighbours.

nest: *see* there are no BIRDS in last year's nest; BIRDS in their little nests agree; it's an ILL bird that fouls its own nest.

In vain the NET is spread in the sight of the bird
  [PROVERBS i. 17 Surely in vaine the net is spread in the sight of any bird.] *c* **1395** WYCLIF *Bible* (1850) Proverbs i. 17 A net is leid in veyn before the ighen[1] of briddis. **1581** G. PETTIE tr. *S. Guazzo's Civil Conversation* I. 20ᵛ In vaine (as the Prouerb sayth) The net is pitcht in the sight of the birdes. **1888** J. E. T. ROGERS *Economic Interpretation of Hist.* xxi. The landowners in Pitt's time foresaw this. . . They would certainly be caught, and the net was spread in vain in sight of the bird. **1941** 'M. COLES' *They tell No Tales* x. 'Keep a good look out.' . . 'In vain is the net spread in the sight of the bird, anyway.'

[1] i.e. eyes.

net: *see also* all is FISH that comes to the net.

If you gently touch a NETTLE it'll sting you for your pains; grasp it like a lad of mettle, an' as soft as silk remains.
  The metaphorical phrase *to grasp the nettle*, to tackle a difficulty boldly, is also found.
  1578 LYLY *Euphues* I. 212 True it is Philautus that he which toucheth ye nettle tenderly, is soonest stoung. **1660** W. SECKER *Nonsuch Professor* I. 156 Sin is like the nettle, that stings when it is gently touched, but doth hurt not when it is ruggedly handled. **1753** A. HILL *Works* IV. 120 Tender-handed stroke a nettle, And it stings you, for your pains: Grasp it like a man of mettle, And it soft

as silk remains. **1830** R. FORBY *Vocabulary of East Anglia* 430 'Nip a nettle hard, and it will not sting you'—i.e. Strong and decided measures prevail best with troublesome people. **1925** S. O'CASEY *Juno & Paycock* I. 35 Be firm, Captain. . . If you gently touch a nettle it'll sting you for your pains; grasp it like a lad of mettle, an' as soft as silk remains.

NEVER is a long time
  c **1390** CHAUCER *Canon's Yeoman's Tale* l. 1411 Nevere to thryve were to long a date. **1721** J. KELLY *Scottish Proverbs* 260 Never is a long Term. . . Spoken to them that say they will never get such a Thing effected. **1887** BLACKMORE *Springhaven* I. xvii. She never could pay her rent. But 'never is a long time' . . and . . she stood clear of all debt now. **1979** H. HOWARD *Sealed Envelope* iii. 'I never reveal my sources.' . . 'Never is a long time.'

It is NEVER to late to learn
  A later variation of the next two proverbs.
  1678 R. L'ESTRANGE *Seneca's Morals* III. xx. It is never too late to learn what it is always necessary to know. **1721** J. KELLY *Scottish Proverbs* 266 Never too late to learn. **1927** E. F. BENSON *Lucia in London* ii. We want to know what the cosmopolitan mind is thinking about. Of course we're old, but it is never too late to learn. **1976** J. I. M. STEWART *Memorial Service* vii. You might not be too bad at it. Why not have a try, eh? Never too late to learn.

It is NEVER too late to mend
  See also the two adjacent proverbs.
  1590 GREENE (*title*) Never too late. **1594** LODGE & GREENE *Looking-Glass for London* I3ᵛ Amends may neuer come too late. *c* **1645** J. HOWELL *Familiar Letters* 9 Nov. (1903) III. 139 We have both of us our failings that way . . but it is never over late to mend. **1856** C. READE (*title*) It is never too late to mend. **1934** H. SPRING *Shabby Tiger* iv. Adolf shrugged a shoulder which suggested that it's never too late to mend. **1961** I. JEFFERIES

It wasn't Me! i. How kind. . . Never too late to mend.

NEVER too old to learn
See also the two preceding proverbs.

[SENECA *Epistle* lxxiv. *tamdiu discendum est, quamdiu nescias: si proverbio credimus, quamdiu vivis*, we must go on learning as long as we are ignorant; or, if we believe the proverb, as long as we live.] **1530** A. BARCLAY *Eclogues* (EETS) II. 538 Coridon thou art not to olde for to lere. **1555** *Institution of Gentleman* B7^v No man can be to olde to learne. **1670** J. RAY *English Proverbs* 112 Never too old to learn. **1712** J. ARBUTHNOT *Law is Bottomless Pit* I. vii. A Lawyer I was born, and a Lawyer I will be; one is never too Old to learn. **1858** TROLLOPE *Dr. Thorne* I. x. One should never be too old to learn—there's always something new worth picking up. **1981** P. O'DONNELL *Xanadu Talisman* ii. I've made sure you were schooled in every possible subject . . but you're never too old to learn.

never: *see also* BETTER late than never.

NEW brooms sweep clean
The phrase *new broom* 'one newly appointed to a position who makes changes in personnel or procedures' derives from this proverb.

1546 J. HEYWOOD *Dialogue of Proverbs* II. i. F3^v Som therto said, the grene new brome swepith cleene. **1578** LYLY *Euphues* I. 232 Ah well I wotte[1] that a newe broome sweepeth cleane. **1616** J. WITHALS *Dict.* (rev. ed.) 569 New bromes sweepe cleane, yet old friendship still retaine. **1776** G. COLMAN *New Brooms!* 15, I am glad he is gone—Glad!—To be sure. *New Brooms*, you know. **1877** J. A. FROUDE *Short Studies* 3rd Ser. 55 New brooms sweep clean. Abbot Thomas, like most of his predecessors, began with attempts at reformation. **1979** F. OLBRICH *Sweet & Deadly* iv. He was all right at first. It was a case of a new broom sweeping clean.

[1] know.

What is NEW cannot be true
1639 J. CLARKE *Parœmiologia Anglo-Latina* 228 The newest things, not always truest. **1791** BOSWELL *Life of Johnson* II. 283, I found that generally what was new was false. **1880** J. NICHOL *Byron* ix. We are told . . that he knew little of art or music. . . It is true but not new. But when Hunt proceeds to say that Byron had no sentiment . . it is new enough, but is manifestly not true. **1928** *Times* 4 Feb. 8 Sir Arthur Evans has fallen a victim . . to the old slogan 'What is new cannot be true.'

NEW lords, new laws
[a **1450** *St. Editha* (1883) 96 Willyham Conquerour was made here kyng, And made newe lordus and eke new lawe.] a **1547** E. HALL *Chronicle* (1548) Hen. VI 169 Tholde spoken prouerbe, here toke place: New Lordes, new lawes. **1824** SCOTT *St. Ronan's Well* II. i. But new lords new laws—naething but fine and imprisonment, and the game no a feather the plentier. **1874** HARDY *Far from Madding Crowd* I. viii. 'I was lately married to a woman, and she's my vocation now.' . . 'New lords new laws, as the saying is.'

You can't put NEW wine in old bottles
The idea is also expressed allusively as a metaphorical phrase.

[MATTHEW ix. 17 Neither doe men put new wine into old bottels: else the bottels breake, and the wine runneth out, and the bottels perish.] **1912** L. STRACHEY *Landmarks in French Literature* vi. The new spirits had animated the prose of Chateaubriand and the poetry of Lamartine; but . . the *form* of both these writers retained most of the important characteristics of the old tradition. It was new wine in old bottles. **1948** A. J. TOYNBEE *Civilization on Trial* vi. The new wines of industrialism and democracy have been poured into old bottles and they have burst the old bottles beyond repair. **1960** I. JEFFERIES *Dignity & Purity* viii. 'I don't think you can put new wine in old bottles.' I looked doubtful. . . 'A lot of this could be rationalized.' **1974** T. SHARPE *Porter-house Blue* x. 'Motives? . . Good old-fashioned lust.' 'That hardly explains the explosive nature of his end.' . . 'You can't put new wine in old bottles.'

new: *see also* there is NOTHING new under the sun; it is best to be OFF with the old love before you are on with the new; you cannot TEACH an old dog new tricks.

news: *see* BAD news travels fast; GO abroad and you'll hear news of home; NO news is good news.

nibble: *see* a BLEATING sheep loses a bite.

night: *see* BARNABY bright, Barnaby bright, the longest day and the shortest night; all CATS are grey in the dark; RED sky at night shepherd's delight; red sky in the morning shepherd's warning.

NINE tailors make a man
The literal meaning is that a gentleman must select his attire from various sources. It is now sometimes specifically associated with bell-ringing (see quots. 1908 and 1934).

   1613 *Tarlton's Jests* C1 Two Taylors goes to a man. **1647** N. WARD *Simple Cobbler* 26 It is a more common then convenient saying, that nine Taylers make a man; it were well if nineteene could make a woman to her minde. **1776** *Poor Robin's Almanack* II. C6ᵛ Do ye know how many Taylors make a Man? Why Nine—Nine Taylors make a Man. **1819** SCOTT *Letter* 26 July (1933) V. 427 They say it takes *nine* tailors to make a man—apparently, *one* is sufficient to ruin him. **1908** H. B. WALTERS *Church Bells* v. When the Knell is rung, it is a frequent practice to indicate the . . sex of the deceased. . . The old saying 'nine tailors make a man' is really 'nine tellers',[1] or three times three. **1912** A. BRAZIL *New Girl at St. Chad's* i. There's a saying that it takes nine tailors to make a man, so if your name is Taylor you can only be the ninth part of a lady!' **1934** D. L. SAYERS *Nine Tailors* IV. iii. The

voice of the bells of Fenchurch St. Paul. . . Nine Tailors Make a Man.
  [1] strokes.

nine: *see also* PARSLEY seed goes nine times to the Devil; POSSESSION is nine points of the law; it is not SPRING until you can plant your foot upon twelve daisies; a STITCH in time saves nine.

NO man can serve two masters
  Cf. *You cannot serve* GOD *and Mammon.*

  [MATTHEW vi. 24 No man can serue two masters.] *c* **1330** in T. Wright *Political Songs* (1839) 325 No man may wel serue tweie lordes to queme.[1] *c* **1477** CAXTON *Jason* (EETS) 57 No man may wel serue two maistres, for that one corumpeth that other. **1642** D. ROGERS *Naaman* vi. You cannot have your will . . and Christ too; no man can serve two Masters. **1853** R. C. TRENCH *On Lessons in Proverbs* v. Our lord . . has said: 'No man can serve two masters.' . . So the Spanish proverb: He who has to serve two masters, has to lie to one. **1979** 'C. AIRD' *Some die Eloquent* v. The Coroner's Officer existed in a sort of leaderless no-man's-land. Hostilities had broken out over this more than once. . . No man can serve two masters.
  [1] please.

NO man is a hero to his valet
Attributed to MADAME CORNUEL (1605–94): *il n'y a pas de héros pour son valet-de-chambre,* no man is a hero to his valet.

  1603 FLORIO tr. *Montaigne's Essays* III. ii. Few men haue beene admired of their familiers. . . In my climate of Gascoigne they deeme it as iest to see mee in print. **1764** S. FOOTE *Patron* II. 31 It has been said . . that no man is a hero to his valet de chambre; now I am afraid when you and I grow a little more intimate . . you will be horribly disappointed in your high expectations. **1910** *Times* 20 Jan. (Literary Supplement) 17 Many men have been heroes to their valets, and most (except Pope and Poe) to their biographers. **1940** A. CHRISTIE *One, Two, buckle My Shoe* i. It has been said that no man is a hero to his valet. To that may be added that few men are heroes to themselves at the moment of visiting their dentist. **1980** 'C. AIRD' *Passing Strange* xiii. Just as no man was a hero to his valet, so no member of a profession was a sea-green incorruptible to a policeman.

NO moon, no man
1878 HARDY *Return of Native* I. i. iii. 'No moon, no man.' 'Tis one of the truest sayings ever spit out. The boy never comes to anything that's born at new moon. **1878** T. F. THISTLETON-DYER *English Folk-Lore* ii. In Cornwall, when a child is born in the interval between an old moon and the first appearance of a new one, it is said that it will never live to reach the age of puberty. Hence the saying 'No moon, no man.'

NO names, no pack-drill
If nobody is named as being responsible for something, then nobody can be blamed and punished. Also used more generally in favour of reticence on a subject. *Pack-drill* is a military punishment in which the offender is compelled to march up and down in full marching order.
1923 O. ONIONS *Peace in our Time* I. ii. Men had a way of omitting the names of those of whom they spoke; no names no pack-drill. **1925** S. O'CASEY *Juno & Paycock* II. 61, I know some as are as sweet as the blossoms that bloom in the May—oh, no names, no pack dhrill. **1955** M. ALLINGHAM *Beckoning Lady* ii. It just means no name, no pack drill, and always speak well of them as has money to sue. **1979** L. MEYNELL *Hooky & Villainous Chauffeur* x. If you gave your word to a girl you'd stick to it. But there as plenty as don't. . . No names, no packdrill, as my old dad used to say.

NO news is good news
1616 JAMES I in *Loseley MSS* (1836) 403 No newis is bettir then evill newis. **1640** J. HOWELL *Familiar Letters* 3 June (1903) II. 144, I am of the Italians mind that said, 'Nulla nuova, buona nuova,' (no news, good news). **1850** F. E. SMED-

LEY *Frank Fairlegh* x. Arguing . . (on the 'no news being good news' system) that I should have heard again if anything had gone wrong, I dismissed the subject from my mind. **1974** T. SHARPE *Porterhouse Blue* xxi. 'He can't reply,' the Senior Tutor pointed out. 'I find that most consoling. After all no news is good news.'

no: *see also* HALF a loaf is better than no bread; there's no great LOSS without some gain; there's no PLACE like home; if there were no RECEIVERS, there would be no thieves; there is no ROYAL road to learning; SEE no evil, hear no evil, speak no evil; SHROUDS have no pockets; no SMOKE without fire; TIME and tide wait for no man; no TIME like the present.

nobody: *see* EVERYBODY's business is nobody's business; it's an ILL wind that blows nobody any good.

A NOD's as good as a wink to a blind horse
A fanciful assertion, often abbreviated as in quot. 1822, that the slightest hint is enough to convey one's meaning in the case.
1794 W. GODWIN *Caleb Williams* I. viii. Say the word; a nod is as good as a wink to a blind horse. **1822** B. MALKIN *Gil Blas* (rev. ed.) I. II. ix. I shall say no more at present; a nod is as good as a wink. **1925** S. O'CASEY *Shadow of Gunman* in *Two Plays* I. 142 You needn't say no more—a nod's as good as a wink to a blind horse. **1979** L. MEYNELL *Hooky & Villainous Chauffeur* vi. The way he behaves . . Other men's wives. Still, I expect you know as much about that as I do. . . They say a nod's as good as a wink to a blind horse.

nod: *see also* (verb) HOMER sometimes nods.

noise: *see* EMPTY vessels make the most sound.

none: *see* a BAD excuse is better than none; there's none so BLIND as those who will not see; none but the BRAVE deserve the fair; there's none so DEAF as those who will not hear; TWO is company, but three is none.

no one: *see* no one should be JUDGE in his own cause.

nose: *see* don't CUT off your nose to spite your face.

NOTHING comes of nothing

[ALCAEUS *Fragment* clxxiii. οὐδεν ἐκ δένος γένοιτο, nothing comes of nothing; L. *ex nihilo nihil fit.*] *c* **1380** CHAUCER *Boethius* v. pr. i. For this sentence is verray and soth, that 'no thing hath his beynge of naught.' **1551** CRANMER *Answer to Gardiner* 369 *Sicut ex nihilo nihil fit, Ita nihil in nihilum rediguitur*, As nothyng can be made of nought, so nothynge can be tourned into nought. **1605–6** SHAKESPEARE *King Lear* I. i. 89 Nothing will come of nothing. Speak again. **1818** SCOTT *Heart of Midlothian* I. i. You are to give me all your business. . . If you have none, the learned gentleman here knows nothing can come of nothing. **1946** E. R. CURTIS *Lady Sarah Lennox* iv. 'What did you think of it?' 'Nothing, Sir.' 'Nothing comes of nothing!' the King exclaimed impatiently. **1979** E. H. GOMBRICH *Sense of Order* viii. Nothing comes out of nothing. The great ornamental styles could no more have been the invention of one man . . than could the organ fugue.

NOTHING for nothing

Cf. *You don't get* SOMETHING *for nothing.*

a **1704** T. BROWN *Works* (1707) I. 131 Thou know'st the proverb, nothing due for nought. **1800** M. EDGEWORTH *Castle Rackrent* 167 Nothing for nothing, or I'm under a mistake with you, Jason. **1858** G. J. WHYTE-MELVILLE *Interpreter* xxv. Sir Harry . . recollected the old-established principle of himself and his clique, 'Nothing for nothing, and very little for a halfpenny'. **1908** A. MACLAREN *Ezekiel* 172 The last touch in the picture is meanness, which turned everything

into money. . . Is not 'nothing for nothing' an approved maxim to-day? **1981** N. FREELING *One Damn Thing after Another* iv. Their heart's not in it. Nothing for nothing and not much for sixpence.

NOTHING is certain but death and taxes

1726 DEFOE *Hist. Devil* II. vi. Not the Man in the Moon, . . not the Inspiration of Mother Shipton, or the Miracles of Dr. Faustus, Things as certain as Death and Taxes, can be more firmly believ'd. **1789** B. FRANKLIN *Letter* 13 Nov. in *Writings* (1907) X. 69 In this world nothing can be said to be certain, except death and taxes. **1912** *Spectator* 18 May 785 It is not merely the . . amount of the taxes. . . It is their compulsory and irresistible incidence. . . 'There are only two evils from which no man can escape—death and the King's taxes.' **1939** L. I. WILDER *By Shores of Silver Lake* xxv. Everything's more or less a gamble. . . Nothing is certain but death and taxes.

NOTHING is certain but the unforeseen

Cf. *The* UNEXPECTED *always happens.*

1886 J. A. FROUDE *Oceana* vii. There is a proverb that 'nothing is certain but the unforeseen,' and in fact few things turn out as we expect them. **1905** A. MACLAREN *Gospel according to St. Matthew* I. 322 There is nothing certain to happen, says the proverb, but the unforeseen. Tomorrow *will have* its cares.

There is NOTHING lost by civility

Cf. CIVILITY *costs nothing.*

1892 G. & W. GROSSMITH *Diary of a Nobody* xviii. I . . assisted several ladies to ices, remembering an old saying that 'There is nothing lost by civility.'

There is NOTHING new under the sun

[ECCLESIASTES i. 9 There is no new thing vnder the sunne.] **1592** G. DELAMOTHE *French Alphabet* II. 7 Under the large Cope of heauen, we see not a new thing. **1664** A. BRADSTREET *Works* (1867) 53 There is no new thing under the sun. **1801** T. JEFFERSON *Writings* (1904)

X. 229 We can no longer say there is nothing new under the sun. **1850** KINGSLEY *Alton Locke* I. xviii. There is nothing new under the sun; all that, is stale and trite to a septuagenarian, who has seen where it all ends. **1979** E. H. GOMBRICH *Sense of Order* i. It rather looks as if there were nothing new under the sun and as if all change was only apparent.

NOTHING should be done in haste but gripping a flea

a **1655** N. L'ESTRANGE in *Anecdotes & Traditions* (1839) I. 55 A grave gentleman in this Kingdome us'd this phrase often: 'Do nothing rashly, but catching of fleas.' **1678** J. RAY *English Proverbs* (ed. 2) 151 Nothing most be done hastily but killing of fleas. **1721** J. KELLY *Scottish Proverbs* 261 Nothing to be done in haste, but gripping of Fleas. . . Spoken when we are unreasonably urged to make haste. **1927** J. BUCHAN *Witch Wood* xii. What's the need o' hurry when the body's leg is still to set. As my auld mither used to say, naething suld be done in haste but grippin' a flea. [Cf. **1981** *Daily Telegraph* 16 July 18 Mr. Whitelaw . . has not taken shelter behind the often disastrously misleading maxim that 'nothing should be done in hot blood'.]

NOTHING so bad but it might have been worse

[**1817** SCOTT *Rob Roy* II. xiii. There's naething sae gude on this side o' time but it might hae been better.] **1876** I. BANKS *Manchester Man* III. xiii. However, there is nothing so bad but it might be worse. **1885** E. J. HARDY *How to be Happy though Married* xxi. Let us resolve to look at the bright side of things. . . 'Nothing so bad but it might have been worse.' **1908** *Times* 5 Oct. 3 Farmers . . will regard the . . meteorological changes as illustrating the ancient axiom to the effect that circumstances are never so bad that they cannot be worse.

NOTHING so bold as a blind mare

a **1628** J. CARMICHAELL *Proverbs in*

Scots no. 1435 The blind horse is hardiest. **1721** J. KELLY *Scottish Proverbs* 266 Nothing so bold as a blind Mare. **1922** J. BUCHAN *Huntingtower* x. He spoke of the still unconquered enemy with . . disrespect, so that Mrs Morran was moved to observe that there was 'naething sae bauld as a blind mear'.

There is NOTHING so good for the inside of a man as the outside of a horse

A proverb asserting the benefit of horse-riding to health.

1906 G. W. E. RUSSELL *Social Silhouettes* xxxii. The Squire will wind up . . with an apocryphal saying which he attributes to Lord Palmerston—'There's nothing so good for the inside of a man as the outside of a horse.' **1946** M. C. SELF *Horseman's Encyclopædia* 338 'There is something about the outside of a horse which is good for the inside of a man.' This adage is wiser than might first appear. But one should not consider just the hours spent in the saddle as beneficial.

NOTHING succeeds like success

[Fr. *rien ne réussit comme le succès*, nothing succeeds like success.] **1867** A. D. RICHARDSON *Beyond Mississippi* xxxiv. 'Nothing succeeds like success.' There *was* much Southern sympathy on the island; now all are our dear friends. **1872** W. BESANT *Ready-Money Mortiboy* I. ix. In Mr. Mortiboy's judgment no proverb could be better than . . 'Nothing succeeds like success.' Success dazzled him. **1980** H. TREVOR-ROPER *History & Imagination* 9 Nothing succeeds like success, and if Hitler had founded his empire . . we can well imagine how later historians would have treated him.

NOTHING venture, nothing gain

A variant of the next proverb.

[**1481** CAXTON *Reynard* (1880) xii. He that will wynne he muste laboure and auenture.] **1624** T. HEYWOOD *Captives* IV. i. I see hee that nought venters, nothinge gaynes. **1668** C. SEDLEY *Mulberry Garden* III. ii. Who ever caught any

thing With a naked hook? nothing venture, nothing Win. **1876** BLACKMORE
Cripps III. iv. We must all have been in
France . . if—well, never mind. Nothing
venture, nothing win. **1967** D. MORRIS
Naked Ape iv. For him [the withdrawn
individual] the old saying: 'Nothing
ventured, nothing gained' has been re-
written: 'Nothing ventured, nothing
lost'. **1979** A. PRICE *Tomorrow's Ghost*
vii. That was decidedly interesting—
'And Pearson Cole?' Nothing venture,
nothing gain.

NOTHING venture, nothing have

c **1385** CHAUCER *Troilus & Criseyde* II.
807 He which that nothing under-
taketh, Nothyng n'acheveth, be hym
looth or deere.[1] **1546** J. HEYWOOD *Dialogue of Proverbs* I. xi. E1 Noght venter
noght haue. **1559** T. COOPER *Bibliotheca*
(ed. 3) s.v. Fortis, Fortune foretherethe[2]
bolde aduenturers, nothyng venture,
nothyng haue. **1791** BOSWELL *Life of
Johnson* II. 166, I am, however, generally for trying 'Nothing venture, nothing
have'. **1841** DICKENS *Old Curiosity Shop*
I. xxix. I'm sorry the gentleman's
daunted—nothing venture, nothing
have—but the gentleman knows best.
1937 A. CHRISTIE *Death on Nile* II. xxix.
Nothing venture nothing have! It's
about the only time in my life I shall be
near to touching a fat lot of money.
    [1] *be hym looth or deere*, whether hateful or
pleasing to him.   [2] i.e. furthers.

nothing: *see also* BELIEVE nothing of what

you hear, and only half of what you see;
BLESSED is he who expects nothing, for
he shall never be disappointed; CIVILITY
costs nothing; there is nothing like
LEATHER; you don't get SOMETHING for
nothing; SOMETHING is better than nothing; the SUN loses nothing by shining
into a puddle.

notice: *see* LONG foretold long last, short
notice soon past.

There's NOWT so queer as folk

1905 *English Dialect Dict.* IV. 304
There's nowt[1] sae queer as folk, *Old
saying*. **1939** J. WOOD *Letter* 30 May in J.
Chambers *Letters* (1979) 186, I trust you
. . find plenty of interest in people &
their doings. Really 'there is nowt so
funny as folk'. **1955** R. E. MEGARRY *Mis-
cellany-at-Law* I. 72 There is the infinite
charm and variety of human nature itself: 'there's nowt so queer as folk.'
1979 L. MEYNELL *Hooky & Villainous
Chauffeur* viii. There's nowt so queer as
folk, people get up to all sorts of odd
things.
    [1] a dialectal variant of 'nought'; nothing.

nowt: *see also* when in DOUBT, do nowt;
HEAR all, see all, say nowt.

number: *see* there is LUCK in odd numbers;
there is SAFETY in numbers.

nut: *see* the GODS send nuts to those who
have no teeth.

O

When the OAK is before the ash, then you will only get a splash; when the ash is before the oak, then you may expect a soak

1852 *Notes & Queries* 1st Ser. V. 581 When the oak comes out before the ash, there will be fine weather in harvest. I . . find it generally correct. **1911** *Times Literary Supplement* 4 Aug. 285 One of the commonest weather rhymes in most parts of England deals with the budding of the oak and the ash:— When the oak's before the ash Then you'll only get a splash, When the ash is before the oak Then you may expect a soak. But in North Germany the signs are exactly inverted, and also in Cornwall.

Beware of an OAK, it draws the stroke; avoid an ash, it counts the flash; creep under the thorn, it can save you from harm

Advice on where to shelter from lightning during a thunderstorm.

1878 *Folk-Lore Record* I. 43 Mothers teach their children to say—Beware of an oak, It draws the stroke; Avoid an ash; It counts the flash; Creep under the thorn, It can save you from harm. **1945** F. THOMPSON *Lark Rise* xvii. Some one would . . warn him to keep away from trees during a thunderstorm. . . Others would quote: Under oak there comes a stroke, Under elm there comes a calm, And under ash there comes a crash.

oak: see also GREAT oaks from little acorns grow; LITTLE strokes fell great oaks; a REED before the wind lives on, while mighty oaks do fall.

obedience: see the FIRST duty of a soldier is obedience.

odd: see there is LUCK in odd numbers.

odious: see COMPARISONS are odious.

It is best to be OFF with the old love before you are on with the new

1801 M. EDGEWORTH *Belinda* I. x. I can give you my advice gratis, in the formula of an old Scotch song. . . "Tis good to be off with the old love, Before you be on with the new.' **1819** SCOTT *Bride of Lammermoor* III. ii. It is best to be off wi' the old love Before you be on wi' the new. **1891** A. LANG *Essays in Little* 6 Dumas . . met the great man at Marseilles, where . . Alexandre chanced to be 'on with the new love' before being completely 'off with the old'. **1923** E. V. LUCAS *Advisory Ben* xxxix. That proverb about being off with the old love is a very sound one. **1980** I. ST. JAMES *Money Stones* III. vi. Off with the old and on with the new. Why not just come out with it? Tell her it's all finished.

OFFENDERS never pardon

1640 G. HERBERT *Outlandish Proverbs* no. 561 The offender never pardons. **1672** DRYDEN *Conquest of Granada* II. I. ii. Forgiveness to the Injur'd does belong; But they ne'r pardon who have done the wrong. **1876** I. BANKS *Manchester Man* III. xiii. He was of Mrs. Ashton's mind, that, 'as offenders never pardon', Augusta needed a friend.

OLD habits die hard

1758 B. FRANKLIN in *London Chronicle* 26–28 Dec. 632, I hear the reader say, Habits are hard to break, and those . . accustomed to idleness or extravagance do not easily change their manners. **1792** J. BELKNAP *Foresters* ix. Old habits are not easily broken, and . . they endeavoured . . to transfer the blame from him to his wife. **1980** I. MITCHELL *Dove of War* iv. 'What brings you to these parts?'. . 'Old habits die hard.'

You cannot put an OLD head on young shoulders

1591 H. SMITH *Preparative to Marriage* 14 It is not good grafting of an olde head vppon young shoulders, for they will neuer beare it willingly but grudgingly. **1794** E. DRINKER *Journal* 31 Dec. (1889) 256 Tis not the way I could wish my children to conclude the year— in parties— but we can't put old heads on young shoulders. **1951** *Sport* 30 Mar.–5 Apr. 11, I no longer believe in the old proverb that you cannot put an old head on young shoulders.

OLD sins cast long shadows

[Cf. **1638** J. SUCKLING *Aglaura* v. in *Plays* (1971) 110 Our sins, like to our shadowes, When our day is in its glorie scarce appear: Towards our evening how great and monstrous they are!] **1940** A. CHRISTIE *Sad Cypress* II. iii. 'You know . . who her real father was?' . . 'I don't *know* anything. I could make a guess. Old sins have long shadows, as they say.' **1957** V. BRITTAIN *Testament of Experience* I. iii. If he hadn't been killed, they would probably never have become what they were. . . Bygone battles, like old sins, cast long shadows. **1973** A. CHRISTIE *Postern of Fate* III. iv. It all took place . . years and years ago. New sins have old shadows. Or is it Old sins make long shadows?

OLD soldiers never die

1920 J. FOLEY (*song-title*) Old soldiers never die. **1930** BROPHY & PARTRIDGE *Songs & Slang of British Soldier 1914–18* II. 67 Old soldiers never die—They simply fade away. **1933** F. RICHARDS *Old Soldiers never Die* xxviii. We generally wound up our evenings with the old song, set to the tune of a well-known hymn, 'Old soldiers never die, they simply fade away'. **1940** *Times* 6 Apr. 4 There is an old saying that 'Old soldiers never die'—but they may starve . . when other State pensioners are receiving increased benefits. **1979** *Daily Telegraph* 31 Oct. 2 Union leaders have complained to the Government that old soldiers never die, but are given jobs in the Civil Service.

old: *see also* BETTER be an old man's darling, than a young man's slave; you cannot CATCH old birds with chaff; there's no FOOL like an old fool; there's many a GOOD tune played on an old fiddle; HANG a thief when he's young and he'll no' steal when he's old; a MAN is as old as he feels, and a woman as old as she looks; NEVER too old to learn; you can't put NEW wine in old bottles; it is best to be OFF with the old love before you are on with the new; an old POACHER makes the best gamekeeper; you cannot SHIFT an old tree without it dying; you cannot TEACH an old dog new tricks; YOUNG folks think old folks to be fools but old folks know young folks to be fools; YOUNG men may die but old men must die.

You cannot make an OMELETTE without breaking eggs

[Fr. *on ne fait pas d'omelette sans casser des œufs*, one does not make an omelette without breaking eggs.] **1859** T. P. THOMPSON *Audi Alteram Partem* II. xc. We are walking upon eggs and . . the omelet will not be made without the breaking of some. **1897** R. L. STEVENSON *St. Ives* viii. You cannot make an omelette without breaking eggs, . . and it is no bagatelle to escape from Edinburgh Castle. One of us, I think, was even killed. **1922** H. WALPOLE *Cathedral* II. iv. He could imagine . . the scene and the reception his advice would have. Nevertheless, how sad that undoubtedly you cannot make an omelette without first breaking the eggs! **1974** J. MANN *Sticking Place* iv. 'In your philosophy, it seems that some men have no right to live at all.' . . 'You can't make an omelette without breaking eggs, Mother.'

ONCE a ——, always a ——

This formula produces a limitless variety of proverbs, many of them depreciatory. Some are of long standing, others appear to be nonce uses. A small selection of such sayings is illustrated below and in the next two entries.

1622 J. MABBE tr. *Aleman's Guzman* I. I. i. Once a knaue, and euer a knaue: . . For he that hath once beene naught, is presumed to bee so still . . not considering . . whether . . hee had reformed his manners or no. **1655** T. FULLER *Church Hist. Britain* VII. xxviii. Latimer by the courtesie of England (once a bishop, and ever a bishop) was in civility saluted Lord. **1659** N. R. *Proverbs* 82 Once a whore and alwaies a whore. **1706** J. STEVENS *Spanish & English Dict.* s.v. Hurtar, Once a Thief, always a Thief. **1820** M. EDGEWORTH *Letter* 15 Nov. in *Maria Edgeworth in France & Switzerland* (1979) 277 She loses her rank . . by marrying one of inferior rank. . . French and Russians are with reason surprised with the superior gallantry of our customs which say once a Lady and always a lady. **1953** R. CHANDLER *Long Good-bye* xliv. I went out of the door and got out of the building fast. Once a patsy,[1] always a patsy.

[1] dupe (slang).

ONCE a priest, always a priest

An act enabling the clergy of the Church of England to unfrock themselves (The Clerical Disabilities Act) was passed on 9 Aug. 1870.

1859 G. A. SALA *Twice round Clock* 290 The great case of Horne Tooke *versus* the House of Commons—'Once a priest forever a priest'. **1865** L. STEPHEN *Life & Letters* (1906) ix. As in this . . country we stick to the maxim, 'once a parson, always a parson,' I could not . . go in for law. **1916** JOYCE *Portrait of Artist* (1967) iv. You must be quite sure, Stephen, that you have a vocation because it would be terrible if you found afterwards that you had none. Once a priest always a priest, remember. **1920** *Bookman* Sept. 192 No former celibate, with Boris's incapacity for blotting out his past, could be happy until he returned to his cell—once a priest always a priest, is a true enough motto so far as he is concerned.

ONCE a whore, always a whore

1613 H. PARROT *Laquei Ridiculosi* II. cxxi. Well you may change your name, But once a Whore, you shall be still the same. **1659** N.R. *Proverbs* 82 Once a whore and alwaies a whore. **1754** *World* 31 Jan. 344 Supposing him to have acquired so much wealth, the proverb of 'Once a whore, and always a whore', is less significant. **1824** H. MARSHALL *Hist. Kentucky* II. v. 'Once a prostitute, and always a prostitute,' is a fair mode of argument—at least, among politicians. **1981** N. LOFTS *Old Priory* v. iii. There is a saying, 'Once a whore, always a whore.'

ONCE bitten, twice shy

Cf. A BURNT *child dreads the fire.*

1853 R. S. SURTEES *Sponge's Sporting Tour* xxxvii. Jawleyford had been bit once, and he was not going to give Mr. Sponge a second chance. **1894** G. F. NORTHALL *Folk-Phrases* 20 Once bitten, twice shy. **1920** CONRAD *Rescue* III. ix. Once bit twice shy. He had no mind to be kidnapped. **1979** S. ALLAN *Mortal Affair* i. The vaguely concealed caution of not wanting to become involved again—once bitten twice shy.

When ONE door shuts, another opens

1586 D. ROWLAND tr. *Lazarillo* D3^v This proverbe was fulfild, when one doore is shut the other openeth. **1620** T. SHELTON tr. *Cervantes' Don Quixote* III. vii. Where one door is shut, another is opened. **1710** S. PALMER *Proverbs* 49 When one Door Shuts another Opens. . . How often does the Divine Bounty surprize us with unthought of Felicity! **1821** J. GALT *Annals of Parish* xxvi. Here was an example . . of the truth of the old proverb that as one door shuts another opens; . . A full equivalent for her [the light-headed Lady Macadam] was given in this hot and fiery Mr. Cayenne. **1925** S. O'CASEY *Juno & Paycock* I. 16 'The job couldn't come at a betther time.'. . 'Ah, God never shut wan door but he opened another.' **1979** D. SANDERS *Queen sends for Mrs Chadwick* 80, I do not accept that I have . . come to the end of the road. . . When one door closes, another one opens.

ONE for sorrow, two for mirth; three for a wedding, four for a birth

A traditional country proverb found in a variety of forms, which refers to the number of magpies seen on a particular occasion.

a **1846** B. HAYDON *Autobiography* (1853) I. v. During the journey four magpies rose . . and flew away. . . I repeated . . the old saw, 'one for sorrow, two for mirth, three for a wedding, and four for death.' **1846** M. A. DENHAM *Proverbs relating to Seasons, &c.* 35 One for sorrow: two for mirth: three for a wedding: four for a birth: five for silver: six for gold: seven for a secret, not to be told: eight for heaven: nine for hell: and ten for the devil's own sel.[1] **1913** A. C. BENSON *Along Road* 162, I never see magpies myself without relating the old rhyme: 'One for sorrow, Two for mirth, Three for a death, Four for a birth; Five, you will shortly be In a great company.' **1981** J. GARDNER *Licence Renewed* ii. Bond thought of the old adage, 'One for sorrow, two for joy.' There were a lot of single magpies swooping near the car.
[1] i.e. self.

ONE for the mouse, one for the crow, one to rot, one to grow

Other forms of this traditional country saying relating to sowing are also illustrated here.

1850 *Notes & Queries* 1st Ser. II. 515 How to sow Beans. 'One for the mouse, One for the crow, One to rot, One to grow.' **1941** L. I. WILDER *Little Town on Prairie* ii. 'Kernels,' said Pa. 'Four kernels. . . One for the blackbird, One for the crow, And that will leave Just two to grow.' **1961** N. LOFTS *House at Old Vine* I. 34 Careful farmers . . sow their seed broadcast, saying: One for wind and one for crow One to die and one to grow.

ONE nail drives out another

[ARISTOTLE *Politics* 1314a ἥλῳ γὰρ ὁ ἧλος, ὥσπερ ἡ παροιμία, one nail knocks out another, according to the proverb.] a **1250** *Ancrene Wisse* (1962) 206 An neil driueth ut then o ther. **1555** J. HEYWOOD *Two Hundred Epigrams* no. 112 One nayle dryueth out an other. **1591** SHAKESPEARE *Two Gentlemen of Verona* II. iv. 189 As one nail by strength drives out another, So the remembrance of my former love Is by a newer object quite forgotten. c **1645** J. HOWELL *Familiar Letters* 17 Sept. (1903) III. 87 Languages and words . . may be said to stick in the memory like nails or pegs in a wainscot door, which used to thrust out one another oftentimes. **1852** E. FITZGERALD *Polonius* cxvii. One nail drives out another. **1979** V. CANNING *Satan Sampler* ix. He needed a home with a woman in it. One nail drove out another.

ONE year's seeding means seven years' weeding

On the danger of allowing weeds to grow and seed themselves: used figuratively.

1889 T. F. THISTLETON-DYER *Folklore of Plants* xi. A weed that runs to seed Is a seven years' weed . . implies that disagreeable actions . . only too frequently cling to a man in after years. **1980** *Daily Telegraph* 19 Jan. 20 My advice to weedpersons is: do not let your weeds grow to maturity and seed—'one year's seeding means seven years' weeding'.

one: *see also* BETTER one house spoiled than two; a BIRD never flew on one wing; two BOYS are half a boy, and three boys are no boy at all; the BUYER has need of a hundred eyes, the seller of but one; every DOG is allowed one bite; don't put all your EGGS in one basket; one ENGLISHMAN can beat three Frenchmen; one FUNERAL makes many; one GOOD turn deserves another; one HALF of the world does not know how the other half lives; one HAND for oneself and one for the ship; one HAND washes the other; one HOUR's sleep before midnight is worth two after; one man's LOSS is another man's gain; one man's MEAT is another man's poison; one PICTURE is worth ten thousand words; one man may STEAL a horse, while another may not look over a hedge; one STEP at a time;

from the SUBLIME to the ridiculous is
only a step; one SWALLOW does not
make a summer; TWO heads are better
than one; if TWO ride on a horse, one
must ride behind; one VOLUNTEER is
worth two pressed men; one WEDDING
brings another.

one-eyed: *see* in the COUNTRY of the blind,
the one-eyed man is king.

open: *see* (adjective) open CONFESSION is
good for the soul; a DOOR must either be
shut or open; (verb) a GOLDEN key can
open any door; when ONE door shuts,
another opens.

**The OPERA isn't over till the fat lady
sings**
   1978 *Washington Post* 13 June B1 The
opera isn't over till the fat lady sings. . .
One day three years ago, Ralph Carpen-
ter, who was then Texas Tech's sports
information director, declared to the
press box contingent in Austin, 'The
rodeo ain't over till the bull riders ride.'
Stirred to top that deep insight, San
Antonio sports editor Dan Cook coun-
tered with, 'The opera ain't over till the
fat lady sings.' **1980** *Record* (OUP) Dec.
8 Do not relax your vigilance in the fight
against the punk paperbacks. . . 'The
opera ain't over 'til the fat lady sings.'

opinion: *see* he that COMPLIES against his
will is of his own opinion still; so many
MEN, so many opinions.

OPPORTUNITY makes a thief
   c **1220** *Hali Meidenhad* (EETS) 23 Man
seith that eise maketh theof. **1387** J.
TREVISA tr. *Higden's Polychronicon* (1879)
VII. 379 At the laste the bischop seide to
hym, 'Me thenke that opportunitie
makethe a thefe'. **1623** W. CAMDEN *Remains concerning Britain* (ed. 3) 275
Opportunity makes the thief. **1670** J.
RAY *English Proverbs* 129 Opportunity
makes the thief. . . Therefore, masters
. . ought to secure their moneys and
goods under lock and key, that they do
not give . . a temptation to steal. **1835**
SOUTHEY *Doctor* III. cv. Opportunity,

which makes thieves, makes lovers
also. **1979** *Daedalus* Summer 107 A
child steals from the sleeping woman's
pocket, acting out the proverb 'Opportunity makes a thief.'

**OPPORTUNITY never knocks twice at
any man's door**
Fortune occurs instead of *opportunity* in
earlier forms of the saying. Several
quotations below represent slightly
different ideas based on the original
proverb. In quots. 1809 and 1981,
Opportunity is said to knock once or
more, but in other quotations, once
only.
   1567 G. FENTON *Bandello* 216 Fortune
once in the course of our life, dothe put
into our handes the offer of a good
torne. **1809** *Port Folio* (Philadelphia)
Nov. 431 Fortune knocks once, at least,
at every man's door. **1889** W. F. BUTLER
C. G. Gordon iii. Fate, it is said, knocks
once at every man's door. . . Gordon
had just passed his thirtieth year when
Fortune . . knocked at . . the door which
was to lead him to fame. **1891** J. J. ING-
ALLS *Opportunity* in *Truth* (NY) 19 Feb.
17, I [Opportunity] knock unbidden
once at every gate! If sleeping, wake: if
feasting rise before I turn away . . [for] I
return no more! **1941** 'P. WENTWORTH'
Unlawful Occasions xxiv. It was an
opportunity with a capital O, and if she
threw it away it would never come back
again. Opportunity never knocks twice
at any man's door. **1981** J. BINGHAM
Brock 178 Though opportunity may
knock twice, there was no guarantee
that it would knock as insistently as
now.

opportunity: *see also* ENGLAND's difficulty
is Ireland's opportunity; MAN's extremity is God's opportunity.

opposite: *see* DREAMS go by contraries.

OTHER times, other manners
The proverb occurs in many forms.
   [PINDAR *Fragment* xliii. ἄλλοτ' ἀλλοῖα
φρόνει, different thoughts at different
times; Fr. *autres temps, autres mœurs*,

other times, other customs.] **1576** G.
PETTIE *Petit Palace* 34 Other times, other
wayes. **1902** A. DOBSON *Samuel Richardson* iv. Notwithstanding the favourite
explanation of 'other times, other manners', contemporary critics of Clarissa
found very much the same fault with
her history as people do to-day. **1945** F.
THOMPSON *Lark Rise* viii. Other days,
other ways. . . The old country mid-
wives did at least succeed in bringing
into the world many generations of our
forefathers. **1978** G. GREENE *Human Factor* IV. ii. 'We used to have better fu-
nerals in Africa.' . . 'Oh well—other
countries, other manners.'

other: *see also* DO unto others as you would
they should do unto you; the GRASS is
always greener on the other side of the
fence; one HALF of the world does not
know how the other half lives; one
HAND washes the other.

An OUNCE of practice is worth a pound of precept

A number of proverbs similar in form
are illustrated below.

[**1567** W. CECIL *Letter* in C. Read *Mr.
Secretary Cecil* (1955) xxi. Marry,[1] an
ounce of advice is more worth to be executed aforehand than in the sight of
perils.] *c* **1576** T. WHYTHORNE *Autobiography* (1961) 142, I ment not to be on
of thoz who waith[2] A chip of chauns[3]
mor then A pownds wurth of witt.
1592 G. DELAMOTHE *French Alphabet* II.
55 An ounce of discretion, is better
worth, then a pound of hardinesse.[4]
1616 T. ADAMS *Sacrifice of Thankfulness* 19
The prouerbe is true; an Ounce of Discretion, is worth a pound of Learning.
1748 J. ELIOT *Essays upon Field Husbandry*
12 It used to be the Saying of an old
Man, That an Ounce of Experience is
better than a Pound of Science. **1866**
BLACKMORE *Cradock Nowell* II. ix. Remember that rigid probity, and the stric-
test punctuality . . are the very soul of
business, and that an ounce of practice
is worth a pound of precept. **1925** J.
GALSWORTHY *Caravan* 667 'Define it
[Beauty], Mr. Vaness.' 'An ounce of fact

is worth a ton of theory.—It stands before me.' **1981** P. O'DONNELL *Xanadu
Talisman* ix. She said rather primly, 'An
ounce of wisdom is worth a pound of
wit.'

[1] why! to be sure!   [2] weigheth, (modern)
weighs; esteems.   [3] luck.   [4] audacity.

OUT of debt, out of danger

1639 J. CLARKE *Parœmiologia Anglo-
Latina* 82 Out of debt and deadly danger. **1667** H. PEACHAM *Worth of Penny*
(ed. 2) 8 How bold, confident, merry,
lively, and ever in humour are moneyed
men (for being out of debt, they are out
of danger). **1710** S. PALMER *Proverbs* 132
Out of Debt out of Danger. . . A Man in
Debt is a Slave, and can't act with
Liberty. **1908** E. M. SNEYD-KYNNERSLEY
H.M.I. xxi. Call it 'distributing capital
expenditure over a term of years', and
even a rural dean succumbs. 'Out of
debt, out of danger,' but 'out of debt,
out of progress.'

OUT of sight, out of mind

c **1250** *Proverbs of Alfred* (1907) 46 He
that is ute bi-loken[1] He is inne sone for-
yeten.[2] *c* **1450** tr. T. à Kempis' *De Imitatione Christi* (EETS) I. xxxiii. Whan Man
is oute of sight, son he passith oute of
mynde.[3] **1545** R. TAVERNER tr. *Erasmus'
Adages* (ed. 2) D6[v] Whereunto also
agreethe oure Englyshe proverbe which
sayethe: Oute of syght, oute of mynde.
1797 A. RADCLIFFE *Italian* III. ix. Old
women now-a-days are not much
thought of; out of sight out of mind with
them. **1863** KINGSLEY *Water Babies* i. Sir
John and the rest saw no more of her;
and out of sight was out of mind. **1979**
'S. WOODS' *This Fatal Writ* 45 The brief . .
was promptly concealed. . . He was
working on the principle 'out of sight,
out of mind'.

[1] shut.   [2] forgotten.   [3] memory (now
common only in standard phrases).

out: *see also* BETTER be out of the world than
out of the fashion; there are as good FISH
in the sea as ever came out of it; when
the GORSE is out of bloom, kissing's out

of fashion; don't HALLOO till you are out
of the wood; out of the MOUTHS of
babes—; MURDER will out; never tell
TALES out of school; TRUTH will out.

outside: *see* there is NOTHING so good for
the inside of a man as the outside of a
horse.

over: *see* the OPERA isn't over till the fat
lady sings; the SHARPER the storm, the
sooner it's over.

own: *see* the DEVIL looks after his own.

owt: *see* HEAR all, see all, say nowt; you
don't get SOMETHING for nothing.

ox: *see* BETTER a dinner of herbs than a
stalled ox where hate is.

P

It is the PACE that kills
 1855 S. A. HAMMETT *Wonderful Adventures of Captain Priest* xv. The well-known sporting maxim, that 'It is the pace that kills'. **1901** S. LANE-POOLE *Sir H. Parkes in China* xx. There is an old proverb about the pace that kills, and . . Sir Harry was killing himself by work at high pressure. **1936** N. MARSH *Death in Ecstasy* II. xvii. Don't overdo it. . . This is the pace that kills.

package: *see* the BEST things come in small packages.

pack-drill: *see* NO names, no pack-drill.

pain: *see* GENIUS is an infinite capacity for taking pains; PRIDE feels no pain.

paint: *see* a BLIND man's wife needs no paint.

painted: *see* the DEVIL is not so black as he is painted.

pan: *see* if IFS and ands were pots and pans, there'd be no work for tinkers' hands.

parcel: *see* the BEST things come in small packages.

pardon: *see* OFFENDERS never pardon.

Parkinson's law: *see* WORK expands so as to fill the time available.

PARSLEY seed goes nine times to the Devil
 1658 in Mennis & Smith *Wit Restored* 35 There is a saying in the North Riding of Yorkshire that The weed [parsley] before it's borne Nine times the devill sees. **1885** *Notes & Queries* 6th Ser. XI. 467 Parsley seed (when it has been sown) goes nine times to the devil. **1908** D. H. LAWRENCE *Letter* 4 May (1962) I. 7 People say parsley seed goes seven times (some are moderate, discarding the holy number as unfit, and say five) to the Old Lad, it is so long a-germinating. **1981** in A. Hewins *Dillen* xiv. It's a queer plant, parsley: 'sow on Good Friday, seven times down to Hell afore it chits.[1]'
 [1] sprouts, germinates (dialect).

parsnip: *see* FINE words butter no parsnips.

part: *see* (noun) DISCRETION is the better part of valour; if you're not part of the SOLUTION, you're part of the problem; (verb) the BEST of friends must part.

parted: *see* a FOOL and his money are soon parted.

Things PAST cannot be recalled
 Cf. *What's* DONE *cannot be undone.*
 a **1500** H. MEDWALL *Nature* H3[v] A thyng don can not be called agayn. *a* **1535** MORE *Edward V* in J. Hardyng *Chronicle* (1543) II. 36 Sith[1] thynges passed cannot be gaine called, muche more ought wee to bee ware. **1616** T. DRAXE *Adages* 151 That that is past, cannot be recalled or helped. **1804** M. EDGEWORTH *Popular Tales* II. 130 Since a thing past can't be recalled . . we may be content. **1979** *Country Life* 24 May 1683 Visually, another period's realities are palpably inaccessible. Things past cannot be—perfectly—recalled.
 [1] since.

pastime: *see* he that would go to SEA for pleasure, would go to hell for a pastime.

PATIENCE is a virtue
 [*Disticha Catonis* I. xxxviii. *maxima enim . . patientia virtus*, patience is the greatest virtue.] **1377** LANGLAND *Piers Plowman* B. XI. 370 Suffraunce[1] is a souereygne vertue. *c* **1390** CHAUCER *Franklin's Tale* l. 773 Pacience is an heigh vertue, certeyn. **1594** LYLY *Mother Bom-*

bie v. iii. Patience is a vertue, but pinching is worse than any vice! **1754** RICHARDSON *Grandison* II. xvii. Aunt Prue in Yorkshire . . will be able to instruct you, that patience is a virtue; and that you ought not to be in haste to take a first offer, for fear you should not have a second. **1858** TROLLOPE *Dr. Thorne* I. xiv. That was only three days ago. Why don't you . . ferret her out? . . Patience is and always was a virtue. **1979** A. FOX *Threat Warning Red* iv. 'I . . can't *wait* for you.' . . 'Patience is said to be a virtue?'

¹ forbearance, long-suffering.

Paul: *see* if SAINT Paul's day be fair and clear, it will betide a happy year.

paved: *see* the ROAD to hell is paved with good intentions.

If you PAY peanuts, you get monkeys
Now commonly used with reference to wage negotiations.

1966 L. COULTHARD in *Director* Aug. 228 Shareholders want the best available businessmen to lead the companies and recognise that you get what you pay for. If you pay in peanuts, you must expect to get monkeys. **1979** P. ALEXANDER *Show me Hero* iii. 'That's forty thousand we're giving away. Seems an awful lot.' 'If you pay peanuts,' said Ashman, 'you get monkeys.' **1979** *Guardian* 11 Sept. 30 The companies' chief negotiator . . was greeted with shouts of 'if you pay peanuts, you get monkeys'.

pay: *see also* they that DANCE must pay the fiddler; DEATH pays all debts; SPEAK not of my debts unless you mean to pay them; the THIRD time pays for all.

He who PAYS the piper calls the tune
Cf. *They that* DANCE *must pay the fiddler*, where the emphasis is reversed.

1895 *Daily News* 18 Dec. 9 Londoners had paid the piper, and should choose the tune. **1910** *Spectator* 22 Oct. 643 Until British electors know that the dollars have been returned they will be wise in placing little trust in Nationalist 'loyalty'. He who pays the piper calls

the tune. **1979** *Daily Telegraph* 7 June 2 We have to find some way of giving them a fair deal since we shall be dependent on them. . . He who pays the piper calls the tune.

You PAYS your money and you takes your choice
Both *pays* and *takes* are non-standard, colloquial forms.

1846 *Punch* X. 16 'Which *is* the Prime Minister?' . . 'Which ever you please, my little dear. You pays your money, and you takes your choice.' **1904** V. S. LEAN *Collectanea* IV. 205 You pays your money and you takes your choice. You pays your moneys and what you sees is A cow or a donkey just as you please. **1979** W. KIENZLE *Rosary Murders* 62 Ah, the plastic church of today. . . You pays your money and you takes your choice.

If you want PEACE, you must prepare for war
[VEGETIUS *Epitoma Rei Militaris* III. Introd. *qui desiderat pacem, praeparet bellum,* he who desires peace must prepare for war.] *a* **1547** E. HALL *Chronicle* (1548) Edw. IV 209 He forgat the olde adage, saynge in tyme of peace prouyde for warre. **1593** M. SUTCLIFFE *Practice of Arms* A2ᵛ He that desireth peace, he must prepare for warres. **1624** BURTON *Anatomy of Melancholy* (ed. 2) II. iii. The Commonwealth of Venice in their Armory haue this inscription, Happy is that Citty which in time of peace thinkes of warre, a fit Motto for euery mans priuate house. **1885** C. LOWE *Prince Bismarck* II. x. Lord Beaconsfield had acted on the maxim that 'if you want peace, you must prepare for war'. **1929** *Listener* 28 Aug. 278 'If you want peace, prepare for peace.' This is the reverse of the old motto, 'If you want peace, prepare for war'.

peanut: *see* if you PAY peanuts, you get monkeys.

pear: *see* WALNUTS and pears you plant for your heirs.

Do not throw PEARLS to swine

Often used allusively, especially in the phrase *to cast pearls before swine.*

[MATTHEW vii. 6 Giue not that which is holy vnto the dogs, neither cast ye your pearles before swine.] **1340** *Ayenbite of Inwit* (EETS) 152 Thet we ne thrauwe naght oure preciouse stones touore the zuyn. **1362** LANGLAND *Piers Plowman* A. xi. *Noli mittere Margeri,* perles Among hogges. **1526** *Pilgrimage of Perfection* II. iii. The holy fathers thought it nat expedient to disclose the secrete misteryes to euery worldly person. . . Cast not your perles before hogges. **1550** R. CROWLEY *Epigrams* A3ᵛ Before suche swyne no pearles maye be caste. **1694** ALLYN & PITKIN *Connecticut Vindicated* 7 Cast not your Pearls before Swine. **1816** S. SMITH *Letter* in S. Holland *Memoir* (1855) II. 134 Elgin has done a very useful thing in taking them [the Elgin Marbles] away from the Turks. Do not throw pearls to swine. **1925** WODEHOUSE *Sam the Sudden* xi. 'Young women do not interest me.' The proverb about casting pearls before swine occurred to Sam. **1967** M. DRABBLE *Jerusalem the Golden* vi. They wrote to each other, long, intimate, witty letters, the kind of letters that Clara fancied she had for years been casting before if not swine at least less than perfect readers.

A PECK of March dust is worth a king's ransom

The month of March is traditionally wet and blustery.

1533 J. HEYWOOD *Play of Weather* D1 One bushell of march dust is worth a kynges raunsome. **1685** R. BOYLE *Discourse of Causes of Insalubrity of Air* 55 It is proverbially said in England, that a Peck[1] of March Dust is worth a King's Ransom: So infrequent is dry Weather, during that Month. **1836** E. HOWARD *Rattlin the Reefer* III. viii. A spoonful of the soup to-day will be worth a king's ransom—a peck of March dust! pooh! **1978** R. WHITLOCK *Calendar of Country Customs* iii. [The farmer] values dry, cold weather, such as often occurs in

late February and March. 'A peck of dust in March is worth a king's ransom,' is still an oft-quoted proverb.

[1] a dry measure of two gallons.

peck: *see also* we must EAT a peck of dirt before we die.

The PEN is mightier than the sword

[Cf. CICERO *De Officiis* I. xxii. *cedant arma togae,* arms give way to persuasion.] **1582** G. WHETSTONE *Heptameron of Civil Discourses* iii. The dashe of a Pen, is more greeuous then the counter use of a Launce. *a* **1712** W. KING *Eagle & Robin* in *Poetical Works* (1781) III. 49 Poor Bob . . A goosequill on for weapon ty'd, Knowing by use that now and then A sword less hurt does than a pen. **1839** BULWER-LYTTON *Richelieu* II. ii. Beneath the rule of men entirely great, The pen is mightier than the sword. **1976** T. SHARPE *Wilt* iii. The man who said the pen was mightier than the sword ought to have tried reading *The Mill on the Floss* to Motor Mechanics.

Take care of the PENCE and the pounds will take care of themselves

1750 LD. CHESTERFIELD *Letter* 5 Feb. (1932) IV. 1500 Old Mr. Lowndes,[1] the famous Secretary of the Treasury, . . used to say . . Take care of the pence, and the pounds will take care of themselves. **1912** G. B. SHAW *Pygmalion* II. 132 Take care of the pence and the pounds will take care of themselves is as true of personal habits as of money. **1979** R. CASSILIS *Arrow of God* IV. xvii. Little things, Master Mally. Look after the pennies, Master Mally, and the pounds will look after themselves.

[1] William Lowndes, 1652–1724.

A PENNY saved is a penny earned

1640 G. HERBERT *Outlandish Proverbs* no. 506 A penny spar'd is twice got. *a* **1661** T. FULLER *Worthies* (Hunts.) 51 By the same proportion that a penny saved is a penny gained, the preserver of books is a Mate for the Compiler of them. **1695** E. RAVENSCROFT *Canterbury*

Guests II. iv. This I did to prevent expences, for . . a penny sav'd, is a penny got. **1748** THOMSON *Castle of Indolence* I. 26 A Penny saved is a Penny got. **1853** DICKENS *Bleak House* ix. I saved five pounds out of the brickmaker's affair. . . It's a very good thing to save one, let me tell you: a penny saved, is a penny got! **1923** P. G. WODEHOUSE *Inimitable Jeeves* xi. I can save money this way; and believe me, laddie, nowadays . . a penny saved is a penny earned. **1979** A. LURIE *Only Children* 5 You can hear his engine saying the same dull things he said in history—A pen–ny saved is a pen–ny earned, he said, over and over again.

PENNY wise and pound foolish

1607 E. TOPSELL *Four-footed Beasts* 609 If by couetousnesse or negligence, one withdraw from them their ordinary foode, he shall be penny wise, and pound foolish. **1712** ADDISON *Spectator* 7 Feb., I think a Woman who will give up herself to a Man in Marriage, where there is the least room for such an Apprehension . . may very properly be accused . . of being Penny wise and Pound foolish. **1864** MRS. H. WOOD *Trevlyn Hold* II. xxi. He never would insure his ricks. . . Miss Diana has often told him he deserved to have his ricks take fire for being penny wise and pound foolish. **1973** G. GREENE *Honorary Consul* II. i. It's only a bachelor's digs. . . No sense of national pride. Penny wise, pound foolish.

penny: *see also* a BAD penny always turns up; IN for a penny, in for a pound.

Like PEOPLE, like priest

[HOSEA iv. 9 Like people, like priest.] **1589** *Pasquil's Return* C3 Like people, like Priest begins now to be verified. **1670** J. RAY *English Proverbs* 114 Like priest, like people. . . Always taken in the worse sense. **1893** R. HEATH *English Peasant* IV. iii. He had so deep a reverence for the clergy, that it never entered into his mind that perhaps, after all, it was 'like people, like priest'.

people: *see also* IDLE people have the least leisure; MORE people know Tom Fool than Tom Fool knows; the VOICE of the people is the voice of God.

perfect: *see* PRACTICE makes perfect.

perish: *see* he who LIVES by the sword dies by the sword.

perjury: *see* JOVE but laughs at lovers' perjury.

PHYSICIAN, heal thyself

[LUKE iv. 23 Yee will surely say vnto me this prouerbe, Physition, heale thy selfe. (VULGATE: *medice cura teipsum*).] *c* **1400** tr. *Honorius of Autun's Elucidarium* (1909) 29 Blynde leches,[1] heeleth first youre silf! **1519** COLET *Sermon to Convocation* B5ᵛ If you wyll ponder and loke vpon oure mottis[2] fyrst take awaye the blockes out of your eies. Hit is an olde Prouerbe: Phisition heale thy selfe. **1780** T. FRANCKLIN tr. *Lucian's Works* I. 320 According to the old adage, 'physician, cure thyself'. **1875** S. SMILES *Thrift* ii. How can a man . . teach sobriety or cleanliness, if he be himself drunken or foul? 'Physician, heal thyself,' is the answer of his neighbours. **1979** G. SWARTHOUT *Skeletons* 150 Tell me the truth. Don't dig yourself in any deeper. Physician, heal thyself.
[1] doctors.   [2] i.e. motes.

pick: *see* HAWKS will not pick out hawks' eyes; *see* a PIN and pick it up, all the day you'll have good luck.

pickle (small quantity): *see* MANY a little makes a mickle.

One PICTURE is worth ten thousand words

There is no foundation to the ascription of Chinese origin in quot. 1927.

1921 *Printers' Ink* 8 Dec. 96 One look is worth a thousand words. **1927** *Ibid.* 10 Mar. 114 *Chinese proverb*, One picture is worth ten thousand words. . . The same influence could not be created even with the same picture in any other advertising medium. **1954** R. HAYDN *Journal of Edwin Carp* 90 'One picture speaks louder than ten thousand words.' Mr. Bovey repeated the adage this morning when . . he handed me my finished portrait. **1979** *Scientific American* Oct. 118 A picture is worth a thousand words.

Every PICTURE tells a story
[**1847** C. BRONTË *Jane Eyre* I. i. The letter-press . . I cared little for. . . Each picture told a story.] **1904** *Daily Mail* 26 Feb. 8 A London woman and Doan's Backache Kidney Pills. . . 'Every picture tells a story.' **1967** E. WILLIAMS *Beyond Belief* IV. xxiii. Every Picture Tells a Story. . . Sergeant Carr stood with his . . back to the fireplace, the lady next to the budgie, the dog next to her.

pie-crust: *see* PROMISES, like pie-crust, are made to be broken.

pig: *see* what can you EXPECT from a pig but a grunt?

See a PIN and pick it up, all the day you'll have good luck; see a pin and let it lie, bad luck you'll have all the day
[Cf. **1668** PEPYS *Diary* 2 Jan. (1976) IX. 7 The King answered to it with great indifferency. Sir W. Coventry answered: 'I see your Majesty doth not remember the old English proverb, 'He that will not stoop for a pin will never be worth a pound'.] **1843** J. O. HALLIWELL *Nursery Rhymes* 120 See a pin and let it lay, Bad luck you'll have all the day! **1883** C. S. BURNE *Shropshire Folklore* xxi. Pins are held . . unlucky . . in the North of England . . but side by side with this we have the thrifty maxim—See a pin and let it lie, You'll want a pin another day; See a pin and pick it up, All the day you'll have good luck. **1935** A. CHRISTIE *Tape-Measure Murder* in *Miss Marple's Final Cases* (1979) 'There's a pin in your tunic.' . . He said, 'They do say, "See a pin and pick it up, all the day you'll have good luck." '

pin: *see also* it's a SIN to steal a pin.

pint: *see* you cannot get a QUART into a pint pot.

piper: *see* he who PAYS the piper calls the tune.

pitch: *see* he that TOUCHES pitch shall be defiled.

The PITCHER will go to the well once too often
1340 *Ayenbite of Inwit* (EETS) 206 Zuo longe geth thet pot to the wetere: thet hit comth to-broke hom. **1584** J. WITHALS *Dict.* (rev. ed.) B1 So oft goeth the pitcher to the well, that at last it commeth broken home. **1777** N. SHAW *Collections of New London County Hist. Society* (1933) I. 223, I shall send down what I have, but dont you think the Pitcher will go to the well once too often? **1880** *Church Times* 30 Apr. 275 Some of Mr. Gladstone's feats in the way of sweeping obstacles out of his path have been wonderful; but the proverb tells us that the pitcher which goes oft to the well will be broken at last. **1962** A. CHRISTIE *Mirror Crack'd* xvii. A phrase came into her mind. . . The pitcher goes to the well once too often. Nonsense. Nobody could suspect that it was she.

pitcher: *see also* LITTLE pitchers have large ears.

pitchfork: *see* you can DRIVE out Nature with a pitchfork, but she keeps on coming back.

pitied: *see* BETTER be envied than pitied.

PITY is akin to love
1601 SHAKESPEARE *Twelfth Night* III. i. 119, I pity you.—That's a degree to love. **1696** T. SOUTHERNE *Oroonoko* II. i. Do, pity me: Pity's a-kin to Love. *a* **1895** F. LOCKER-LAMPSON *My Confidences* (1896) 95 They say that Pity is akin to Love, though only a Poor Relation; but Amy did not even pity me.

A PLACE for everything, and everything in its place
1640 G. HERBERT *Outlandish Proverbs* no. 379 All things have their place, knew wee how to place them. **1842** MARRYAT *Masterman Ready* II. i. In a well-conducted man-of-war . . every thing in its place, and there is a place for every thing. **1855** T. C. HALIBURTON *Nature & Human Nature* I. vi. There was a place for everything, and everything was in its place. **1928** D. L. SAYERS *Lord Peter views Body* x. 'I thought you were rather partial to anatomical specimens.' 'So I am, but not on the breakfast-table. "A place for everything and everything in its place," as my grandmother used to say.' **1968** P. DICKINSON *Skin Deep* vii. Do you run your whole life like that? . . A place for everything and everything in its place, and all in easy reach.

There's no PLACE like home
Cf. EAST, *west, home's best.*
1571 T. TUSSER *Husbandry* (rev. ed.) H1ᵛ Though home be but homely, yet huswife is taught, That home hath no fellow to such as haue aught. **1823** J. H. PAYNE *Clari* I. i. 'Mid pleasures and palaces though we may roam, Be it ever so humble, there's no place like home. **1939** E. F. BENSON *Trouble for Lucia* xi. 'What a joy to have it back at Mallards again!' . . 'No place like home is there, dear?'

place: *see also* LIGHTNING never strikes the same place twice; there is a TIME and place for everything; a WOMAN's place is in the home.

plague: *see* PLEASE your eye and plague your heart.

plant: *see* it is not SPRING until you can plant your foot upon twelve daisies; WAL-NUTS and pears you plant for your heirs.

Those who PLAY at bowls must look out for rubbers
Rubber is apparently an alteration of *rub*, an obstacle or impediment to the course of a bowl.

[**1595** SHAKESPEARE *Richard II* III. iv. 4 Madam, we'll play at bowls.—'Twill make me think the world is full of rubs.] **1762** SMOLLETT *Sir Launcelot Greaves* I. x. (*heading*) Which sheweth that he who plays at bowls, will sometimes meet with rubbers. **1824** SCOTT *Redgauntlet* III. vi. 'And how if it fails?' said Darsie. 'Thereafter as it may be—' said Nixon; 'they who play at bowls must meet with rubbers.' **1874** L. STEPHEN *Hours in Library* I. 384 De Quincey . . admits . . that the fanaticism of the rub was 'much more reasonable' than the fanaticism of Priestly; and that those who play at bowls must look out for rubbers. **1907** F. W. HACKWOOD *Old English Sports* xi. Another term used in common speech and derived from this game [bowls] is 'rub'; as when we say . . 'he who plays at bowls must look out for rubs'—that is, he must consider the inequalities of the ground, and . . make due allowance for them.

If you PLAY with fire you get burnt
The metaphorical phrase *to play with fire*, to tinker with something potentially dangerous, is also commonly found. Cf. *He that* TOUCHES *pitch shall be defiled.*
[**1655** H. VAUGHAN *Silex Scintillans* II. 15, I played with fire, did counsell spurn, . . But never thought that fire would burn, Or that a soul could ake.] **1884** R. H. THORPE *Fenton Family* xiv. If people will play with fire, they must expect to be burned by it some time. If I had not learned the game, and thought myself a good player, I'd never have lost Mother's money. **1980** P. KINSLEY *Vatchman Switch* xxiv. If you play with fire you get burnt. Shouldn't mess around in Crown Colonies.

play: *see also* (noun) FAIR play's a jewel; GIVE and take is fair play; TURN about is fair play; all WORK and no play makes Jack a dull boy; (verb) when the CAT's away, the mice will play.

You can't PLEASE everyone
1472 E. PASTON *Letter* 16 May in *Paston Letters* (1971) I. 635, I am in serteyn the

contrary is true—yt is nomore but that he can not plese all partys. **1616** T. DRAXE *Adages* 45 One can hardly please all men. **1844** RUSKIN *Journal* 30 Apr. in *Diaries 1835–47* (1956) 274 At Ward's about window—nothing done. Gastineau came up and don't like mine: can't please everybody. **1981** *Daily Telegraph* 16 May 18 The old adage, 'you can't please everyone', holds good.

PLEASE your eye and plague your heart
  c **1617** A. BREWER *Lovesick King* (1655) III. E3ᵛ She may please your eye a little . . but vex your heart. **1748** SMOLLETT *Roderick Random* II. xl. Many a substantial farmer . . would be glad to marry her; but she was resolved to please her eye, if she should plague her heart. **1829** COBBETT *Advice to Young Men* III. cxxic. 'Please your eye and plague your heart' is an adage that want of beauty invented, I dare say, more than a thousand years ago. **1876** I. BANKS *Manchester Man* III. vi. But I *will* marry him, mamma—I'll please my eye, if I plague my heart.

please: *see also* LITTLE things please little minds.

pleasure: *see* BUSINESS before pleasure; he that would go to SEA for pleasure, would go to hell for a pastime.

plum: *see* a CHERRY a year a merry year.

An old POACHER makes the best gamekeeper
  Cf. *Set a* THIEF *to catch a thief.*
  c **1390** CHAUCER *Physician's Tale* l. 83 A theef of venysoun, that hath forlaft His likerousnesse[1] and al his olde craft, Kan kepe a forest best of any man. **1695** T. FULLER *Church Hist. Britain* IX. iii. Always set a—to catch a—; and the greatest dear-stealers, make the best Parke-keepers. **1878** R. JEFFERIES *Gamekeeper at Home* ix. There is a saying that an old poacher makes the best gamekeeper, on the principle of setting a thief to catch a thief. **1970** V. CANNING *Great Affair* iii. What the Church needed,

possibly, was a good leavening of sinners in its ministry, on the principle that poachers make the best gamekeepers.
  [1] depravity.

pocket: *see* SHROUDS have no pockets.

point: *see* POSSESSION is nine points of the law.

poison: *see* one man's MEAT is another man's poison.

poke: *see* you should KNOW a man seven years before you stir his fire.

policy: *see* HONESTY is the best policy.

politeness: *see* CIVILITY costs nothing; PUNCTUALITY is the politeness of princes.

POLITICS makes strange bedfellows
  Politics has long been considered a plural noun; its use with a singular verb is comparatively recent. Cf. ADVERSITY *makes strange bedfellows.*
  1839 P. HONE *Diary* 9 July (1927) I. 404 Party politics, like poverty, bring men 'acquainted with strange bedfellows'. **1870** C. D. WARNER *My Summer in Garden* (1871) 187 The Doolittle raspberries have sprawled all over the strawberry-beds: so true is it that politics makes strange bed-fellows. **1936** M. MITCHELL *Gone with Wind* lviii. Ashley Wilkes and I are mainly responsible. Platitudinously but truly, politics make strange bedfellows. **1980** P. VAN GREENAWAY *Dissident* vii. Even enemies have something in common. Statecraft produces strange bedfellows.

It is a POOR heart that never rejoices
  1834 MARRYAT *Peter Simple* I. v. 'Well,' continued he, 'it's a poor heart that never rejoiceth.' He then poured out half a tumbler of rum. **1841** DICKENS *Barnaby Rudge* iv. What happened when I reached home you may guess. . . Ah! Well, it's a poor heart that never rejoices. **1935** E. F. BENSON *Lucia's Progress* viii. They were all men together, he said, and it was a sad heart that never

rejoiced. **1979** J. SCOTT *Clutch of Vipers* iv. 'It's a poor heart', Frankie told him, 'that never rejoices.'

poor: *see also* one LAW for the rich and another for the poor; the RICH man has his ice in the summer and the poor man gets his in the winter.

Pope: *see* it is ill SITTING at Rome and striving with the Pope.

port: *see* ANY port in a storm.

POSSESSION is nine points of the law
There is no specific legal ruling which supports this proverb—though the concept is widely acknowledged—but in early use the satisfaction of ten (sometimes twelve) points was commonly asserted to attest full entitlement or ownership. Possession, represented by nine (or eleven) points, is therefore the closest substitute for this.

[**1595** *Edward III* E3 Tis you are in possession of the Crowne, And thats the surest poynt of all the Law.] **1616** T. DRAXE *Adages* 163 Possession is nine points in the Law. **1659** J. IRETON *Oration* 5 This Rascally–devill . . denys to pay a farthing of rent. Tis true, possession is nine points of the Law, Yet give Gentlemen, right's right. **1709** O. DYKES *English Proverbs* 213 Possession is a mighty Matter indeed; and we commonly say, 'tis eleven Points of the Law. It goes a great Way to the giving of Security, but not any Right. **1822** T. L. PEACOCK *Maid Marian* v. In those days possession was considerably more than eleven points of the Law. The baron was therefore convinced that the earl's outlawry was infallible. **1920** J. GALSWORTHY *In Chancery* II. xiv. We're the backbone of the country. They [Leftists] won't upset us easily. Possession's nine points of the Law. **1980** M. BABSON *Queue here for Murder* xviii. I'm in the Penthouse Suite and I'm staying there. 'Possession', he added . . 'is nine points of the law.' 'Eleven,' Belva Barrie said automatically.

possible: *see* ALL things are possible with God; all's for the BEST in the best of all possible worlds.

post: *see* the post of HONOUR is the post of danger.

A POSTERN door makes a thief
Cf. OPPORTUNITY *makes a thief.*

c **1450** *Proverbs of Good Counsel* in *Book of Precedence* (EETS) 69 A nyse wyfe, & a back dore, Makyth often tymus A ryche man pore. **1573** J. SANFORDE *Garden of Pleasure* 107 The posterne dore[1] destroyeth the house. **1611** J. DAVIES *Scourge of Folly* 146 The Posterne doore makes theefe and whore. But, were that dam'd with Stone, or Clay, Whoores and Theeues would find a way. **1732** T. FULLER *Gnomologia* no. 6176 The Postern Door Makes Thief and Whore. **1977** J. AIKEN *Five-Minute Marriage* xi. 'I shall never be able to sleep securely in this room, if thieves are to be always breaking in and waking me up!' 'A postern door do always make a thief.'

[1] *posterne door*, back door.

pot: *see* if IFS and ands were pots and pans, there'd be no work for tinkers' hands; a LITTLE pot is soon hot; you cannot get a QUART into a pint pot; a WATCHED pot never boils.

pound: *see* IN for a penny, in for a pound; an OUNCE of practice is worth a pound of precept; take care of the PENCE and the pounds will take care of themselves; PENNY wise and pound foolish.

pour: *see* it never RAINS but it pours.

When POVERTY comes in at the door, love flies out of the window
[**1474** CAXTON *Game of Chess* III. iii. Herof men saye a comyn prouerbe in englond that loue lastest as longe as the money endurith.] **1631** R. BRATHWAIT *English Gentlewoman* vi. It hath been an old Maxime; that as pouerty goes in at one doore, loue goes out at the other. **1639** J. CLARKE *Parœmiologia Anglo-Latina* 25 When povertie comes in at

doores, love leapes out at windowes.
1790 *Universal Asylum* Aug. 84, I hope,
ladies, none of you may ever experience, that 'when poverty comes in
at the door, love flies out at the
windows'. **1894** J. LUBBOCK *Use of Life*
iii. It is a mean proverb that, 'When
poverty comes in at the door, love flies
out of the window'.

POVERTY is no disgrace, but it is a great inconvenience

1591 J. FLORIO *Second Fruits* 105
Neuer be ashamed of thy calling, for
Pouertie is no vice, though it be an
inconvenience. **1721** J. KELLY *Scottish
Proverbs* 278 Poortha[1] is a Pain, but no
Disgrace. Unless it be the Effects of Laziness, and Luxury. **1945** F. THOMPSON
Lark Rise i. 'Poverty's no disgrace, but
'tis a great inconvenience' was a common saying among the Lark Rise
people; but . . their poverty was no less
than a hampering drag upon them.

[1] Poverty.

POVERTY is not a crime

1591 J. FLORIO *Second Fruits* 105
Pouertie is no vice. **1640** G. HERBERT
Outlandish Proverbs no. 844 Poverty is
no sinne. **1785** C. MACKLIN *Man of
World* IV. 56 Her Poverty is not her
crime, Sir, but her misfortune. **1839**
DICKENS *Nicholas Nickleby* lv. 'Remember how poor we are.' Mrs. Nickleby . .
said through her tears that poverty was
not a crime. **1945** F. THOMPSON *Lark
Rise* ii. There's nothing the matter with
Lark Rise folks but poverty, and that's
no crime. If it was, we should likely be
hung ourselves.

poverty: *see also* ADVERSITY makes strange
bedfellows.

powder: *see* put your TRUST in God, and
keep your powder dry.

POWER corrupts

The proverb is now commonly used in
allusion to quot. 1887.
   1876 TROLLOPE *Prime Minister* IV. viii.
We know that power does corrupt, and

that we cannot trust kings to have loving hearts. **1887** LD. ACTON *Letter* in *Life
& Letters of Mandel Creighton* (1904) I.
xiii. Power tends to corrupt, and absolute power corrupts absolutely. Great
men are almost always bad men, even
when they exercise influence and not
authority. **1957** V. BRITTAIN *Testament
of Experience* II. ix. The processes by
which 'power corrupts' are perhaps
inevitable. **1979** C. McCARRY *Better
Angels* IV. xii. He doesn't *know* that
power corrupts; there's nothing dark in
him.

power: *see also* KNOWLEDGE is power;
MONEY is power.

PRACTICE makes perfect

1553 T. WILSON *Art of Rhetoric* 3 Eloquence was vsed, and through practise
made parfect. **1599** H. PORTER *Two
Angry Women of Abingdon* l. 913 Forsooth as vse makes perfectnes, so seldome seene is soone forgotten. **1761**
J. ADAMS *Diary* (1961) I. 192 Practice
makes perfect. **1863** C. READE *Hard
Cash* III. iv. He lighted seven fires, skillfully on the whole, for practice makes
perfect. **1979** D. LESSING *Shikasta* 185 It
is like playing the piano or riding a
bicycle. Practice makes perfect.

practice: *see also* an OUNCE of practice is
worth a pound of precept.

PRACTISE what you preach

1377 LANGLAND *Piers Plowman* B. XIII.
79 This goddes gloton . . Hath no pyte
on vs pore. He performeth yuel[1], That he
precheth he preueth[2] nought. **1639**
T. FULLER *Holy War* I. xxiii. The Levites
. . had 48 cities . . being better provided
for then many English ministers, who
may preach of hospitalitie to their
people, but cannot go to the cost to
practice their own doctrine. **1678**
R. L'ESTRANGE *Seneca's Morals* II. ii. We
must practise what we preach. **1854**
THACKERAY *Newcomes* I. xiv. Take counsel by an old soldier, who fully practises
what he preaches, and beseeches you to
beware of the bottle. **1945** F. THOMPSON

Lark Rise iv. Songs of a high moral tone, such as: Waste not, want not . . And practise what you preach.

¹ *perforneth yuel*: i.e. performeth evil.
² demonstrates.

PRAISE the child, and you make love to the mother

1829 COBBETT *Advice to Young Men* IV. clxxxi. It is an old saying, 'Praise the child, and you make love to the mother'; and it is surprising how far this will go. **1885** E. J. HARDY *How to be Happy though Married* xix. 'Praise the child, and you make love to the mother,' and it is a thing that no husband ought to overlook.

praise: *see also* (noun) SELF-praise is no recommendation.

pray: *see* the FAMILY that prays together stays together.

precept: *see* EXAMPLE is better than precept; an OUNCE of practice is worth a pound of precept.

prepare: *see* HOPE for the best and prepare for the worst; if you want PEACE, you must prepare for war.

present: *see* no TIME like the present.

preservation: *see* SELF-preservation is the first law of nature.

pressed: *see* one VOLUNTEER is worth two pressed men.

PREVENTION is better than cure

[*c* **1240** BRACTON *De Legibus* v. x. *melius & utilius [est] in tempore occurrere, quam post causam vulneratam quaerere remedium*, it is better and more useful to meet a problem in time than to seek a remedy after the damage is done.] **1618** T. ADAMS *Happiness of Church* 146 Preuention is so much better then healing, because it saues the labour of being sicke. **1732** T. FULLER *Gnomologia* no. 3962 Prevention is much preferable to Cure. **1826** J. PINTARD *Letter* 19 Apr.

(1940) II. 257 Prevention is better than cure. . . With perseverance we shall save numbers of little Devils from becoming big ones. **1954** R. HAYDN *Journal of Edwin Carp* 148 'Why do you wear those old galoshes when the sun's shining?' . . 'Prevention's better than Cure.' **1981** M. SELLARS *From Eternity to Here* vi. Coates . . addressed a meeting of private detectives and tried to convince us that prevention was better than cure.

price: *see* EVERY man has his price.

PRIDE feels no pain

1614 T. ADAMS *Devil's Banquet* II. 73 Pride is neuer without her own paine, though shee will not feele it: be her garments what they will, yet she will neuer be too hot, nor too colde. **1631** JONSON *New Inn* II. i. Thou must make shift with it. Pride feeles no pain. Girt thee hard, Pru. **1721** J. KELLY *Scottish Proverbs* 277 Pride finds no cold. Spoken . . to Beaus¹ with their open Breasts, and Ladies with their extravagant Hoops.² **1837** T. HOOK *Jack Brag* III. iii. Truly, indeed, does the proverb say that 'pride knows no pain'. One fiftieth part of the turmoil and exertion which Jack underwent . . would in all probability have secured him ease and competence. **1981** *Radio Times* 28 Feb.–6 Mar. 43 (Advt.), Pride feels no pain, the saying goes. Thankfully, with Clarks [shoes] it doesn't have to.

¹ well-dressed men; fops or dandies.
² wide skirts worn over a framework of hoops.

PRIDE goes before a fall

[PROVERBS xvi. 18 Pride goeth before¹ destruction: and an hautie² spirit before a fall.] *c* **1390** GOWER *Confessio Amantis* I. 3062 Pride . . schal doun falle. **1509** A. BARCLAY *Ship of Fools* 195ᵛ First or last foule pryde wyll haue a fall. **1784** JOHNSON *Letter* 2 Aug. (1952) III. 191, I am now reduced to think . . of the weather. Pride must have a fall. **1856** MELVILLE *Piazza Tales* 431 The bell's main weakness was where man's blood had flawed it. And so pride went before the fall.

1930 W. S. MAUGHAM *Cakes & Ale* v. I
suppose he thinks he'd be mayor
himself. . . Pride goeth before a fall.
1980 M. L. WEST in K. J. Dover *Ancient
Greek Literature* iii. The spectacle of
Xerxes' defeat tremendously reinforced
the traditional conviction that pride
goes before a fall.
 [1] *goeth before*, precedes.   [2] i.e., haughty.

priest: *see* ONCE a priest, always a priest;
like PEOPLE, like priest.

prince: *see* whosoever DRAWS his sword
against the prince must throw the scabbard away; PUNCTUALITY is the politeness of princes.

problem: *see* if you're not part of the SOLU-
TION you're part of the problem.

PROCRASTINATION is the thief of time
1742 YOUNG *Night Thoughts* I. 18 Procrastination is the Thief of Time; Year
after year it steals, till all are fled. 1850
DICKENS *David Copperfield* xii. Never do
to-morrow what you can do to-day.
Procrastination is the thief of time. 1935
O. NASH *Primrose Path* 100 Far from
being the thief of Time, procrastination
is the king of it. 1980 P. VAN GREEN-
AWAY *Dissident* ii. What once could
leisurely be dwelt on . . must go by the
board, speed being the thief of time in
our neck of the century.

PROMISES, like pie-crust, are made to be broken
1681 *Heraclitus Ridens* 16 Aug., He
makes no more of breaking Acts of Par-
liaments, than if they were like Promises and Pie-crust made to be broken.
1871 TROLLOPE *Ralph the Heir* II. iv.
'Promises like that are mere pie-crust,'
said Ralph. 1948 F. THOMPSON *Still
glides Stream* v. Ain't you larned yet that
a gal's promises be but piecrust, made to
be broke? 1981 *Family Circle* Feb. 66
Promises, like pie-crusts, they say, are
made to be broken. Not at Sainsbury's.
Every single pie they sell lives up to the
promise of its famous name.

The PROOF of the pudding is in the eating
Proof means 'test' rather than the more
normal 'verification, proving to be true'.
c 1300 *King Alisaunder* (EETS) l. 4038
Jt is ywrite that euery thing Hymself sheweth in the tastyng.   1623
W. CAMDEN *Remains concerning Britain*
(ed. 3) 266 All the proofe of a pudding, is
in the eating. 1666 G. TORRIANO *Italian
Proverbs* 100 (*note*) As they say at the
winding up, or the proof of the pudding
is in the eating. 1738 SWIFT *Polite Conversation* II. 132 The Proof of the Pudden
is in the Eating. 1842 R. H. BARHAM
Ingoldsby Legends 2nd Ser. 25 With respect to the scheme . . I've known soldiers adopt a worse stratagem. . .
There's a proverb however, I've always
thought clever . . The proof of the Pudding is found in the eating. 1924
J. GALSWORTHY *White Monkey* III. xii. Let
us . . look at the thing more widely. The
proof of the pudding is in the eating.
1979 *Daily Telegraph* 27 Oct. 3 The proof
of the pudding is in the eating. The villa-
gers eat their own vegetables and suffer
no illness.

A PROPHET is not without honour save in his own country
[MATTHEW xiii. 57 A Prophet is not
without honour, saue in his owne coun-
trey, and in his owne house.] *a* 1485
CAXTON in Malory *Works* (1967) I. p.cxlv,
The word of God . . sayth that no man
is accept for a prophete in his owne
contreye. 1603 FLORIO tr. *Montaigne's
Essays* III. ii. No man hath beene a
Prophet . . in his owne country, saith
the experience of histories. 1771 SMOL-
LETT *Humphry Clinker* III. 92 The captain, like the prophets of old, is but little
honoured in his own country. 1946
W. S. MAUGHAM *Then & Now* xxx. In
Florence . . they had no great confidence in his judgment and never followed his advice. 'A prophet is not
without honour save in his own country.'

propose: *see* MAN proposes, God
disposes.

prosper: *see* CHEATS never prosper.

protect: *see* HEAVEN protects children, sailors, and drunken men.

prove: *see* the EXCEPTION proves the rule.

provide: *see* TAKE the goods the gods provide.

PROVIDENCE is always on the side of the big battalions
    [**1673** MME. DE SÉVIGNÉ *Letter* 22 Dec. *la fortune est toujours, comme disait le pauvre M. de Turenne, pour les gros bataillons,* fortune is always, as poor Mr. de Turenne used to say, for the big battalions.] **1822** A. GRAYDON *Memoirs* v. Heaven was ever found favourable to strong battalions. **1842** A. ALLISON *Hist. Europe* X. lxxviii. Providence was always on the side of dense battalions. **1904** 'SAKI' *Reginald* 63 Someone has observed that Providence is always on the side of the big dividends. **1943** R. A. J. WALLING *Corpse by any Other Name* iii. Our statesmen . . ought to have learned years ago that Providence is always on the side of the big battalions. **1979** *Guardian* 9 July 9 Many thousands more voices now are raised in the name of sanity. But I dare say God is still on the side of the big battalions.

public: *see* one does not WASH one's dirty linen in public.

Any PUBLICITY is good publicity
    1974 P. CAVE *Dirtiest Picture Postcard* xiv. Haven't you ever heard the old adman's adage . . 'any publicity is good publicity'? **1975** 'E. LATHEN' *By Hook or by Crook* xiii. It's hard to say how that will affect the turnout. They say that any publicity is good publicity. **1980** R. HILL *Killing Kindness* iii. There are all kinds of gain. . . In the entertainment world, there's no such thing as bad publicity.

pudding: *see* the PROOF of the pudding is in the eating.

puddle: *see* the SUN loses nothing by shining into a puddle.

It is easier to PULL down than to build up
    1577 R. STANYHURST *Hist. Ireland* in Holinshed *Chronicles* 89 It is easie to raze, but hard to buylde. **1587** J. BRIDGES *Defence of Government in Church of England* vi. 518 We may quicklier pull downe with one hande, than wee can easilie builde againe with both. **1644** J. HOWELL *Dodona's Grove* 134 In politicall affaires, as well as mechanicall, it is farre easier to pull downe, then build up. **1909** *Times* 29 Apr. 9 Turkey and her new rulers . . have astonished those who thought they knew the Turks best by . . the vigour . . with which the great change has been conducted. . . But it is easier always and everywhere to pull down than to build up.

PUNCTUALITY is the politeness of princes
    [Fr. *l'exactitude est la politesse des rois,* punctuality is the politeness of kings (attributed to Louis XVIII, 1755–1824).] **1834** M. EDGEWORTH *Helen* II. ix. 'Punctuality is the virtue of princes.'. . Mr. Harley . . would have ridiculed so antiquated a notion. **1854** R. S. SURTEES *Handley Cross* (ed. 2) xli. Punctuality is the purlitness o' princes, and I doesn't like keepin' people waitin'. **1940** C. DICKSON *And so to Murder* 181 Punctuality . . has been called the politeness of kings. It's more than that: it's plain good business. **1981** P. MCCUTCHAN *Shard calls Tune* xv. One should never keep people waiting; punctuality was the politeness of princes.

PUNCTUALITY is the soul of business
    1853 T. C. HALIBURTON *Wise Saws* I. iii. 'Punctuality,' sais I, 'my lord, is the soul of business.' **1911** W. CROSSING *Folk Rhymes of Devon* 16 Punctuality is the soul of business, and in these days of cheap watches there can be no excuse for anybody failing to cultivate the habit. **1940** C. DICKSON *And so to Murder* 181 Punctuality . . has been called the

politeness of kings. It's more than that: it's plain good business.

punished: *see* CORPORATIONS have neither bodies to be punished nor souls to be damned.

To the PURE all things are pure

[TITUS i. 15 Vnto the pure all things are pure, but vnto them that are defiled, and vnbeleeuing, is nothing pure.] **1854** S. M. HAYDEN *Early Engagements* ii. Would that our earth were more frequently brightened and purified by such spirits. . . 'To the pure all things are pure.' **1895** G. ALLEN *Woman who Did* vii. Herminia, for her part, never discovered she was talked about. To the pure all things are pure. **1941** N. MARSH *Death in Ecstasy* xvi. How dare you suggest such a thing? . . Maurice Pringle burst out laughing. . . Look. To the pure all things are pure. **1979** C. CURZON *Leaven of Malice* iv. Perhaps I had been oversensitive. . . To the pure, after all, all things are pure.

purpose: *see* the DEVIL can quote Scripture for his own ends.

purse: *see* you can't make a SILK purse out of a sow's ear.

Never PUT off till tomorrow what you can do today

The proverb is often humorously reversed.

c **1390** CHAUCER *Tale of Melibee* l. 2984 An olde proverbe . . seith that 'the goodnesse that thou mayst do this day, do it, and abide nat ne delaye it nat till to-morwe'. **1616** T. DRAXE *Adages* 42 Deferre not vntill to morrow, if thou canst do it to day. **1633** J. HOWELL *Familiar Letters* 5 Sept. (1903) II. 140 Secretary Cecil . . would ofttimes speak of himself, 'It shall never be said of me that I will defer till to-morrow what I can do to-day.' **1749** CHESTERFIELD *Letter* 26 Dec. (1932) IV. 1478 No procrastination; never put off till to-morrow what you can do to-day. **1869** C. H. SPURGEON *John Ploughman's Talk* vii. These slow coaches think that to-morrow is better than to-day, and take for their rule an old proverb turned topsy-turvy— 'Never do to-day what you can put off till tomorrow.' **1980** J. LEES-MILNE *Harold Nicolson* xv. Lord Sackville was . . a lovable, easy-going but indolent peer whose philosophy is best summarized in one of his pet sayings: 'Never do to-day what you can possibly put off until tomorrow.'

put: *see also* don't put all your EGGS in one basket; you can't put NEW wine in old bottles; you cannot put an OLD head on young shoulders; put a STOUT heart to a stey brae; put your TRUST in God, and keep your powder dry.

Q

The QUARREL of lovers is the renewal of love

[TERENCE *Andria* l. 555 *amantium irae amoris integratiost,* lovers' quarrels are a strengthening of love.] *c* **1520** *Terence in English* C1 The angers of louers renew love agayn. **1576** R. EDWARDES *Paradise of Dainty Devises* 42 Now haue I founde, the prouerbe true to proue, The fallyng out of faithfull frends, is the renuyng of love. **1624** BURTON *Anatomy of Melancholy* (ed. 2) III. ii. She would . . picke quarrells vpon no occasion, because she would be reconciled to him againe. . . The falling out of lovers is the renuing of loue. **1754** RICHARDSON *Grandison* III. xviii. The falling out of Lovers . . is the renewal of Love. Are we not now better friends than if we had never differed? **1874** TROLLOPE *Phineas Redux* II. xxix. She knew that 'the quarrel of lovers is the renewal of love'. At any rate, the woman always desires that it may be so, and endeavours to reconcile the parted ones. **1905** *Graphic* (Christmas) 14 (*caption*) The quarrel of lovers is the renewal of love. **1980** M. GILBERT *Death of Favourite Girl* ii. Bear in mind, ladies, that a lovers' quarrel sometimes signifies the rebirth of love.

quarrel: *see also* (noun) it takes TWO to make a quarrel; (verb) a BAD workman blames his tools.

You cannot get a QUART into a pint pot
The metaphorical phrase *to get* (or *put*) *a quart*[1] *into a pint pot* is also used.

1896 *Daily News* 23 July 4 They had been too ambitious. They had attempted what he might describe in homely phrase as putting a quart into a pint pot. **1948** P. M. WARNER *Embroidery Mary* xi. When they . . got down to . . packing it was found to be a case of 'quarts into pint pots will not go.' **1967** RIDOUT & WITTING *English Proverbs Explained* 170 You can't get a quart into a pint pot, so we'll have to make do with much less luggage. **1974** W. FOLEY *Child in Forest* I. 101 A quart may not go into a pint pot, but my feet had to go into those boots.
[1] an imperial measure of two pints.

queer: *see* there's NOWT so queer as folk.

quench: *see* DIRTY water will quench fire.

question: *see* ASK no questions and hear no lies; FOOLS ask questions that wise men cannot answer; there are TWO sides to every question.

QUICKLY come, quickly go
Cf. EASY *come, easy go.*

1583 B. MELBANCKE *Philotimus* 151 Quickly spent, thats easely gotten. **1631** J. MABBE tr. *F. de Rojas' Celestina* I. 8 Quickly be wonne, and quickly be lost. **1869** W. C. HAZLITT *English Proverbs* 322 Quickly come, quickly go. **1979** N. GOLLER *Tomorrow's Silence* iv. 'Was he alright when you came home?' . . 'Yes, what comes quickly must go quickly, that's what I say.'

quickly: *see also* he GIVES twice who gives quickly.

quiet: *see* the best DOCTORS are Dr Diet, Dr Quiet, and Dr Merryman.

quote: *see* the DEVIL can quote Scripture for his own ends.

R

The RACE is not to the swift, nor the battle to the strong

[ECCLESIASTES ix. 11 The race is not to the swift, nor the battle to the strong.]
1632 BURTON *Anatomy of Melancholy* (ed. 4) II. iii. It is not honesty, learning, worth, wisdome, that preferres men, The race is not to the swift, nor the battell to the stronger [1638 strong]. **1873** C. M. YONGE *Pillars of House* III. xxxii. Poor child! she lay . . trying to work out . . why the race is not to the swift, nor the battle to the strong. **1901** G. B. SHAW *Caesar & Cleopatra* in *Three Plays for Puritans* 96 The descendants of the gods did not stay to be butchered, cousin. The battle was not to the strong; but the race was to the swift. **1976** T. SHARPE *Wilt* xix. The race was not to the swift after all, it was to the indefatigably inconsequential and life was random.

ragged: *see* there's many a GOOD cock come out of a tattered bag.

RAIN before seven: fine before eleven

1853 *Notes & Queries* 1st Ser. VIII. 218 Weather Proverbs. . . Rain before seven, fine before eleven. **1909** *Spectator* 20 Mar. 452 'Rain before seven, shine before eleven,' is one of the most trustworthy of all country saws. **1940** B. DE VOTO (*title*) Rain before seven.

rain: *see also* blessed are the DEAD that the rain rains on; if in FEBRUARY there be no rain, 'tis neither good for hay nor grain; SAINT Swithun's day if thou be fair for forty days it will remain.

It never RAINS but it pours

1726 J. ARBUTHNOT (*title*) It cannot rain but[1] it pours. **1771** T. GRAY *Letter* 2 Feb. (1971) III. 1164 It never rains, but it pours, my dear Doctor. . . Mr. Br: has added to his Mastership . . a living hard by Cambridge. **1857** TROLLOPE *Barchester Towers* III. xii. A wife with a large fortune too. It never rains but it pours,

does it, Mr. Thorne? **1979** L. BARNEA *Reported Missing* vii. I listened to the radio. Ben Gurion had suffered a stroke. . . It never rains but it pours.

[1] used (now somewhat archaically) to introduce an inevitable accompanying circumstance.

It is easier to RAISE the Devil than to lay him

1655 T. FULLER *Church Hist. Britain* X. iv. The Boy having gotten a habit of counterfeiting . . would not be undeviled by all their Exorcisms, so that the Priests raised up a Spirit which they could not allay. **1725** N. BAILEY tr. *Erasmus' Colloquies* 202 'Tis an old Saying and a true, 'Tis an easier Matter to raise the Devil, than 'tis to lay him. **1845** MACAULAY *Works* (1898) XII. 136 Did you think, when, to serve your turn, you called the Devil up, that it was as easy to lay him as to raise him? **1890** 'R. BOLDREWOOD' *Miner's Right* II. viii. Exorcists of all lands . . have ever found the fiend more easy to invoke than to lay.

ransom: *see* a PECK of March dust is worth a king's ransom.

rat: *see* the CAT, the rat, and Lovell the dog, rule all England under the hog.

reach: *see* STRETCH your arm no further than your sleeve will reach.

reap: *see* as you sow, so you reap; they that sow the wind shall reap the whirlwind.

There is REASON in the roasting of eggs
There is reason behind every action, however odd it may seem.

1659 J. HOWELL *Proverbs* (English) 12 Ther's reason in rosting of Eggs. **1785** BOSWELL *Journal of Tour of Hebrides* 24 (*note*) Every man whatever is more or less a cook, in seasoning what he him-

self eats.—Your definition is good, said
Mr. Burke, and I now see the full force of
the common proverb, 'There is *reason* in
roasting of eggs'. **1867** TROLLOPE *Last
Chronicle of Barset* II. lxxv. But there's
reason in the roasting of eggs, and . .
money is not so plentiful . . that your
uncle can afford to throw it into the
Barchester gutters. **1915** SOMERVILLE &
ROSS *In Mr. Know's Country* ix. I seemed
to myself merely an imbecile, sitting in
heavy rain, staring at a stone wall. Half
an hour, or more, passed. 'I'm going out
of this,' I said to myself defiantly;
'there's reason in the roasting of eggs.'
[**1949** D. SMITH *I capture Castle* iii. Mother
said: 'There's reason in everything and
Thomas ought to be in bed.']

recalled: *see* things PAST cannot be recalled.

receive: *see* it is BETTER to give than to
receive.

**If there were no RECEIVERS, there
would be no thieves**
    c **1390** CHAUCER *Cook's Tale* l. 4415
Ther is no theef with-oute a lowke,[1] That
helpeth hym to wasten and to sowke.[2]
1546 J. HEYWOOD *Dialogue of Proverbs* I.
xii. F1 This prouerbe preeues, Where be
no receyuers, there be no theeues. **1614**
T. ADAMS *Devil's Banquet* II. 67 The
Calumniator is a wretched Thiefe, and
robs man of the best thing he hath. . .
But if there were no receiuer, there
would be no Thiefe. **1884** R. JEFFERIES
Red Deer v. No one would buy a stolen
deer, knowing the inevitable conse-
quences, and as there are no receivers . .
there are no thieves. **1926** *Times* 22
Nov. 11 It had often been said in those
Courts that if there were no receivers
there would be no thieves.
    [1] accomplice.   [2] cheat.

reckoning: *see* SHORT reckonings make
long friends.

recommendation: *see* SELF-praise is no recommendation.

**RED sky at night, shepherd's delight;
red sky in the morning, shepherd's
warning**
    c **1395** WYCLIF *Bible* Matthew xvi. 2
The eeuenynge maad, ye seien, It shal
be cleer, for the heuene is lijk to reed;
and the morwe, To day tempest, for
heuen shyneth heuy, or sorwful.
c **1454** R. PECOCK *Follower to Donet* (EETS)
54 We trowen[1] that this day schal be a
reyny day for that his morownyng was
reed, or that to morow schal be a fayre
day for that his euentide is reed. **1592–3**
SHAKESPEARE *Venus & Adonis* l. 453 Like
a red morn, that ever yet betoken'd
Wreck to the seaman . . Sorrow to
shepherds. **1611** BIBLE *Matthew* xvi. 2
When it is euening, yee say, It will bee
faire weather: for the skie is red. And in
the morning, It will be foule weather to
day: for the skie is red and lowring.
1893 R. INWARDS *Weather Lore* 53 Sky red
in the morning shepherd's warning.
1979 P. ALEXANDER *Show me Hero* xxv.
delight. **1920** *Punch* 14 July 36 Red sky
at night shepherd's delight. . . Red sky
in the morning shepherd's warning.
1979 P. ALEXANDER *Show me Hero* xxv.
'Going to be a fine day,' he said at last.
'Red sky in the morning, shepherd's
warning,' Ashman said.
    [1] believe.

redressed: *see* a FAULT confessed is half
redressed.

**A REED before the wind lives on, while
mighty oaks do fall**
    c **1385** CHAUCER *Troilus & Criseyde* II.
1387 And reed that boweth down for
every blast, Ful lightly, cesse wynd, it
wol aryse. **1621** BURTON *Anatomy of
Melancholy* II. iii. Though I liue obscure,
yet I liue cleane and honest, and when
as the lofty oake is blowne downe, the
silly[1] reed may stand. **1732** T. FULLER
Gnomologia no. 3692 Oaks may fall,
when Reeds stand the Storm. **1954** R.
HAYDN *Journal of Edwin Carp* 20 Remembering that 'a reed before the wind
lives on—while mighty oaks do fall,' I
attempted to remove the pencil marks
with my pocket eraser.
    [1] frail, insignificant.

regulated: see ACCIDENTS will happen (in the best-regulated families).

rejoice: see it is a POOR heart that never rejoices.

There is a REMEDY for everything except death

[Med. L. *contra malum mortis, non est medicamen in hortis,* against the evil of death there is no remedy in the garden.] *c* **1430** LYDGATE *Dance of Machabree* (EETS) l. 432 Agens deeth is worth[1] no medicine. **1573** J. SANFORDE *Garden of Pleasure* 52 There is a remedie for all things, sauing for death. **1620** T. SHELTON tr. *Cervantes' Don Quixote* II. lxiv. There is a remedy for everything but death, said Don Quixote; for tis but hauing a Barke ready at the Sea side, and in spite of all the world we may embarke our selues. *a* **1895** F. LOCKER-LAMPSON *My Confidences* (1896) 95 There is a remedy for everything except Death . . so the bitterness of this disappointment has long passed away.

[1] useful, valuable.

remedy: see also DESPERATE diseases must have desperate remedies.

removal: see THREE removals are as bad as a fire.

renewal: see the QUARREL of lovers is the renewal of love.

repair: see a WOMAN and a ship ever want mending.

repeat: see HISTORY repeats itself.

repent: see MARRY in haste and repent at leisure.

rest: see (noun) a CHANGE is as good as a rest: (verb) AFTER dinner rest a while, after supper walk a mile.

return: see a BAD penny always turns up; the DOG returns to his vomit.

REVENGE is a dish that can be eaten cold

Vengeance need not be exacted immediately.

[**1620** T. SHELTON tr. *Cervantes' Don Quixote* II. lxiii. Reuenge is not good in cold bloud.] **1885** C. LOWE *Prince Bismarck* I. iv. He had defended Olmütz, it is true, but . . with a secret resolution to 'eat the dish of his revenge cold instead of hot'. **1895** J. PAYN *In Market Overt* xvii. Invective can be used at any time; like vengeance, it is a dish that can be eaten cold. **1975** J. O'FAOLAIN *Women in Wall* iii. Revenge . . is a meal that's as tasty cold as hot. Tastier cold sometimes.

REVENGE is sweet

[Cf. HOMER *Iliad* XVIII. 109 χόλος . . ὅς τε πολὺ γλυκίων μέλιτος καταλειβομένοιο ἀνδρῶν ἐν στήθεσσιν ἀέξεται ἠΰτε καπνός, anger . . that far sweeter than trickling honey wells up like smoke in the breasts of men.] **1566** W. PAINTER *Palace of Pleasure* 300 Vengeance is sweete vnto him, which in place of killing his enemy, giueth life to a perfect frende. **1609** JONSON *Silent Woman* IV. v. O reuenge, how sweet art thou! **1658** *Whole Duty of Man* XVI. 346 'Tis a devilish phrase in the mouth of men, that revenge is sweet. . . Is it possible there can be any such sweetnesse in it? **1775** SHERIDAN *St. Patrick's Day* II. 22 'Revenge is sweet' . . and though disappointed of my designs upon your daughter, . . I'm revenged on her unnatural father. **1861** H. KINGSLEY *Ravenshoe* II. x. Revenge is sweet—to some. Not to him. **1980** J. PORTER *Dover beats Band* xv. He came to the conclusion that though revenge may be sweet, knowledge . . is better than money in the bank.

revenue: see THRIFT is a great revenue.

REVOLUTIONS are not made with rosewater

[**1789** CHAMFORT in Marmontel *Works* (1818) II. 294 *voulez-vous qu'on vous fasse des révolutions à l'eau rose,* do you require that revolutions be made with rose-

water?] **1819** BYRON *Letter* 3 Oct. (1976)
VI. 226 On either side harm must be
done before good can accrue—revolutions are not to be made with rose
water. **1894** J. LUBBOCK *Use of Life* xi. It
is sometimes said that Revolutions are
not made with rose-water. Greater
changes, however, have been made in
the constitution of the world by argument than by arms.

reward: *see* VIRTUE is its own reward.

**The RICH man has his ice in the summer
and the poor man gets his in the winter**
    [**1721** J. KELLY *Scottish Proverbs* 335
There is nothing between a poor Man
and a rich but a piece of an ill Year.
Because, in that space, many things
may fall out, that may make a rich Man
poor.] **1921** W. B. MASTERSON in *Morning Telegraph* (NY) 27 Oct. 7 There are
those who argue that everything breaks
even in this old dump of a world of
ours. . . These ginks who argue that
way hold that because the rich man gets
ice in the Summer and the poor man
gets it in the winter things are breaking
even for both. *a* **1957** L. I. WILDER *First
Four Years* (1971) ii. Everything evens
up in the end. . . The rich man has his
ice in the summer and the poor man gets
his in the winter.

rich: *see also* it is BETTER to be born lucky
than rich; one LAW for the rich and
another for the poor.

**If you can't RIDE two horses at once, you
shouldn't be in the circus**
    1935 G. MCALLISTER *James Maxton* xiv.
Maxton[1] made a brief intervention in the
debate to say . . that he did not believe it
was necessary to pass a resolution for
disaffiliation [of the I.L.P. from the
Labour Party]. He had been told that he
could not ride two horses. 'My reply to
that is', he said . . 'that if my friend
cannot ride two horses—what's he
doing in the bloody circus?' **1979** *Daily
Telegraph* 15 Mar. 15 A producer who
'can't ride two horses at the same time
shouldn't be in the circus.'. . Current

affairs televison should be both serious
and entertaining. **1979** *New Society* 27
Sept. 666 If you can't ride two horses at
once, you shouldn't be in the bloody
circus.
    [1] Independent Labour Party MP, 1932–46.

ride: *see also* set a BEGGAR on horseback,
and he'll ride to the Devil; if TWO ride on
a horse one must ride behind; if WISHES
were horses beggars would ride.

**He who RIDES a tiger is afraid to dis-
mount**
Once a dangerous or troublesome venture is begun, the safest course is to
carry it through to the end.
    1875 W. SCARBOROUGH *Collection of
Chinese Proverbs* no. 2082 He who rides a
tiger is afraid to dismount. . . *Ch'i 'hu
nan hsia pei.* **1902** A. R. COLQUHOUN *Mas-
tery of Pacific* xvi. These colonies are . .
for her [France] the tiger which she has
mounted (to use the Chinese phrase)
and which she can neither manage nor
get rid of. **1962** S. E. FINER *Man on Horseback* viii. [The Military] must still fear
the results of a fall from power. . . 'Who
rides the tiger can never dismount.'
'Whosoever draws his sword against
the prince must throw the scabbard
away.' These two proverbs pithily express the logic of the situation.

ridiculous: *see* from the SUBLIME to the
ridiculous is only a step.

right: *see* the CUSTOMER is always right; DO
right and fear no man; GOD's in his
heaven, all's right with the world;
MIGHT is right; TWO wrongs don't make
a right.

ring: *see* GIVE a thing, and take a thing, to
wear the Devil's gold ring.

ripe: *see* SOON ripe, soon rotten.

rise: *see* EARLY to bed and early to rise,
makes a man healthy, wealthy, and
wise; a STREAM cannot rise above its
source.

The ROAD to hell is paved with good intentions

Earlier forms of the proverb omit the first three words.

[Cf. ST. FRANCIS DE SALES *Letter* lxxiv. *le proverbe tiré de notre saint Bernard, 'L'enfer est plein de bonnes volontés ou désirs,'* the proverb taken from our St. Bernard, 'Hell is full of good intentions or desires.'] **1574** E. HELLOWES tr. *Guevara's Epistles* 205 Hell is full of good desires. **1654** R. WHITLOCK *Observations on Manners of English* 203 It is a saying among Divines, that Hell is full of good Intentions, and Meanings. **1736** J. WESLEY *Journal* 10 July (1910) I. I. 246 It is a true saying, 'Hell is paved with good intentions'. **1847** J. A. FROUDE *Shadows of Clouds* ix. I shall have nothing to hand in, except intentions,—what they say the road to the wrong place is paved with. **1855** H. G. BOHN *Hand-Book of Proverbs* 514 The road to hell is paved with good intentions. **1980** *Guardian* 8 Jan. 11 The educational road to hell is, it seems, as paved with good intentions as any other.

road: *see also* there is no ROYAL road to learning.

All ROADS lead to Rome

[Med. Lat. *mille vie ducunt hominem per secula Romam*, a thousand roads lead man for ever towards Rome. *c* **1391** CHAUCER *Astrolabe* Prologue l. 40 Right as diverse pathes leden diverse folk the righte way to Rome.] **1806** R. THOMSON tr. *La Fontaine's Fables* IV. XII. xxiv. All roads alike conduct to Rome. **1872** W. BLACK *Strange Adventures of Phaeton* vi. You know all roads lead to Rome, and they say that Oxford is half-way to Rome. **1912** J. S. HUXLEY *Individual in Animal Kingdom* vi. All roads lead to Rome: and even animal individuality throws a ray on human problems. **1936** W. HOLTBY *South Riding* I. v. Her official 'subjects' were History and Civics, but all roads led her to Rome—an inexhaustible curiosity about the contemporary world and its inhabitants. **1980** F. BUECHNER *Godric* (1981) 60 All roads lead to Rome, they say, and ours leads us a crooked way.

roasting: *see* there is REASON in the roasting of eggs.

robbery: *see* a fair EXCHANGE is no robbery.

The ROBIN and the wren are God's cock and hen; the martin and the swallow are God's mate and marrow

The rhyme is found in a variety of forms.

[*a* **1508** SKELTON *Poems* (1969) 45 The prety wren . . is our Ladyes hen.] **1787** F. GROSE *Provincial Glossary* (Popular Superstitions) 64 There is a particular distich in favour of the robin and wren: A robin and a wren Are God Almighty's cock and hen. Persons killing [them] . . or destroying their nests, will infallibly, within the course of a year, break a bone, or meet with some other dreadful misfortune. On the contrary, it is deemed lucky to have martins and swallows build their nests in the eaves of a house. **1826** R. WILBRAHAM *Cheshire Glossary* (ed. 2) 105 The following metrical adage is common in Cheshire: The Robin and the Wren Are God's cock and hen, The Martin and the Swallow are God's mate and marrow.[1] **1945** F. THOMPSON *Lark Rise* ix. No boy would rob a robin's or a wren's nest . . for they believe that: The robin and the wrens Be God Almighty's friends. And the martin and the swallow Be God Almighty's birds to follow.

[1] companion.

ROBIN Hood could brave all weathers but a thaw wind

1855 W. NEVILLE *Life & Exploits of Robin Hood* ii. Every one, at least every Yorkshireman, is familiar with the observation that Robin Hood could brave all weathers but a thaw wind.[1] **1931** J. BUCHAN *Blanket of Dark* xii. I dread the melting wind which makes seas of rivers and lakes of valleys. Robin Hood feared little above ground, but he feared the thaw-wind.

[1] *thaw wind:* 'a cold piercing wind from the

S. or SE. which often accompanies the breaking up of a long frost' (J. Bridge, *Cheshire Proverbs*).

rock: *see* the HAND that rocks the cradle rules the world.

rod: *see* SPARE the rod and spoil the child.

A ROLLING stone gathers no moss

[ERASMUS *Adages* III. iv. λίθος κυλινδόμενος τὸ φῦκος οὐ ποιεῖ, a rolling stone does not gather sea-weed; *musco lapis volutus haud obducitur*, a rolling stone is not covered with moss.] **1362** LANGLAND *Piers Plowman* A. x. 101 Selden Moseth[1] the Marbelston that men ofte treden. **1546** J. HEYWOOD *Dialogue of Proverbs* I. xi. D2 The rollyng stone neuer gatherth mosse. **1579** GOSSON *Ephemerides of Phialo* 5[v] A rowling stone gathers no mosse, & a running hed wil neuer thriue. **1710** A. PHILIPS *Pastorals* II. 8 A Rolling Stone is ever bare of Moss. **1841** DICKENS *Old Curiosity Shop* II. xlviii. Your popular rumour, unlike the rolling stone of the proverb, is one which gathers a deal of moss in its wanderings up and down. **1979** *Listener* 5 July 16 A roadside notice . . said in one long line: *Loose stones travel slowly*. Well, I dare say they do: rolling stones, we know, gather no moss.

[1] i.e. mosses: becomes covered by moss.

When in ROME, do as the Romans do

[Med. Lat. *si fueris Romae, Romano vivito more; si fueris alibi, vivito sicut ibi*, if you are at Rome, live after the Roman fashion; if you, are elsewhere, live as they do there (attributed to St. Ambrose).] *c* **1475** in *Modern Philology* (1940) XXXVIII. 122 Whan tho herd that Rome Do so of ther the dome [when you are at Rome do as they do there]. **1552** R. TAVERNER tr. *Erasmus' Adages* (ed. 3) 51[v] That which is commonly in euery mans mouth in England Whan you art at Rome, do as they do at Rome. **1766** in L. H. Butterfield et al. *Adams Family Correspondence* (1963) I. 55 My advice to you is among the Romans, do as the romans do. **1836** E. HOWARD *Rattlin the Reefer* I.

xxii. 'Do at Rome as the Romans do,' is the essence of all politeness. **1960** N. MITFORD *Don't tell Alfred* viii. 'I thought the English never bothered about protocol?' 'When in Rome, however, we do as the Romans do.'

ROME was not built in a day

[Med. Fr. *Rome ne fut pas faite toute en un jour*, Rome was not made in one day.] **1545** R. TAVERNER tr. *Erasmus' Adages* (ed. 2) D1[v] Rome was not buylt in one daye. **1546** J. HEYWOOD *Dialogue of Proverbs* I. xi. D4 Rome was not bylt on a daie (quoth he) & yet stood Tyll it was fynysht. **1646** in *Publications of Prince Society* (1865) I. 236 Rome was not built in a day. . . Let them produce any colonie . . where more hath been done in 16 yeares. **1849** C. BRONTË *Shirley* I. vi. As Rome . . had not been built in a day, so neither had Mademoiselle Gerard Moore's education been completed in a week. **1941** P. CHEYNEY *Trap for Bellamy* iv. Life is what you make it. Rome wasn't built in a day. **1979** *Daily Telegraph* 16 Nov. 18 Rome was not built in a day and artistic enterprise must for the moment remain the privilege of those who live in major centres.

Rome: *see also* all ROADS lead to Rome; it is ill SITTING at Rome and striving with the Pope.

There is always ROOM at the top

Commonly used to encourage competition.

1900 W. JAMES *Letter* 2 Apr. (1920) II. 121 Verily there is room at the top. S— seems to be the only Britisher worth thinking of. **1914** A. BENNETT *Price of Love* vii. The Imperial had set out to be the most gorgeous cinema in the Five Towns; and it simply was. Its advertisements read: 'There is always room at the top.' **1933** W. S. MAUGHAM *Sheppey* III. 89 You have to be pretty smart with all the competition there is nowadays. . . There's always room at the top. **1957** J. BRAINE *Room at Top* xxviii. You're the sort of young man we want. There's always room at the top. **1980** M.

DRABBLE *Middle Ground* 140 There's room at the top, maybe, but only for the clever ones.

roost: *see* CURSES, like chickens, come home to roost.

root: *see* IDLENESS is the root of all evil; MONEY is the root of all evil.

Give a man ROPE enough and he will hang himself

Rope is used both literally, and figuratively—'licence, freedom'.

1639 T. FULLER *Holy War* v. vii. They were suffered to have rope enough, till they had haltered themselves. **1670** J. RAY *English Proverbs* 148 Give a thief rope enough, and he'll hang himself. **1698** in *William & Mary College Quarterly* (1950) VII. 106 The Kings prerogative . . will be hard for his Successor to retrieve, though there's a saying give Men Rope enough, they will hang themselves. **1876** TROLLOPE *Prime Minister* II. xvii. Give Sir Orlando rope enough and he'll hang himself. **1941** G. BAGBY *Red is for Killing* x. 'I like to build a pretty complete case before making an arrest.' . . 'If you give a man enough rope he hangs himself.'

rose-water: *see* REVOLUTIONS are not made with rose-water.

rot: *see* ONE for the mouse, one for the crow.

The ROTTEN apple injures its neighbour

The proverb is also found in a number of variant forms, some of which are illustrated below.

[L. *pomum compunctum cito corrumpit sibi junctum*, a rotten apple quickly infects its neighbour.] **1340** *Ayenbite of Inwit* (EETS) 205 A roted eppel amang the holen,[1] maketh rotie the yzounde.[2] **1577** J. NORTHBROOKE *Treatise against Dicing* 95 A peny naughtily[3] gotten, sayth Chrysostome, is like a rotten apple layd among sounde apples, which will rot all the rest. **1736** B. FRANKLIN *Poor Richard's Almanack* (July)

The rotten apple spoils his companion. **1855** H. G. BOHN *Hand-Book of Proverbs* 514 The rotten apple injures its neighbour. **1979** D. MacKENZIE *Raven feathers his Nest* 19 The police . . have a deserved reputation for uprightness. . . But one bad apple can spoil the whole barrel.

[1] i.e. whole (ones). [2] i.e. sound.
[3] dishonestly.

rotten: *see also* SOON ripe, soon rotten.

roundabout: *see* what you LOSE on the swings you gain on the roundabouts.

There is no ROYAL road to learning

[PROCLUS *Commentary on Euclid* (Friedlein) 68 μὴ εἶναι βασιλικὴν ἀτραπὸν ἐπὶ γεωμετρίαν, there is no royal short cut to geometry (quoting Euclid). **1745** E. STONE tr. *Euclid's Elements* (ed. 2) II. A2ᵛ There is no other Royal Way or Path to Geometry.] **1824** R. W. EMERSON *Journal* (1961) II. 268 There is no royal road to Learning. **1857** TROLLOPE *Barchester Towers* II. i. There is no royal road to learning; no short cut to the acquirement of any valuable art. **1941** H. G. WELLS *You can't be too Careful* II. vi. 'There's no Royal Road to Learning,' said Mr. Myame. 'No. "Thorough" has always been my motto.'

rubber: *see* those who PLAY at bowls must look out for rubbers.

rue: *see* MARRY in May, rue for aye.

rule: *see* (noun) the EXCEPTION proves the rule; there is an EXCEPTION to every rule; (verb) DIVIDE and rule; the HAND that rocks the cradle rules the world.

If you RUN after two hares you will catch neither

[ERASMUS *Adages* III. ccxxxvii. *duos insequens lepores, neutrum capit*, he who chases two hares catches neither.] **1509** A. BARCLAY *Ship of Fools* H5 A fole is he . . Whiche with one haunde tendyth[1] to take two harys in one instant. **1580** LYLY *Euphues* II. 157, I am redie to take potions . . yet one thing maketh to feare, that in running after two Hares, I catch neither. **1732** T. FULLER *Gnomologia* no. 2782 If you run after two Hares, you will catch neither. **1880** C. H. SPURGEON *John Ploughman's Pictures* 24 If we please one we are sure to get another grumbling. We shall be like the man who hunted many hares at once and caught none. **1981** P. O'DONNELL *Xanadu Talisman* v. Let's take things a step at a time. You know what they say. If you run after two hares you will catch neither.

[1] intends.

You cannot RUN with the hare and hunt with the hounds

Also used in the metaphorical phrase *to run with the hare and hunt with the hounds*.

a **1449** LYDGATE *Minor Poems* (EETS) 821 He . . holdeth bothe with hounde and hare. **1546** J. HEYWOOD *Dialogue of Proverbs* I. x. C3 There is no mo[1] suche tytifils[2] in Englands grounde, To holde with the hare, and run with the hounde. **1694** *Trimmer's Confession of Faith* 1, I can hold with the Hare, and run with the Hound: Which no Body

can deny. **1896** M. A. S. HUME *Courtships of Queen Elizabeth* xii. Leicester, as usual, tried to run with the hare and hunt with the hounds, to retain French bribes and yet to stand in the way of French objects. **1975** J. O'FAOLAIN *Women in Wall* v. Clotair's henchmen say: 'You cannot *run* with the hare and hunt with the hounds.' The peasants have an even clearer way of putting this: 'You cannot', they say, 'side with the cow and the clover'.

[1] more. [2] scoundrels, knaves; from Titivil, formerly a common name for a demon.

run: *see also* he who FIGHTS and runs away, may live to fight another day; the LAST drop makes the cup run over; STILL waters run deep; we must learn to WALK before we can run.

rush: *see* FOOLS rush in where angels fear to tread.

Russian: *see* SCRATCH a Russian and you find a Tartar.

rust: *see* BETTER to wear out than to rust out.

S

Sabbath: *see* Monday's CHILD is fair of face.

sack: *see* EMPTY sacks will never stand upright.

SAFE bind, safe find

1546 J. HEYWOOD *Dialogue of Proverbs* I. iii. A4 Than catche & hold while I may, fast bind, fast fynde. **1573** T. TUSSER *Husbandry* (rev. ed.) II. 8 Drie sunne, drie winde, safe bind, safe find. **1655** T. FULLER *Church Hist. Britain* IV. iv. Because sure binde, sure finde, he [Richard III] is said, and his Queen, to be Crowned again in York with great solemnity. **1890** D. C. MURRAY *John Vale's Guard* I. vi. 'Safe bind, safe find,' said Uncle Robert, locking the door and pocketing the key. **1937** D. L. SAYERS *Busman's Honeymoon* xx. As I says to Frank Crutchley, safe bind, safe find, I says.

safe: *see also* it is BEST to be on the safe side; BETTER be safe than sorry.

There is SAFETY in numbers

[PROVERBS xi. 14 In the multitude of counsellors there is safetie.] **1680** BUNYAN *Mr. Badman* 133, I verily think, (since in the multitude of Counsellors there is safety) that if she had acquainted the Congregation with it, . . she had had more peace. **1816** J. AUSTEN *Emma* II. i. She determined to call upon them and seek safety in numbers. **1914** T. DREISER *Titan* xvii. He was beginning to run around with other women. There was safety in numbers. **1979** 's. WOODS' *This Fatal Writ* 82 Julian isn't married. . . As far as women are concerned he seems to believe there's safety in numbers.

said: *see* LEAST said, soonest mended.

sailor: *see* HEAVEN protects children, sailors, and drunken men.

If SAINT Paul's day be fair and clear, it will betide a happy year

[*c* **1340** ROBERT OF AVESBURY *Hist.* (1720) 266 *clara dies Pauli bona tempora denotat anni*, a clear St. Paul's day[1] denotes good times for the year.] **1584** R. SCOT *Discovery of Witchcraft* XI. xv. If Paule th'apostles daie be cleare, it dooth foreshew a luckie yeare. **1687** J. AUBREY *Gentilism & Judaism* (1881) 94 The old verse so much observed by Countrey-people: 'If Paul's day be faire and cleare It will betyde a happy yeare.' **1846** M. A. DENHAM *Proverbs relating to Seasons, &c.* 24 If St. Paul's day be fine and clear, It doth betide a happy year; But if by chance it then should rain, It will make dear all kinds of grain. **1975** M. KILLIP *Folklore of Isle of Man* xiii. In January the testing day was . . the 25th: St. Paul's Day stormy and windy, Famine in the world and great death of mankind, Paul's day fair and clear, Plenty of corn and meal in the world.

[1] the Conversion of St. Paul is traditionally celebrated on 25 January.

SAINT Swithun's day, if thou be fair, for forty days it will remain; Saint Swithun's day, if thou bring rain, for forty days it will remain

St. Swithun (or Swithin) was a bishop of Winchester. He died in 862. The form of the rhyme varies.

1600 JONSON *Every Man out of Humour* I. iii. O, here, S. Swithin's,[1] the xv day, variable weather, for the most part raine. . . Why, it should raine fortie daies after, now, or lesse, it was a rule held afore I was able to hold a plough. **1697** *Poor Robin's Almanack* July B2ᵛ In this month is St. Swithin's day; On which, if that it rain, they say, Full forty days after it will, Or more or less some rain distill. **1846** M. A. DENHAM *Proverbs relating to Seasons, &c.* 52 St. Swithin's day, if thou dost rain, For forty days it will remain: St. Swithin's day, if thou be fair, For forty days 'twill rain na mair. **1892** C. M. YONGE *Old Woman's Outlook* 169 St. Swithin's promise is by no means infallible, whether for wet or fair weather. In . . Gloucestershire, they prefer a shower on his day, and call it christening the apples; but Hampshire . . hold[s] that— If Swithun's day be fair and clear, It betides a happy year; If Swithun's day be dark with rain, Then will be dear all sorts of grain. **1978** R. WHITLOCK *Calendar of Country Customs* viii. Even today innumerable people take note of the weather on St. Swithun's Day, 15 July. . . St. Swithun's Day, if thou be fair, For forty days it will remain. St. Swithun's Day, if thou bring rain, For forty days it will remain.

[1] 15 July.

On SAINT Thomas the Divine kill all turkeys, geese, and swine.

1742 *Agreeable Companion* 59 Thomas Divine,[1] Brewing and Baking, and Killing of Swine. **1846** M. A. DENHAM *Proverbs relating to Seasons, &c.* 64 The day of St. Thomas, the blessed divine, Is good for brewing, baking, and killing fat swine. **1979** C. MORSLEY *News from English Countryside* 164 This couplet reminded farmers of the day on which they should make their last slaughters for the Christmas table. On St. Thomas the Divine Kill all turkeys, geese and swine.

[1] St. Thomas the Apostle, whose feast-day has been traditionally celebrated on 21 December in the West.

saint: *see also* the DEVIL was sick, the Devil a saint would be, the Devil was well, the devil a saint was he!; the GREATER the sinner, the greater the saint.

Help you to SALT, help you to sorrow

1666 G. TORRIANO *Italian Proverbs* 245 At table, one ought not to present any one, either salt, or the head of any creature. **1872** J. GLYDE *Norfolk Garland* i. The spilling of salt is very ominous, and the proverb is well known: Help me to salt, Help me to sorrow. **1945** F. THOMPSON *Lark Rise* xxxvi. No one

would at table spoon salt on to another person's plate, for 'Help you to salt, help you to sorrow'.

Saturday: *see* Monday's CHILD is fair of face.

What's SAUCE for the goose is sauce for the gander

What is suitable for a woman is suitable for a man. The proverb is also occasionally used in wider contexts.

1670 J. RAY *English Proverbs* 98 That that's good sawce for a goose, is good for a gander. . . This is a woman's Proverb. **1692** R. L'ESTRANGE *Fables of Aesop* cccii. Sauce for a Goose is Sauce for a Gander. **1894** BLACKMORE *Perlycross* III. v. A proverb of large equity . . declares . . that 'sauce for the goose is sauce for the gander'. This maxim is pleasant enough to the goose. **1979** *Daily Telegraph* 20 Oct. 1 They upheld a complaint that . . only men had to work in part of the factory making coloured bursting shells. 'What is sauce for the goose is sauce for the gander nowadays,' said Lord Denning.

sauce: *see also* HUNGER is the best sauce.

SAVE us from our friends

Now often used in this abbreviated form (see also earlier quotations).

1477 A. WYDEVILLE *Dicts of Philosophers* 127 Ther was one that praied god to kepe him from the daunger of his frendis. **1585** Q. ELIZABETH in J. E. Neale *Elizabeth I & Her Parliament* (1957) iv. There is an Italian proverb which saith, From my enemy let me defend myself; but from a pretensed friend, good Lord deliver me. **1604** J. MARSTON *Malcontent* IV. ii. Now, God deliver me from my friends . . for from mine enemies I'll deliver myself. **1884** *Railway Engineer* V. 265 The old proverb, 'Save us from our friends', may be well applied to the diligent gentlemen who . . toiled through labyrinths of reports since 1877, to dress up a few exaggerated cases against the . . brake. **1979** 's.

woods' *Proceed to Judgement* 140 Heaven save us from our friends!

save: *see also* a STITCH in time saves nine.

saved: *see* a PENNY saved is a penny earned.

say: *see* DO as I say, not as I do; when in DOUBT, do nowt; what EVERYBODY says must be true; HEAR all, see all, say nowt; what MANCHESTER says today, the rest of England says tomorrow; *also* SAID.

scabbard: *see* whosoever DRAWS his sword against the prince must throw the scabbard away.

scarce: *see* GOOD men are scarce.

scarlet: *see* an APE's an ape, a varlet's a varlet, though they be clad in silk or scarlet.

scheme: *see* the BEST-laid schemes of mice and men gang aft agley.

school: *see* EXPERIENCE keeps a dear school; never tell TALES out of school.

scorned: *see* HELL hath no fury like a woman scorned.

SCRATCH a Russian and you find a Tartar
The proverb is also used allusively, especially of other nationalities.

[Fr. *grattez le Russe et vous trouverez le Tartare*, scratch the Russian and you will find the Tartar (attributed to Napoleon).] **1823** J. GALLATIN *Diary* 2 Jan. (1914) 229 Very true the saying is, 'Scratch the Russian and find the Tartar.' *c* **1863** J. R. GREEN in *Notes & Queries* (1965) CCX. 348 They say, if you scratch a Russian you always find the Tartar beneath. **1899** F. A. OBER *Puerto Rico* xii. Scratch a Puerto Rican and you find a Spaniard underneath, so the language and home customs of Spain prevail here. **1911** *Spectator* 2 Dec. 964 Until a short time ago the aphorism, 'Scratch a Russian and you find a Tar-

tar,' was the sum of British comprehension of the Russian character. **1947** J. FLANNER in *New Yorker* 31 May 6 Scratch a Pole and you find a Pole, even if he is a Communist. **1955** D. GARNETT *Flowers of Forest* viii. 'Scratch a Russian and you find a Tartar.' . . It seemed to me then, and often since, that the virtues are, by their nature, superficial.

Scripture: *see* the DEVIL can quote Scripture for his own ends.

He that would go to SEA for pleasure, would go to hell for a pastime
A sailors' proverb.

1899 A. J. BOYD *Shellback* viii. Shentlemens vot goes to sea for pleasure vould go to hell for pastime. **1910** D. W. BONE *Brassbounder* xxvi. He gave a half-laugh and muttered the old formula about 'the man who would go to sea for pleasure, going to hell for a pastime!' **1924** R. CLEMENTS *Gipsy of Horn* iii. 'He who would go to sea for pleasure, would go to hell for a pastime' is an attempt at heavy satire. **1933** M. LOWRY *Ultramarine* i. 'What made you come to sea anyway?' 'Search me. . . To amuse myself, I suppose.' 'Well, a man who'd go to sea for fun'd go to hell for a pastime. . . It's an old sailor expression.'

sea: *see also* there are as good FISH in the sea as ever came out of it.

search: *see* on the FIRST of March, the crows begin to search.

SECOND thoughts are best
[EURIPIDES *Hippolytus* l. 436 αἱ δεύτεραί πως φροντίδες σοφώτεραι, the second thoughts are invariably wiser.] **1577** HOLINSHED *Chronicles* 438 Oftentymes it chaunceth, that latter thoughts are better aduised than the first. **1581** G. PETTIE tr *S. Guazzo's Civil Conversation* I. 23ᵛ I finde verified that Prouerbe, That the second thoughts are euer the best. **1681** DRYDEN *Spanish Friar* II. 22 Second thoughts, they say, are best: I'll consider of it once again. **1813** BYRON *Letter* 11

Dec. (1974) III. 196 In composition I do not think *second* thoughts are best, though *second* expressions may improve the first ideas. **1908** C. FITCH *Beau Brummel* I. i. Second thoughts seem to be always the best. **1981** P. O'DONNELL *Xanadu Talisman* v. That was my first thought. . . But second thoughts are always best.

secret: *see* THREE may keep a secret, if two of them are dead.

SEE no evil, hear no evil, speak no evil
The proverb is conventionally represented by three monkeys covering their eyes, ears, and mouth respectively with their hands.

1926 *Army & Navy Stores Catalogue* 197 The three wise monkeys. 'Speak no evil, see no evil, hear no evil.' **1973** O. SELA *Portuguese Fragment* xxviii. Like one of the proverbial three monkeys he was seeing no evil, and even if he did, he was certainly not speaking about it. **1978** T. L. SMITH *Money War* III. 233 It's the sort of thing they want done but do not want to know about. See no evil, hear no evil, speak no evil.

see: *see also* BELIEVE nothing of what you hear, and only half of what you see; there's none so BLIND as those who will not see; what the EYE doesn't see, the heart doesn't grieve over; HEAR all, see all, say nowt; they that LIVE longest, see most; LOOKERS-ON see most of the game; see a PIN and pick it up, all the day you'll have good luck; *also* SEEING, SEEN.

Good SEED makes a good crop
Cf. *As you* SOW, *so you reap.*

[**1492** *Dialogue of Salomon & Marcolphus* (1892) 5 He that sowyth chaf shall porely mowe.] **1569** W. WAGER *Longer Thou Livest* A2 To be a good man it is also expedient Of good Parents to be begotten and borne. . . Commonly of good Seed procedeth good Corne. **1700** T. TRYON *Letters* i. If the Seed he Sowes be good . . his Crop is according; . . if he Sows Tares . . will he expect Wheat?

1940 L. I. WILDER *Long Winter* xvii. Seedtime's pretty sure to come around. . . And good seed makes a good crop.

seed: *see also* the BLOOD of the martyrs is the seed of the Church; PARSLEY seed goes nine times to the Devil.

seeding: *see* ONE year's seeding means seven years' weeding.

SEEING is believing
1609 S. HARWARD *MS* (Trinity College, Cambridge) 85 Seeing is leeving. **1639** J. CLARKE *Parœmiologia Anglo-Latina* 90 Seeing is beleeving. **1712** J. ARBUTHNOT *Lewis Baboon* iv. There's nothing like Matter of Fact; Seeing is Believing. **1848** J. C. & A. W. HARE *Guesses at Truth* (ed. 2) 2nd Ser. 497 Seeing is believing, says the proverb. . . Though, of all our senses, the eyes are the most easily deceived, we believe them in preference to any other evidence. **1975** A. PRICE *Our Man in Camelot* v. 'Show him the stuff. . . Let him make his own mind up.' 'Okay. Maybe you're right at that. Seeing is believing, I guess.'

SEEK and ye shall find
[SOPHOCLES *Oedipus Tyrannus* l. 110 τὸ δὲ ζητούμενον ἁλωτόν, ἐκφεύγειν δὲ τἀμελούμενον, what is sought is found; what is neglected evades us; MATTHEW vii. 7 Aske, and it shalbe giuen you: seeke, and ye shall finde.] **1530** in J. Palsgrave *L'éclaircissement de la Langue Française* A5 He that wyll seke may fynde And in a brefe tyme attayne to his utterest desire. *c* **1538** J. BALE *King Johan* (1931) l. 192 Serche & ye shall fynd, in every congregacyn that long[1] to the pope. **1783** J. JAY *Letter* 14 Nov. (1891) III. 95 'Seek and you shall find' does not, it seems, always extend to that [health] of the body. **1980** R. COLLINS *Case of Philosopher's Ring* xiii. There is danger in the saying, 'Seek and ye shall find'.
[1] belongs.

seem: *see* BE what you would seem to be.

seen: *see* CHILDREN should be seen and not heard.

SELF-praise is no recommendation
[Cf. L. *laus in proprio ore sordescit,* praise in one's own mouth is offensive.] **1826** COBBETT *Weekly Register* 17 June 743 In general it is a good rule . . that self-praise is no commendation. **1853** DICKENS *Bleak House* lv. Self-praise is no recommendation, but I may say for myself that I am not so bad a man of business. **1967** RIDOUT & WITTING *English Proverbs Explained* 137 'I admit I didn't score any of the goals, but it was largely due to me that we won the game.' 'Self-praise is no recommendation.'

SELF-preservation is the first law of nature
a **1631** DONNE *Biathanatos* (1646) I. ii. It is onely upon this reason, that selfe-preservation is of Naturall Law. **1675** [MARVELL] *Complete Poems* (1872) I. 439 Self-preservation, Nature's first great Law. **1681** DRYDEN *Spanish Friar* IV. ii. Self-preservation is the first of Laws: . . When Subjects are oppress'd by Kings, They justifie Rebellion by that Law. **1821** SCOTT *Pirate* I. v. Triptolemus . . had a reasonable share of that wisdom which looks towards self-preservation as the first law of nature. **1952** 'A. A. FAIR' *Top of Heap* xvii. Loyalty is a fine thing . . but self-preservation is the first law of nature.

sell: *see* BUY in the cheapest market and sell in the dearest.

seller: *see* the BUYER has need of a hundred eyes, the seller of but one.

send: *see* GOD never sends mouths but He sends meat; GOD sends meat, but the Devil sends cooks.

SEPTEMBER blow soft, till the fruit's in the loft
1571 T. TUSSER *Husbandry* (rev. ed.) F2 Septembre blowe soft, Till fruite be in loft. **1732** T. FULLER *Gnomologia* no. 6214

September, blow soft, Till the Fruit's in the Loft. **1906** E. HOLDEN *Country Diary of Edwardian Lady* (1977) 121 September blow soft,—Till the fruit's in the loft. **1928** *Daily Mail* 3 Sept. 10 'September blow soft till the apple's in the loft' is what we desire of this traditionally beautiful month.

serve: *see* you cannot serve GOD and Mammon; NO man can serve two masters.

If you would be well SERVED, serve yourself
Cf. *If you* WANT *a thing done well, do it yourself.*
1659 G. TORRIANO *English & Italian Dict.* 39 Who hath a mind to any thing let him go himself. **1706** J. STEVENS *Spanish & English Dict.* s.v. Querer, If you would be well serv'd, serve your self. **1871** J. E. AUSTEN-LEIGH *Memoir of Jane Austen* (ed. 2) ii. 'If you would be well served, serve yourself.' Some gentlemen took pleasure in being their own gardeners. **1981** *Times* 28 Apr. 15 Absurd that the important things in one's life should be made by another person—'One is never so well served as by oneself.'

served: *see also* FIRST come, first served; YOUTH must be served.

session: *see* HOME is home, as the Devil said when he found himself in the Court of Session.

set: *see* set a BEGGAR on horseback, and he'll ride to the Devil; set a THIEF to catch a thief.

seven: *see* KEEP a thing seven years and you'll always find a use for it; you should KNOW a man seven years before you stir his fire; ONE year's seeding means seven years' weeding; PARSLEY seed goes nine times to the Devil; RAIN before seven, fine before eleven; SIX hour's sleep for a man, seven for a woman, and eight for a fool.

shadow: *see* COMING events cast their

shadows before; OLD sins cast long shadows.

shame: *see* TELL the truth and shame the Devil.

shared: *see* a TROUBLE shared is a trouble halved.

The SHARPER the storm, the sooner it's over

[SENECA *Natural Questions* VII. ix. *procellae, quanto plus habent virium, tanto minus temporis,* the harder storms are, the shorter they last.] **1872** F. KILVERT *Diary* 9 June (1977) II. 207 Mrs. Vaughan will have a good family soon. Her children come fast. But the harder the storm the sooner 'tis over. **1913** *Folk-Lore* XXIV. 76 The sharper the storm, the sooner it's over.

sheep: *see* a BLEATING sheep loses a bite; one might as well be HANGED for a sheep as a lamb.

shepherd: *see* RED sky at night shepherd's delight, red sky in the morning shepherd's warning.

You cannot SHIFT an old tree without it dying

c **1518** A. BARCLAY tr. *Mancinus' Mirror of Good Manners* G4ᵛ An old tre transposed shall fynde smal auauntage. **1670** J. RAY *English Proverbs* 22 Remove an old tree, and it will wither to death. **1721** J. KELLY *Scottish Proverbs* 284 Remove an old Tree, and it will wither. Spoken by a Man who is loth to leave a Place in his advanc'd years, in which he has long lived. **1831** W. M. PRAED *Political & Occasional Poems* (1888) 166 I'm near threescore; you ought to know You can't transplant so old a tree. **1906** KIPLING *Puck of Pook's Hill* 259 'You've cleaved to your own parts pretty middlin' close, Ralph.' 'Can't shift an old tree 'thout it dyin'.'

shine: *see* happy is the BRIDE that the sun shines on; MAKE hay while the sun shines.

shining: *see* the SUN loses nothing by shining into a puddle.

Do not spoil the SHIP for a ha'porth of tar

Ship is a dialectal pronunciation of *sheep,* and the original literal sense of the proverb was 'do not allow sheep to die for the lack of a trifling amount of tar', tar being used to protect sores and wounds on sheep from flies. The current form was standard by the mid-nineteenth century. The metaphorical phrase *to spoil the ship for a ha'porth of tar* is also found.

1623 W. CAMDEN *Remains concerning Britain* (ed. 3) 265 A man will not lose a hog, for a halfeperth[1] of tarre. **1631** J. SMITH *Advertisements for Planters* XIII. 30 Rather . . lose ten sheepe, than be at the charge of halfe penny worth of Tarre. **1670** J. RAY *English Proverbs* 103 Ne're lose a hog for a half-penny-worth of tarre [(ed. 2) 154 Some have it, lose not a sheep, &c. Indeed tarr is more used about sheep than swine]. **1861** C. READE *Cloister & Hearth* I. i. Never tyne[2] the ship for want of a bit of tar. **1869** W. C. HAZLITT *English Proverbs* 432 To spoil the ship for a halfpennyworth of tar. In Cornwall, I heard a different version, which appeared to me to be more consistent with probability: 'Don't spoil the sheep for a ha'porth of tar.' **1910** *Spectator* 19 Feb. 289 The ratepayers . . are accused of . . cheeseparing, of spoiling the ship for a ha'p'orth of tar. **1980** 'J. MARCUS' *Marsh Blood* iii. Well, he says, don't want to spoil the ship for a ha'porth of tar and could come in useful having the extra private bath.

[1] i.e. halfpennyworth. [2] lose.

ship: *see also* one HAND for oneself and one for the ship; a WOMAN and a ship ever want mending.

shirt: *see* NEAR is my shirt, but nearer is my skin.

From SHIRTSLEEVES to shirtsleeves in three generations

Shirtsleeves denote the need to work hard for one's living. This saying has been attributed to A. Carnegie (1835–1919), manufacturer and philanthropist, but is not found in his published writings. Cf. *From* CLOGS *to clogs is only three generations.*

1907 N. M. BUTLER *True & False Democracy* ii. No artificial class distinctions can long prevail in a society like ours [in the US] of which it is truly said to be often but three generations 'from shirtsleeves to shirt-sleeves'. **1957** J. S. BRUNER in *Psychological Review* LXIV. 125 From shirtsleeves to shirtsleeves in three generations: we are back with the founding and founded content of the pre-Gestalt Gestalters. **1980** J. KRANTZ *Princess Daisy* xvii. What's this? Shirtsleeves to shirtsleeves in three generations.

shoe: *see* it's ILL waiting for dead men's shoes; for WANT of a nail the shoe was lost.

The SHOEMAKER's son always goes barefoot

A skilled or knowledgeable person commonly neglects to give his own family the benefit of his expertise. Also used in wider contexts.

1546 J. HEYWOOD *Dialogue of Proverbs* I. xi. E1[v] But who is wurs shod, than the shoemakers wyfe, With shops full of newe shapen shoes all hir lyfe? **1773** R. GRAVES *Spiritual Quixote* I. III. ii. The Shoe-maker's wife often goes in ragged shoes. . . Although there had been a [Methodist] Society begun here by Mr. Whitfield, yet . . the people of Gloucester are not much the better for having had so great a Prophet born amongst them. **1876** S. SMILES *Life of Scotch Naturalist* xviii. His large family . . were all . . well shod, notwithstanding the Scottish proverb to the contrary. 'The Smith's meer[1] and the shoemaker's bairns are aye the worst shod.' **1981** 'E. PETER' *Saint Peter's Fair* 30 Spruce in his dress, but down at heel, Cadfael noticed—proof of the old saying that the shoemaker's son is always the one who goes barefoot!

[1] i.e. mare.

shop: *see* KEEP your shop and your shop will keep you.

shopkeeper: *see* the ENGLISH are a nation of shopkeepers.

shorn: *see* GOD tempers the wind to the shorn lamb; many go out for WOOL and come home shorn.

A SHORT horse is soon curried

A slight task is soon completed.

c **1350** *Douce MS* 52 no. 17 Short hors is son j-curryed.[1] *a* **1530** R. HILL *Commonplace Book* (EETS) 128 A shorte hors is son curried. **1732** T. FULLER *Gnomologia* no. 395 A short Horse is soon curried. **1820** SCOTT *Abbot* I. xi. A short tale is soon told—and a short horse soon curried. **1939** L. I. WILDER *By Shores of Silver Lake* xxx. A short horse is soon curried. This is our tightest squeeze yet, . . but it's only a beginning. **1948** F. P. KEYES *Dinner at Antoine's* xx. That's a short horse and soon curried. Let's go see this Captain Murphy and put an end to it.

[1] rubbed down with a curry-comb.

SHORT reckonings make long friends

1530 R. WHITFORDE *Work for Householders* A4 The commune prouerbe is that ofte rekenynge[1] holdest longe felawship. **1641** D. FERGUSSON *Scottish Proverbs* (STS) no. 668 Oft compting makes good friends. **1673** J. DARE *Counsellor Manners* xciii. Short reckonings (we say) make long friends. **1842** S. LOVER *Handy Andy* viii. There must be no nonsense about the wedding. . . Just marry her off, and take her home. Short reckonings make long friends. **1918** BARONESS ORCZY *Man in Grey* 15 Short reckonings make long friends. I'll have a couple of hundred francs now.

[1] settling (also, settlement) of accounts.

short: *see also* ART is long and life is short; LONG foretold long last, short notice soon past.

shortest: *see* BARNABY bright, Barnaby bright, the longest day and the shortest night; the LONGEST way round is the shortest way home.

shoulder: *see* you cannot put an OLD head on young shoulders.

show: *see* TIME will tell.

shower: *see* APRIL showers bring forth May flowers.

SHROUDS have no pockets

    1854 R. C. TRENCH *On Lessons in Proverbs* (ed. 2) v. With an image Dantesque in its vigour, that 'a man shall carry nothing away with him when he dieth', take this Italian, *Our last robe, that is our winding sheet, is made without pockets.* **1909** A. MACLAREN *Epistle to Ephesians* 41 There is nothing that is truly our wealth which remains outside of us, and can be separated from us. 'Shrouds have no pockets.' **1961** M. KELLY *Spoilt Kill* II. 20 'He had a win on the pools and it's burning him.' 'Shrouds don't need pockets, love,' he said with a grin.

shut: *see* a DOOR must either be shut or open; when ONE door shuts, another opens; it is too late to shut the STABLE-door after the horse has bolted.

shy: *see* ONCE bitten, twice shy.

sick: *see* the DEVIL was sick, the Devil a saint would be, the Devil was well, the devil a saint was he!; HOPE deferred makes the heart sick.

side: *see* it is BEST to be on the safe side; the BREAD never falls but on its buttered side; the GRASS is always greener on the other side of the fence; PROVIDENCE is always on the side of the big battalions; there are TWO sides to every question.

sight: *see* in vain the NET is spread in the sight of the bird; OUT of sight, out of mind.

SILENCE is a woman's best garment

    [Cf. SOPHOCLES *Ajax* l. 293 γυναιξὶ κόσμον ἡ σιγὴ φέρει, it is fitting for a woman to be silent; I CORINTHIANS xiv. 34 Let your women keepe silence in the Churches, for it is not permitted vnto them to speake.] **1539** R. TAVERNER tr. *Erasmus' Adages* 50 *Mulierem ornat silentium.* Silence garnysheth a woman . . whych thynge also the Apostle Paule requyreth. **1659** J. HOWELL *Proverbs* (English) 11 Silence the best ornament of a woman. **1732** T. FULLER *Gnomologia* no. 4166 Silence is a fine Jewel for a Woman; but it's little worn. **1977** J. AIKEN *Five-Minute Marriage* iv. Quiet, miss! Silence is a woman's best garment.

SILENCE is golden

See also SPEECH *is silver, but silence is golden.*

    1865 W. WHITE *Eastern England* II. ix. Silence is golden, says the proverb. We apprehend the full significance . . in some lone hamlet situate amid a 'thousand fields'. **1923** A. HUXLEY *Antic Hay* xx. Silence is golden, as her father used to say when she used to fly into tempers and wanted to say nasty things to everybody within range. **1980** J. O'NEILL *Spy Game* xxv. 'I'll tell you the rest . . on the way back.' He sealed her lips with a finger. 'Meanwhile, silence is golden.'

SILENCE means consent

    [L. *qui tacet consentire videtur*, he who is silent seems to consent.] *c* **1380** WYCLIF *Select English Works* (1871) III. 349 Oo[1] maner of consent is, whanne a man is stille and tellith not. **1591** LYLY *Endymion* v. iii. Silence, Madame, consents. *c* **1616–30** *Partial Law* (1908) v. iv. 'I will nothing say.'. . 'Then silence gives consent.' **1847** A. HELPS *Friends in Council* ix. I have known a man . . bear patiently . . a serious charge which a few lines would have entirely answered.'. . 'Silence does not give consent in these cases.' **1914** L. WOOLF *Wise Virgins* v. He . . did not speak. 'I assume that silence means consent,' said Arthur. **1980** *Guardian* 28 Apr. 9 As Krushchev once put it: 'Silence means consent.'

    [1] one.

silence: *see also* SPEECH is silver, but silence is golden.

You can't make a SILK purse out of a sow's ear

1518 A. BARCLAY *Eclogues* (EETS) v. 360 None can . . make goodly silke of a gotes flece. **1579** GOSSON *Ephemerides of Phialo* 62ᵛ Seekinge . . too make a silke purse of a Sowes eare, that when it shoulde close, will not come togeather. **1672** W. WALKER *English & Latin Proverbs* 44 You cannot make a . . silk purse of a sows ear; a scholar of a blockhead. **1834** MARRYAT *Peter Simple* I. xii. The master . . having been brought up in a collier, he could not be expected to be very refined. . . 'It was impossible to make a silk purse out of a sow's ear.' **1915** D. H. LAWRENCE *Rainbow* i. You can't make a silk purse out of a sow's ear, as he told his mother very early, with regard to himself. **1959** M. BRADBURY *Eating People is Wrong* ii. For the mass of men there is not too much to be said or done; you can't make a silk purse out of a sow's ear.

silk: *see also* an APE's an ape, a varlet's a varlet, though they be clad in silk or scarlet.

silver: *see* every CLOUD has a silver lining; SPEECH is silver, but silence is golden.

It's a SIN to steal a pin

1875 A. B. CHEALES *Proverbial Folk-Lore* 129 It is a sin To steal a pin, as we, all of us, used to be informed in the nursery. **1945** F. THOMPSON *Lark Rise* xiii. Children were taught to 'know it's a sin to steal a pin' . . when they brought home some doubtful finding.

sin: *see also* OLD sins cast long shadows.

sincerest: *see* IMITATION is the sincerest form of flattery.

sing: *see* little BIRDS that can sing and won't sing must be made to sing; the OPERA isn't over till the fat lady sings.

singly: *see* MISFORTUNES never come singly.

sinner: *see* the GREATER the sinner, the greater the saint.

sit: *see* where MACGREGOR sits is the head of the table.

It is ill SITTING at Rome and striving with the Pope

Cf. *When in* ROME, *do as the Romans do.*

a **1628** J. CARMICHAELL *Proverbs in Scots* no. 1847 Ye may not sit in Rome and strive with the Pape. **1721** J. KELLY *Scottish Proverbs* 194 It is hard to sit in Rome, and strive against the Pope. It is foolish to strive with our Governours, Landlords, or those under whose Distress we are. **1908** A. MACLAREN *Ezekiel* 58 'It is ill sitting at Rome and striving with the Pope.' Nebuchadnezzar's palace was not precisely the place to dispute with Nebuchadnezzar.

sitting: *see also* it is as CHEAP sitting as standing.

situation: *see* DESPERATE diseases must have desperate remedies.

SIX hours' sleep for a man, seven for a woman, and eight for a fool

1623 J. WODROEPHE *Spared Hours of Soldier* 310 The Student sleepes six Howres, the Traueller seuen; the Workeman eight, and all Laizie Bodies sleepe nine houres and more. **1864** J. H. FRISWELL *Gentle Life* 259 John Wesley . . considered that five hours' sleep was enough for him or any man. . . The old English proverb, so often in the mouth of George III, was 'six hours for a man, seven for a woman, and eight for a fool'. **1908** *Spectator* 19 Dec. 1047 Is there not a proverb that a man requires six hours' sleep, a woman seven, a child eight and only a fool more? If this be true, thousands of great men were, and are, fools.

skin: *see* NEAR is my shirt, but nearer is my skin.

skin-deep: *see* BEAUTY is only skin-deep.

skittle: *see* LIFE isn't all beer and skittles.

If the SKY falls we shall catch larks
'In ridicule of those who talk of doing many things, if certain other things, not likely, were to happen': Fielding, *Proverbs of all Nations* (1824) 22.

c **1445** *Peter Idley's Instructions to his Son* (1935) I. 178 We shall kacche many larkis whan heuene doith falle. **1546** J. HEYWOOD *Dialogue of Proverbs* I. iv. B1ᵛ When the sky falth we shal haue larks. **1670** J. RAY *English Proverbs* 143 If the sky falls we shall catch larks. **1721** J. KELLY *Scottish Proverbs* 343 What if the Lift¹ fall, you may gather Laverocks.² **1914** G. B. SHAW *Misalliance* p.xxx. I cannot be put off by the news that our system would be perfect if it were worked by angels . . just as I do not admit that if the sky fell we should all catch larks.
¹ sky. ² larks.

sky: *see also* RED sky at night shepherd's delight, red sky in the morning shepherd's warning.

slave: *see* BETTER be an old man's darling, than a young man's slave.

sleep: *see* one HOUR's sleep before midnight is worth two after; SIX hours' sleep for a man, seven for a woman, and eight for a fool.

Let SLEEPING dogs lie
[Cf. early 14th-cent. Fr. *n'esveillez pas lou chien qui dort*, wake not the sleeping dog.] c **1385** CHAUCER *Troilus & Criseyde* III. 764 It is nought good a slepyng hound to wake. **1546** J. HEYWOOD *Dialogue of Proverbs* I. x. D1ᵛ It is euill wakyng of a slepyng dog. **1681** S. COLVIL *Whigs' Supplication* II. 27 It's best To let a sleeping mastiff rest. **1824** SCOTT *Redgauntlet* I. xi. Take my advice, and speer¹ as little about him as he does about you. Best to let sleeping dogs lie. **1976** T. SHARPE *Wilt* xx. He would be better off sticking to indifference and undisclosed affection. 'Let sleeping dogs lie,' he muttered.
¹ ask.

sleeve: *see* STRETCH your arm no further than your sleeve will reach.

A SLICE off a cut loaf isn't missed
1592 SHAKESPEARE *Titus Andronicus* II. i. 87 More water glideth by the mill Than wots¹ the miller of; and easy it is Of a cut loaf² to steal a shive.³ **1639** J. CLARKE *Parœmiologia Anglo-Latina* 118 'Tis safe taking a shive of a cut loafe. **1732** T. FULLER *Gnomologia* no. 3012 It is safe taking a slice off a Cut Loaf. **1901** F. E. TAYLOR *Wit & Wisdom of South Lancashire Dialect* 11 A shoive off a cut loaf's never miss't. (A satirical remark). **1981** N. LOFTS *Old Priory* v. iii. I went into this with my eyes open and a slice off a cut loaf ain't missed.
¹ knows. ² *cut loaf*: a loaf from which some slices have been cut. ³ slice.

slip: *see* there's MANY a slip 'twixt cup and lip.

SLOW but sure
Sure means properly 'sure-footed, deliberate' and is frequently contrasted with *slow*. The related saying *slow and steady wins the race* is also illustrated here.

[**1562** G. LEGH *Accidence of Armoury* 97 Although the Asse be slowe, yet is he sure.] **1692** R. L'ESTRANGE *Fables of Aesop* ccclxix. Slow and sure in these cases, is good counsel. **1762** R. LLOYD *Poems* 38 You may deride my awkward pace, But slow and steady wins the race. **1859** S. SMILES *Self-Help* xi. Provided the dunce has persistency and application, he will inevitably head the cleverer fellow without these qualities. Slow but sure, wins the race. **1894** G. F. NORTHALL *Folk-Phrases* 22 Slow and steady wins the race. **1947** M. PENN *Manchester Fourteen Miles* xvii. No dressmaker . . ever learnt her trade in a hurry. 'Slow but sure' was the beginner's motto. **1979** D. HARTLEY *Land of England* 82 Three miles an hour was fairly good going. . . Slow . . but sure.

slowly: *see* make HASTE slowly; the MILLS of God grind slowly, yet they grind exceeding small.

SMALL is beautiful

1973 E. F. SCHUMACHER *(title)* Small is beautiful. **1975** *Country Life* 25 Dec. 1784 Adapting Schumacher's phrase, we decide that not only small but piecemeal is beautiful. **1977** D. JAMES *Spy at Evening* xxiv. Small Is Beautiful—but big pays more.

small: *see also* the BEST things come in small packages; LITTLE things please little minds; there's no great LOSS without some gain; the MILLS of God grind slowly, yet they grind exceeding small.

smell: *see* MONEY has no smell.

smock: *see* NEAR is my kirtle, but nearer is my smock.

No SMOKE without fire

[Cf. late 13th-cent. Fr. *nul feu est sens fumee ne fumee sens feu,* no fire is without smoke, nor smoke without fire. *c* **1375** J. BARBOUR *Bruce* (EETS) IV. 81 And thair may no man fire sa covir, [Bot] low or reyk[1] sall it discovir.[2]] *c* **1422** HOCCLEVE *Works* (EETS) I. 134 Wher no fyr maad is may no smoke aryse. **1592** G. DE-LAMOTHE *French Alphabet* II. 39 No smoke without fire. **1655** T. FULLER *Church Hist. Britain* II. x. There was no Smoak but some Fire: either he was dishonest, or indiscreet. **1869** TROLLOPE *He knew He was Right* II. lii. He considered that . . Emily Trevelyan had behaved badly. He constantly repeated . . the old adage, that there was no smoke without fire. **1948** 'M. INNES' *Night of Errors* iv. 'Chimneys! . . Who the deuce cares whether there's smoke from every chimney in the house.' 'I do. No smoke without fire.' **1976** D. STOREY *Saville* IV. xix. 'There's no smoke without fire.' 'And the way you're going about it you'll have it like a furnace when there's really nothing there at all.'
[1] *low or reyk,* flame or smoke. [2] reveal.

smooth: *see* the COURSE of true love never did run smooth.

so: *see* SO many MEN, so many opinions.

sober: *see* WANTON kittens make sober cats.

A SOFT answer turneth away wrath

c **1395** WYCLIF *Bible* (1850) Proverbs xv. 1 A soft answere brekith ire. *c* **1445** *Peter Idley's Instructions to his Son* (1935) I. 84 A softe worde swageth[1] Ire. **1611** BIBLE *Proverbs* xv. 1 A soft answere turneth away wrath. **1693** C. MATHER *Wonders of Invisible World* 60 We would use to one another none but the Soft Answers, which Turn away Wrath. **1826** SOUTHEY *Letter* 19 July (1912) 414 A soft answer turneth away wrath. There is no shield against wrongs so effectual as an unresisting temper. **1979** J. SCOTT *Clutch of Vipers* vi. 'Yes, sir!' . . Soft answer, no wrath.
[1] assuages.

soft: *see also* SEPTEMBER blow soft, till the fruit's in the loft.

SOFTLY, softly, catchee monkey

1907 G. BENHAM *Cassell's Book of Quotations* 849 (Proverbs), 'Softly, softly,' caught the monkey—(Negro). **1942** N. BALCHIN *Darkness falls from Air* x. Softly catch monkey. . . That's the answer. **1962** P. BRICKHILL *Deadline* xiii. I didn't pursue it any further then. Softly, softly, catchee monkey—and I hated that phrase. **1978** E. ST. JOHN-STON *One Policeman's Story* vii. They took with them the unique motto of the Lancashire Constabulary Training School, 'Softly, Softly, Catchee Monkey' which inspired the new programme's title, 'Softly, Softly'.

softly: *see also* FAIR and softly goes far in a day.

soldier: *see* the FIRST duty of a soldier is obedience; OLD soldiers never die.

If you're not part of the SOLUTION, you're part of the problem

1975 M. BRADBURY *History Man* v. 'If you're not the solution,' says Peter Madden, 'you're part of the problem.' 'It would be terribly arrogant of me to believe I was the solution to anything.' **1977** C. McFADDEN *Serial* xxvi. Listen, don't you realize if you're not part of the solution you're part of the problem.

some: *see* you WIN a few, you lose a few.

You don't get SOMETHING for nothing

A variant of this, originally from the north country, runs *you don't get owt[1] for nowt[2]* (see below). Cf. NOTHING *for nothing*.

[**1845** DISRAELI *Sybil* I. i. v. To do nothing and get something formed a boy's ideal of a manly career.] **1870** P. T. BARNUM *Struggles & Triumphs* viii. When people expect to get 'something for nothing' they are sure to be cheated. **1947** M. PENN *Manchester Fourteen Miles* xiii. No stranger, she declared emphatically, ever sent to another stranger 'summat for nowt'. It would . . be against nature. **1952** F. PRATT *Double Jeopardy* i. You don't get something for nothing, even in medicine. Perizone has a peculiar secondary effect. It releases all inhibitions. **1979** *Guardian* 18 June 10 Stravinsky and Auden . . [are] saying 'You don't get something for nothing.' If you want the lovely things . . you can't have them unless you're prepared to pay for them. [**1979** *Church Times* 29 June 13 You don't get owt for nowt.]

[1] i.e. ought, aught; anything.
[2] i.e. nought, naught; nothing.

SOMETHING is better than nothing

Cf. HALF *a loaf is better than no bread*.

1546 J. HEYWOOD *Dialogue of Proverbs* I. ix. D1 And by this prouerbe appereth this o[1] thyng, That alwaie somwhat is better than nothyng. **1612** T. SHELTON tr. *Cervantes' Don Quixote* III. vii. I will weare it as I may: for something is better then nothing. **1842** J. T. IRVING *Attorney* xvii. Something is better than no-

thing – nothing is better than starving. **1980** *Country Life* 24 Apr. 1283 Mrs Smith worked out her own charitable rules: give what can be given in kind (for something is better than nothing) but never give money.

[1] one.

something: *see also* if ANYTHING can go wrong, it will.

My SON is my son till he gets him a wife; but my daughter's my daughter all the days of her life

1670 J. RAY *English Proverbs* 53 My son's my son, till he hath got him a wife, But my daughter's my daughter all days of her life. **1863** C. READE *Hard Cash* I. v. 'Oh, mamma,' said Julia warmly, 'and do you think all the marriage in the world . . can make me lukewarm to my . . mother? . . It's a son who is a son only till he gets him a wife: but your daughter's your daughter, all-the-days-of her life. **1943** A. THIRKELL *Growing Up* iii. She doesn't hear from him for months at a time now of course and then it's only a wire as often as not, but your son's your son till he gets him a wife, as the saying is. **1981** *Listener* 27 Aug. 206 There's a very old-fashioned sort of saying we have in the North which goes, 'My son is my son till he finds him a wife, but my daughter is my daughter the rest of her life.'

son: *see also* CLERGYMEN's sons always turn out badly; the DEVIL's children have the Devil's luck; like FATHER, like son; the SHOEMAKER's son always goes barefoot.

SOON ripe, soon rotten

[L. *cito maturum cito putridum*, quickly ripe, quickly rotten.] **1393** LANGLAND *Piers Plowman* C. XIII. 233 And that that rathest[1] rypeth, roteth most saunest. **1546** J. HEYWOOD *Dialogue of Proverbs* I. x. C4[v] In youth she was towarde[2] and without euill. But soone rype sone rotten. **1642** D. ROGERS *Naaman* x. Some indeed . . are moved to . . disdaine by their inferiours forwardnesse, called them hastings, soone ripe, soone

rotten. **1887** s. SMILES *Life & Labour* vi.
Very few prize boys and girls stand the
test of wear. Prodigies are almost always uncertain; they illustrate the
proverb of 'soon ripe, soon rotten'.
[1] most early. [2] hopeful, promising.

sorrow: *see* ONE for sorrow, two for mirth;
help you to SALT, help you to sorrow.

sorrowing: *see* he that GOES a-borrowing
goes a-sorrowing.

sorry: *see* BETTER be safe than sorry.

sort: *see* it takes ALL sorts to make a world.

soul: *see* BREVITY is the soul of wit; open
CONFESSION is good for the soul; COR-
PORATIONS have neither bodies to be
punished nor souls to be damned; the
EYES are the window of the soul; PUNC-
TUALITY is the soul of business.

sound: *see* EMPTY vessels make the most
sound.

source: *see* a STREAM cannot rise above its
source.

**A SOW may whistle, though it has an ill
mouth for it**
    1802 M. EDGEWORTH *Letter* 19 Oct. in
Maria Edgeworth in France & Switzerland
(1979) 10 He waddles on dragging his
boots along in a way that would make a
pig laugh. As Lord Granard[1] says, a pig
may whistle though he has a bad mouth
for it. **1846** J. GRANT *Romance of War* I.
xii. 'I dare say the Spanish sounds very
singular to your ear.' 'Ay, sir; it puts me
in mind o' an auld saying o' my faither
the piper. "A soo may whussle, but its
mouth is no made for't."' **1927**
J. BUCHAN *Witch Wood* xvii. Ye say he
has the speech o' a guid Christian?
Weel–a–weel, a soo may whistle,
though it has an ill mouth for it.
    [1] George Forbes (1760–1837), sixth Earl and
first Baron Granard.

As you SOW, so you reap
    [GALATIANS vi. 7 Whatsoeuer a man

soweth, that shall he also reape.] *a* **900**
CYNEWULF *Christ* in *Anglo-Saxon Poetic
Records* (1936) III. 5 Swa eal manna
bearn sorgum sawath, swa eft ripath
[just as each son of man sows in grief, so
he also reaps]. *c* **1470** *Mankind* in *Macro
Plays* (1962) l. 180 Such as thei haue
sowyn, such xall thei repe. **1664** BUTLER
Hudibras II. ii. And look before you ere
you leap; For as you sow, you are like to
reap. **1871** J. A. FROUDE *Short Studies*
2nd Ser. 10 As men have sown they
must still reap. The profligate . . may
recover . . peace of mind . . but no
miracle takes away his paralysis. **1978**
F. WELDON *Praxis* xxiv. 'You should
never have left them,' said Irma. 'As
you sow, Praxis, so you reap.'

**They that SOW the wind shall reap the
whirlwind**
The proverb is also used as a metaphorical phrase *to sow the wind (and reap the
whirlwind)*.
    [HOSEA viii. 7 They haue sowen the
winde, and they shall reape the
whirlewinde.] **1583** J. PRIME *Fruitful &
Brief Discourse* II. 203 They who sowed a
winde, shall reap a whirlewind, but
they that sowed in iustice shall reape
mercie. **1853** G. W. CURTIS in *Putnam's
Magazine* Apr. 386 Ask the Rev. Cream
Cheese to . . preach from this text: 'They
that sow the wind shall reap the
whirlwind.' **1923** O. DAVIS *Icebound* III.
98 Well—what's passed is passed.
Folks that plant the wind reap the
whirlwind! **1981** J. STUBBS *Ironmaster*
xvii. I know that he who sows the wind
shall reap the whirlwind. I dislodge a
clod of earth, and start a landslide.

sow: *see also* (noun) you can't make a SILK
purse out of a sow's ear.

span: *see* when ADAM delved and Eve
span, who was then the gentleman?

**SPARE at the spigot, and let out at the
bung-hole**
    1642 G. TORRIANO *Select Italian Prov-
erbs* 50 He holdeth in at the spicket,[1] but
letteth out at the bunghole.[2] **1670** J. RAY

English Proverbs 193 Spare at the spig-get, and let it out at the bung-hole. **1721**
J. KELLY *Scottish Proverbs* 299 Spare at the Spiggot, and let out at the Bung Hole. Spoken to them who are careful and penurious in some trifling Things, but neglective in the main Chance. **1885** E. J. HARDY *How to be Happy though Married* xiii. People are often saving at the wrong place. . . They spare at the spigot, and let all run away at the bung-hole.

¹ i.e. spigot; a peg or pin used to regulate the flow of liquid through a tap on a cask.

² a hole through which a cask is filled or emptied, and which is closed by a bung.

SPARE the rod and spoil the child

And introduces a consequence.

c **1000** AELFRIC *Homilies* (1843) II. 324 Se the sparath his gyrde,¹ he hatath his cild. **1377** LANGLAND *Piers Plowman* B. v. 41 Salamon seide . . *Qui parcit virge, odit filium.* The Englich of this latyn is . . Who-so spareth the sprynge,² spilleth³ his children. **1560** *Nice Wanton* A1ᵛ He that spareth the rod, the chyld doth hate. **1611** BIBLE *Proverbs* xiii. 24 He that spareth his rod, hateth his sonne. **1639** J. CLARKE *Parœmiologia Anglo-Latina* 161 Spare the rod and spoyle the child. **1876** I. BANKS *Manchester Man* II. vii. 'Spare the rod and spoil the child' had not been abolished from the educational code fifty-five years back. **1907** E. GOSSE *Father & Son* ii. This action [caning] was justified, as everything he did was justified, by reference to Scripture—'Spare the rod and spoil the child'. **1980** H. BIANCHIN *Devil in Command* vii. Brute force isn't always the answer. . . Have you not heard of 'spare the rod and spoil the child'?

¹ stick, rod.   ² rod, switch.   ³ ruins.

SPARE well and have to spend

1541 M. COVERDALE tr. *H. Bullinger's Christian State of Matrimony* xix. Spare as though thou neuer shuldest dye & yet as mortall spend mesurably. To spare that thou mayest haue to spend in honestye for goodes sake. **1635** J. GORE *Way to Well-doing* 25 A good sparer

makes a good spender. **1721** J. KELLY *Scottish Proverbs* 297 Spare when you're young, and spend when you're old. . . He that saveth his Dinner will have the more for his Supper. **1832** A. HENDERSON *Scottish Proverbs* 16 Spare weel and hae weel. **1977** J. AIKEN *Five-Minute Marriage* x. 'I've given them a polish and they've come up real tiptop! Spare well and have to spend, I allus say.

Never SPEAK ill of the dead

[Gr. τὸν τεθνηκότα μὴ κακολογεῖν, speak no evil of the dead (attributed to the Spartan ephor or civil magistrate Chilon, 6th cent. BC); L. *de mortuis nil nisi bonum*, say nothing of the dead but what is good.] **1540** R. TAVERNER tr. *Erasmus' Flores Sententiarum* A6 Rayle not vpon him that is deade. **1609** S. HARWARD *MS* (Trinity College, Cambridge) 81ᵛ Speake not evill of the dead. **1682** W. PENN *No Cross, No Crown* (ed. 2) xix. Speake well of the dead. **1783** JOHNSON *Lives of Poets* (rev. ed.) IV. 381 He that has too much feeling to speak ill of the dead . . will not hesitate . . to destroy . . the reputation . . of the living. **1945** F. THOMPSON *Lark Rise* xiv. 'Never speak ill of the dead' was one of their maxims. **1959** J. CARY *Captive & Free* xlvii. 'Nothing but good about the dead.' What a lie. Of a piece with all this English softness, sentiment, slop.

SPEAK not of my debts unless you mean to pay them

1640 G. HERBERT *Outlandish Proverbs* no. 998 Speake not of my debts, un-lesse you meane to pay them. **1875** A. B. CHEALES *Proverbial Folk-Lore* 88 Special proverbs supply us with some excellent admonitions. . . Dont talk of my debts unless you mean to pay them. **1981** *Times* 2 Jan. 10 An old proverb recommends you not to speak of my debts unless you mean to pay them.

speak: *see also* ACTIONS speak louder than words; SEE no evil, hear no evil, speak no evil; THINK first and speak afterwards; *also* SPOKEN.

speaking: *see* it's ill speaking between a FULL man and a fasting.

Everyone SPEAKS well of the bridge which carries him over

1678 J. RAY *English Proverbs* (ed. 2) 106 Let every man praise the bridge he goes over. *i.e.* Speak not ill of him who hath done you a courtesie, or whom you have made use of to your benefit; or do commonly make use of. **1797** F. BAILY *Journal* 11 May (1856) 279 Let every one speak well of the bridge which carries him safe over. **1850** KINGSLEY *Alton Locke* I. x. Every one speaks well of the bridge which carries him over. Every one fancies the laws which fill his pockets to be God's laws. **1886** G. DAWSON *Biographical Lectures* i. Our love of compromise . . has also been our great strength. . . We speak well of the bridge that carries us over.

species: *see* the FEMALE of the species is more deadly than the male.

SPEECH is silver, but silence is golden

See also the abbreviated form SILENCE *is golden*.

1834 CARLYLE in *Fraser's* June 668 As the Swiss Inscription says: *Sprechen ist silbern, Schweigen ist golden* (Speech is silvern, Silence is golden). **1865** A. RICHARDSON *Secret Service* ii. A taciturn but edified listener, I pondered upon . . 'speech is silver, while silence is golden'. **1936** W. HOLTBY *South Riding* I. iv. She will give a pound note to the collection if I would cut my eloquence short, so in this case, though speech is silver, silence is certainly golden. **1961** M. SPARK *Prime of Miss Jean Brodie* i. Speech is silver but silence is golden. Mary, are you listening?

speed: *see* more HASTE, less speed.

What you SPEND, you have

c **1300** in M. R. James *Catalogue of Library Pembroke College* (1905) 35 That ich et[1] that ich hadde. That ich gaf that ich habbe. That ich ay held that i nabbe.[2] **1579** SPENSER *Shepherd's Calendar* (May) 56 (Glossary) Ho, ho, who lies here? I the good Earle of Deuonshere, And Maulde my wife, that was ful deare. . . That we spent, we had: That we gaue, we haue: That we lefte we lost. **1773** JOHNSON *Letter* 12 Aug. (1952) I. 338 The monument of Robert of Doncaster . . says . . something like this. What I gave, that I have; what I spent, that I had; what I left that I lost. **1862** *Times* 15 Dec. 8 The most common maxim of the rank and file of British industry is that what you spend you have for it alone cannot be taken away from you.

[1] *ich et*, I ate. [2] do not have.

spend: *see also* SPARE well and have to spend.

spent: *see* what is GOT over the Devil's back is spent under his belly.

sphere: *see* a WOMAN'S place is in the home.

spice: *see* VARIETY is the spice of life.

spigot: *see* SPARE at the spigot, and let out at the bung-hole.

spilt: *see* it is no use CRYING over spilt milk.

spite: *see* don't CUT off your nose to spite your face.

splash: *see* when the OAK is before the ash then you will only get a splash.

spoil: *see* do not spoil the SHIP for a ha'porth of tar; SPARE the rod and spoil the child; TOO many cooks spoil the broth.

spoiled: *see* BETTER one house spoiled than two.

spoken: *see* many a TRUE word is spoken in jest.

spoon: *see* he who SUPS with the Devil should have a long spoon.

spot: *see* the LEOPARD does not change his spots.

spread: *see* in vain the NET is spread in the sight of the bird.

It is not SPRING until you can plant your foot upon twelve daisies

1863 R. CHAMBERS *Book of Days* I. 312 We can now plant our 'foot upon nine daisies' and not until that can be done do the old-fashioned country people believe that spring is really come. 1878 T. F. THISELTON-DYER *English Folk-Lore* i. 'It ain't spring until you can plant your foot upon twelve daisies,' is a proverb still very prevalent. 1910 *Spectator* 26 Mar. 499 Spring is here when you can tread on nine daisies at once on the village green; so goes one of the country proverbs. 1972 CASSON & GRENFELL *Nanny Says* 52 When you can step on six daisies at once, summer has come.

spring: *see* (verb) HOPE springs eternal.

The SQUEAKING wheel gets the grease

Attention is only given to a troublesome person or thing.

a 1937 in J. Bartlett *Familiar Quotations* 518 The wheel that squeaks the loudest Is the one that gets the grease. 1948 in B. Stevenson *Home Book of Proverbs* 2483, I hate to be a kicker, I always long for peace, But the wheel that does the squeaking is the one that gets the grease. 1967 A. F. BARTSCH in Olson & Burgess *Pollution & Marine Ecology* VI. 293 It is a tragedy that we devote more study effort to disturbed areas than to reasonably natural ones. This . . is simply a matter of the 'squeaking wheel getting the grease'. 1969 H. LAMARSH *Memoirs of Bird in Gilded Cage* x. Mrs. Kinnear did . . put on a most relentless campaign over the years in her own interest and in line with the best of political principles—'the squeaky wheel got the grease'. 1974 *Hansard* (Commons) 17 Oct. 502 It is the old story: the squeaky wheel gets the grease.

stable: *see* the MAN who is born in a stable is a horse.

It is too late to shut the STABLE-door after the horse has bolted

In early use the proverb referred to horse-stealing: *has bolted* is a modern substitution for the traditional *is stolen*.

[Med. Fr. *a tart ferme on l'estable, quant li chevaux est perduz*, the stable is shut too late, when the horse is lost.] *c* 1350 *Douce MS* 52 no. 22 When the hors is stole, steke[1] the stabull-dore. *c* 1490 in *Anglia* (1918) XLII. 204 Whan the stede ys stole, than shytte the stable-dore. 1578 LYLY *Euphues* I. 188 It is to late to shutte the stable doore when the steede is stolen: The Trojans repented to late when their towne was spoiled. 1719 DEFOE *Robinson Crusoe* II. 92 A dead Bush was cram'd in [the hedge] to stop them [the Spaniards] out for the present, but it was only shutting the Stable-door after the Stead was stolen. 1886 R. L. STEVENSON *Kidnapped* xiv. A guinea-piece . . fell . . into the sea. . . I now saw there must be a hole, and clapped my hand to the place. . . But this was to lock the stable door after the steed was stolen. 1940 N. MARSH *Death of Peer* x. The horse having apparently bolted, I shall be glad to assist at the ceremony of closing the stable-door. 1979–80 *Verbatim* Winter 2 It is too late . . to shut the stable door after the horse has bolted.

[1] shut, lock.

stalled: *see* BETTER a dinner of herbs than a stalled ox where hate is.

stand: *see* EMPTY sacks will never stand upright; if you don't like the HEAT, get out of the kitchen; every TUB must stand on its own bottom; UNITED we stand, divided we fall.

standing: *see* it is as CHEAP sitting as standing.

starve: *see* FEED a cold and starve a fever; while the GRASS grows, the steed starves.

stay: *see* the FAMILY that prays together stays together.

steady: *see* FULL cup, steady hand.

One man may STEAL a horse, while another may not look over a hedge
People may take different degrees of liberty depending on our opinion of them.

1546 J. HEYWOOD *Dialogue of Proverbs* II. ix. K4 This prouerbe .. saith, that some man maie steale a hors better, Than some other maie stande and loke vpone. **1591** LYLY *Endymion* III. iii. Some man may better steale a horse, then another looke ouer the hedge. **1670** J. RAY *English Proverbs* 128 One man may better steal a horse, then another look over the hedge. If we once conceive a good opinion of a man, we will not be perswaded he doth any thing amiss; but him whom we have a prejudice against, we are ready to suspect on the sleightest occasion. **1894** J. LUBBOCK *Use of Life* ii. 'One man may steal a horse, while another may not look over a hedge' .. because the one does things pleasantly, the other disagreeably. **1921** A. BENNETT *Things that have interested Me* 315 Strange how one artist may steal a horse while another may not look over a hedge. **1957** R. WEST *Fountain Overflows* xi. Fancy him caring for her after all these years. Particularly when she treated him the way she did. But there, some people can steal a horse, and others aren't allowed to look over the gate.

steal: *see also* HANG a thief when he's young, and he'll no' steal when he's old; it's a SIN to steal a pin.

steed: *see* while the GRASS grows, the steed starves.

One STEP at a time
1901 KIPLING *Kim* vi. It's beyond me. We can only walk one step at a time in this world. **1919** J. BUCHAN *Mr. Standfast* xvi. I did not allow myself to think of ultimate escape. .. One step at a time was enough. **1979** V. CANNING *Satan Sampler* ix. The stuff he had found .. took precedence over everything. One step at a time.

step: *see also* it is the FIRST step that is difficult; from the SUBLIME to the ridiculous is only a step.

A STERN chase is a long chase
1823 J. F. COOPER *Pilot* xv. 'If we can once get him in our wake I have no fears of dropping them all.' 'A stern chace[1] is a long chase.' **1919** J. A. BRIDGES *Victorian Recollections* xiv. English poetry has had a start of some centuries, and a stern chase is proverbially a long one.

[1] a chase in which the pursuing ship follows directly in the wake of the pursued.

stey (steep): *see* put a STOUT heart to a stey brae.

It is easy to find a STICK to beat a dog
i.e. an excuse to justify a harsh action or opinion.

1564 T. BECON *Works* I. C5ᵛ Howe easye a thyng it is to fynde a staffe if a man be mynded to beate a dogge. **1581** G. PETTIE tr. *S. Guazzo's Civil Conversation* III. 50 It is an easie matter to finde a staffe to beate a dog. **1782** F. HOPKINSON *Miscellaneous Essays* I. 266 A proverb .. naturally occurs on this occasion: It is easy to find a stick to beat a dog. **1875** S. SMILES *Thrift* xiv. Excuses were abundant. .. It is easy to find a stick to beat a sick dog. **1908** *Times Literary Supplement* 6 Nov. 391 The reviewer seems .. predisposed to the view that any stick is good enough to beat a dog with.

stick: *see* (verb) let the COBBLER stick to his last; throw DIRT enough, and some will stick.

STICKS and stones may break my bones, but words will never hurt me
Cf. HARD *words break no bones.*
1894 G. F. NORTHALL *Folk-Phrases* 23 Sticks and stones will break my bones, but names will never hurt me! Said by one youngster to another calling names. **1980** *Cosmopolitan* Dec. 137 'Sticks and stones may break my bones,' goes the children's rhyme, 'but words will never hurt me.' One wonders

whether the people on the receiving end
. . would agree.

A STILL tongue makes a wise head

1562 J. HEYWOOD *Works* Dd3ᵛ Hauyng
a styll toung he had a besy head. **1776**
T. COGAN *John Buncle, Junior* I. 238
Mum's the word. . . A quiet tongue
makes a wise head, says I. **1869**
W. C. HAZLITT *English Proverbs* 35 A still
tongue makes a wise head. **1937**
J. WORBY *Other Half* iv. 'I believe in the
old saying "A still tongue keeps a wise
head".' 'I guess you're right. . . It's no
business of mine.'

STILL waters run deep

Now commonly used to assert that a
placid exterior hides a passionate nature.

c **1400** *Cato's Morals* in *Cursor Mundi*
(EETS) 1672 There the flode is deppist
the water standis stillist. *c* **1410** LYD-
GATE *Minor Poems* (EETS) 476 Smothe
waters ben ofte sithes[1] depe. **1616**
T. DRAXE *Adages* 178 Where riuers runne
most stilly, they are the deepest.
1721 J. KELLY *Scottish Proverbs* 287
Smooth Waters run deep. **1858** D. M.
MULOCK *Woman's Thoughts about Women*
xii. In maturer age . . the fullest, ten-
derest tide of which the loving heart is
capable may be described by those
'still waters' which 'run deep'. **1979**
M. UNDERWOOD *Victim of Circumstances*
II. 86 As for her, still waters run deep, it
seems. She always looked so solemn. . .
Fancy her shooting him!

[1] *ofte sithes*, oftentimes, often.

sting: *see* if you gently touch a NETTLE it'll
sting you for your pains.

stink: *see* the FISH always stinks from the
head downwards; FISH and guests stink
after three days.

stir: *see* you should KNOW a man seven
years before you stir his fire.

A STITCH in time saves nine

The proverb was originally a couplet.

The number *nine* was apparently intro-
duced fancifully for the sake of asson-
ance.

1732 T. FULLER *Gnomologia* no. 6291 A
Stitch in Time May save nine. **1797**
F. BAILY *Journal* 30 Apr. (1856) 268 After
a little while we acquired a method of
keeping her [a boat] in the middle of the
stream, by watching the moment she
began to vary, and thereby verifying the
vulgar proverb, 'A stitch in time saves
nine'. **1868** READE & BOUCICAULT *Foul
Play* I. ix. Repairing the ship. Found a
crack or two in her inner skin. . . A
stitch in time saves nine. **1979** *Homes &
Gardens* June 105 Looking after oneself
is like looking after a house: a stitch in
time. . .

STOLEN fruit is sweet

The proverb is used in a variety of
forms, principally in allusion to the
temptation of Eve (Genesis iii. 6). See
also the next entry.

[*c* **1390** CHAUCER *Parson's Tale* l. 332
The fleesh hadde delit in the beautee of
the fruyt defended.[1]] **1614** T. ADAMS
Devil's Banquet III. 98 But as the Proverbe hath it . . Apples are sweet, when
they are plucked in the Gardiners
absence. Eve liked no Apple in the Garden so well as the forbidden. **1668** F.
KIRKMAN *English Rogue* II. B1ᵛ So eager
are these sort of people to buy any thing
that is unlicensed, following the
Proverb, that stollen meat is sweetest.
1855 E. GASKELL *North & South* II. i. I
can remember . . your being in some
disgrace . . for stealing apples. . . Some
one had told you that stolen fruit tasted
sweetest. **1935** H. SPRING *Rachel Rosing*
xxiv. He knew that he did not love
her. . . What else, then? . . He was not
going to pretend that this stolen fruit
was not sweet. **1961** N. LOFTS *House at
Old Vine* II. 137 Old men are like children, of whom they say 'Stolen apples
are sweetest'.

[1] forbidden.

STOLEN waters are sweet

See also the preceding entry.

c **1395** WYCLIF *Bible* Proverbs ix. 17

Stoln watris ben swettere. *c* **1548** *Will of Devil* (1863) 9 This saiyng of the retcheles[1] woman in Salomon (Stollen waters ar sweete). **1611** BIBLE *Proverbs* ix. 17 Stollen waters are sweet. **1614** T. ADAMS *Devil's Banquet* I. 3 Sinne shewes you a faire Picture—Stollen waters are sweet. **1721** J. KELLY *Scottish Proverbs* 298 Stoln Waters are sweet. People take great Delight in that which they can get privately. **1976** A. J. RUSSELL *Pour Hemlock* ii. Lucarelli, fond of quoting scripture, ended the memo with 'Stolen waters are sweet'.

[1] i.e. reckless.

stomach: *see* an ARMY marches on its stomach; the WAY to a man's heart is through his stomach.

stone: *see* you cannot get BLOOD from a stone; you BUY land you buy stones, you buy meat you buy bones; CONSTANT dropping wears away a stone; DRIVE gently over the stones; those who live in GLASS houses shouldn't throw stones; a ROLLING stone gathers no moss; STICKS and stones may break my bones, but words will never hurt me.

STONE-dead hath no fellow

Predominantly used by advocates of the death penalty.

c **1633** *Soddered Citizen* (1936) l. 2618 'Is your ffather dead?' . . 'Laid with both Leggs Sir, in one lynnen bootehose That has noe fellowe,[1] stone dead. *c* **1641** CLARENDON *Hist. Rebellion* (1702) I. III. 191 The Earl of Essex . . answer'd, 'Stone Dead hath no Fellow.' **1828** MACAULAY *Essays* (1843) I. 144 Stone-dead hath no fellow. **1926** *Times* 27 Aug. 11 The execution of the death sentence had been postponed for a week, an unusual period in a country where the adage 'stone-dead hath no fellow' wins general support. **1980** G. BLAKISTON *Woburn & Russells* v. Bedford, who was against the death penalty for Stratford, sought to moderate the violent opinions of some of his fellow peers, the Earl of Essex being heard to declare vehemently: 'stone dead hath no fellow'.

[1] companion; equal, counterpart.

stool: *see* BETWEEN two stools one falls to the ground.

storm: *see* AFTER a storm comes a calm; ANY port in a storm; the SHARPER the storm, the sooner it's over.

story: *see* every PICTURE tells a story.

Put a STOUT heart to a stey brae

A Scottish proverb also used as a metaphorical phrase.

a **1585** A. MONTGOMERIE *Cherry & Sloe* (1821) xxxvi. So gets ay, that sets ay, Stout stomackis to the brae.[1] **1721** J. KELLY *Scottish Proverbs* 287 Set a stout Heart to a stay[2] Brea. Set about a difficult Business with Courage and Constancy. **1821** J. GALT *Annals of Parish* i. I began a round of visitations; but oh, it was a steep brae that I had to climb, and it needed a stout heart. For I found the doors . . barred against me. **1916** J. BUCHAN *Greenmantle* xii. He . . shouted to me . . to 'pit a stoot hert tae a stey brae'.

[1] slope or hill-side.   [2] steep.

strange: *see* ADVERSITY makes strange bedfellows; POLITICS makes strange bedfellows.

stranger: *see* FACT is stranger than fiction; TRUTH is stranger than fiction.

straw: *see* you cannot make BRICKS without straw; a DROWNING man will clutch at a straw; it is the LAST straw that breaks the camel's back.

STRAWS tell which way the wind blows

The phrase 'a straw in the wind', a sign of the prevailing opinion, action, etc., is also found.

a **1654** J. SELDEN *Table-Talk* (1689) 31 Take a straw and throw it up into the Air, you shall see by that which way the Wind is. . . More solid things do not shew the Complexion of the times so

well, as Ballads and Libels. **1799** COB-
BETT *Porcupine's Works* (1801) X. 161
'Straws' (to make use of Callender's old
hackneyed proverb) . . 'served to show
which way the wind blows.' **1927** A.
ADAMS *Ranch on Beaver* vii. As straws tell
which way the wind blows . . this day's
work gives us a clean line on these company cattle.

A STREAM cannot rise above its source
1663 S. TUKE *Adventures of Five Hours*
Prologue, He would be ever w'you, but
wants force; The Stream will rise no
higher than the Source. **1732** T. FULLER
Gnomologia no. 4771 The Stream can
never rise above the Spring-head. **1905**
H. A. VACHELL *Hill* 84 Clever chap. . .
But one is reminded that a stream can't
rise higher than its source. **1921** T. R.
GLOVER *Pilgrim* 125 It is held that a
stream cannot rise above its source; but
. . [a] river may have many tributaries,
and one of them may change the character of what we call the main stream.

stream: *see also* don't CHANGE horses in
mid-stream.

strength: *see* UNION is strength.

strengthen: *see* as the DAY lengthens, so
the cold strengthens.

STRETCH your arm no further than your
sleeve will reach
Do not spend more than you can afford.
Cf. CUT *your coat according to your cloth.*
  1541 M. COVERDALE tr. H. Bullinger's
Christian State of Matrimony xix. Strech
out thine arme no farther then thy sleve
wyll retche. **1639** J. CLARKE *Parœmio-*
logia Anglo-Latina 211 Stretch your arme
no further than your sleeve will reach.
1721 J. KELLY *Scottish Proverbs* 277 Put
your Hand no farther than your Sleeve
will reach. That is, spend no more than
your Estate will bear. **1881** W. WESTALL
Old Factory II. ii. It would leave me short
of working capital, and . . I mustn't
stretch my arm further than th'coat-
sleeve will reach.

Everyone STRETCHES his legs according
to the length of his coverlet
Similar in form to the preceding
proverb.
  a **1300** WALTER OF HENLEY *Husbandry*
(1890) 4 Wo that stretchet fortherre than
his wytel[1] wyle reche, in the straue[2] his
fet he mot streche. **1550** W. HARRYS
Market D5[v] Then must many a man . .
stretche out his legges accordynge to the
length of his coverlet. **1640** G. HERBERT
Outlandish Proverbs no. 147 Everyone
stretcheth his legges according to his
coverlet. **1897** 'H. S. MERRIMAN' *In*
Kedar's Tents iv. 'The English . . travel
for pleasure.' . . 'Every one stretches his
legs according to the length of his coverlet,' he said.
  [1] blanket.   [2] i.e. straw.

STRIKE while the iron is hot
A proverb originally alluding to the
blacksmith's art.
  [Late 13th-cent. Fr. *len doit batre le*
fer tandis cum il est chauz, one must
strike the iron while it is hot.] *c* **1390**
CHAUCER *Melibee* l. 2226 Whil that iren is
hoot, men sholden smyte. **1546** J.
HEYWOOD *Dialogue of Proverbs* I. iii. A4
And one good lesson to this purpose I
pyke[1] From the smiths forge, whan
thyron is hote stryke. **1576** G. PETTIE
Petit Palace 181, I think it wisdome to
strike while the iron is hot. **1682** BUN-
YAN *Holy War* 18 Finding . . the affections of the people warmly inclining to
him, he, as thinking 'twas best striking
while the iron is hot, made this . .
speech. **1771** SMOLLETT *Humphry Clinker*
III. 242 If so be as how his regard be
the same, why stand shilly shally? Why
not strike while the iron is hot, and
speak to the 'squire without loss of
time? **1848** THACKERAY *Vanity Fair* xxi.
Let George cut in directly and win
her. . . Strike while the iron's hot. **1974**
T. SHARPE *Porterhouse Blue* xx. 'It seems
an inopportune moment,' said the
Senior Tutor doubtfully. . . 'We must
strike while the iron is hot,' said the
Dean.
  [1] i.e. pick.

strike: *see also* LIGHTNING never strikes the same place twice.

striving: *see* it is ill SITTING at Rome and striving with the Pope.

stroke: *see* LITTLE strokes fell great oaks; beware of an OAK it draws the stroke.

strong: *see* GOOD fences make good neighbours; the RACE is not to the swift, nor the battle to the strong; YORKSHIRE born and Yorkshire bred, strong in the arm and weak in the head.

stronger: *see* a CHAIN is no stronger than its weakest link.

stubborn: *see* FACTS are stubborn things.

The STYLE is the man

[L. *stylus virum arguit*, the style shows the man; G. L. LE CLERC, Conte de BUFFON in *Histoire Naturelle* VII. p. xvii. *Le style est l'homme même* [the style is the man himself].] **1901** A. WHYTE *Bible Characters* V. civ. If the style is the man in Holy Scripture also . . we feel a very great liking for Luke. **1942** H. F. HEARD *Reply Paid* ix. Usually I don't like to have my style modified. 'The style is the man.'

From the SUBLIME to the ridiculous is only a step

[Fr. *du sublime au ridicule il n'y a qu'un pas*, from the sublime to the ridiculous there is only one step (attributed to Napoleon).] **1879** M. PATTISON *Milton* 116 The Hague tittle-tattle . . is set forth in the pomp of Milton's loftiest Latin. . . The sublime and the ridiculous are here blended without the step between. **1909** *Times Literary Supplement* 17 Dec. 492 In the case of Louis XVIII, indeed, the ridiculous was, as it is commonly said to be, only a step removed from the sublime. **1940** W. & E. MUIR tr. *L. Feuchtwanger's Paris Gazette* II. xxxviii. From the sublime to the ridiculous is only a step, but there's no road that leads back from the ridiculous to the sublime.

If at first you don't SUCCEED, try, try, try again

The short poem *Try (try) again* was often quoted in nineteenth-century children's literature, especially in the United States (see quot. 1840). It is popularly attributed to W. E. Hickson, who quoted it (with three *try*'s) in his *Moral Songs* (1857) p. 8, but Palmer's use is earlier. The saying was soon used independently as a proverb.

1840 T. H. PALMER *Teacher's Manual* 223 'Tis a lesson you should heed, Try, try again. If at first you do n't succeed, Try, try again. **1915** E. B. HOLT *Freudian Wish* iii. The child is frustrated, but not instructed; and it is in the situation where, later on in life, we say to ourselves, 'If at first you don't succeed, Try, try, try again!' *a* **1979** A. CHRISTIE *Miss Marple's Final Cases* 39 You mustn't give up, Mr. Rossiter, 'If at first you don't succeed, try, try, try again.'

succeed: *see also* NOTHING succeeds like success.

suck: *see* don't TEACH your grandmother to suck eggs.

Never give a SUCKER an even break

This saying has been attributed to various people, including E. F. Albee and W. C. Fields. It was popularized by Fields, who is said to have used it in the musical comedy *Poppy* (1923), though it does not occur in the libretto. *Poppy* was made into a silent film in 1925 and called *Sally of the Sawdust*. This was in turn remade as a 'talkie' in 1936 (see quot. 1936). The proverb means that one should not allow a fair chance to a fool, or one who may be easily deceived.

1925 *Collier's* 28 Nov. 26 'That line of mine that brings down the house always was true, wasn't it?' 'Which line?' I asked. 'Never give a sucker an even break' he [W. C. Fields] answered. **1936** *N.Y. Herald Tribune* 15 Mar. v. 1 Wasn't it 'Poppy' that provided him with his immortal motto, 'Never give a sucker an even break'? **1940** WODEHOUSE *Eggs, Beans & Crumpets* 158 Never give a sucker an even break. . . But your sermon has made me see that there is something higher and nobler than a code of business ethics. **1979** *Daily Telegraph* 3 Nov. 24 The basic American business philosophy of 'never give a sucker an even break' runs rampant in those [money] markets.

suckling: *see* out of the MOUTHS of babes—.

SUFFICIENT unto the day is the evil thereof

[MATTHEW vi. 34 Sufficient vnto the day is the euill thereof.] **1766** in L. H. Butterfield et al. *Adams Family Correspondence* (1963) I. 56 Sufficient to the Day is the Evil thereof. **1836** J. CARLYLE *Letter* 1 Apr. in *Letters & Memorials* (1893) I. 57 In the meanwhile there were no sense in worrying over schemes for a future, which we may not live to see. 'Sufficient for the day is the evil thereof.' **1979** M. BABSON *So soon done For* vii. 'I'll deal with these [bills] later.' . . 'Sufficient unto the day,' Kay agreed.

sufficient: *see also* a WORD to the wise is enough.

summer: *see also* the RICH man has his ice in the summer and the poor man gets his in the winter; one SWALLOW does not make a summer.

The SUN loses nothing by shining into a puddle

[Gr. ὁ ἥλιος εἰς τοὺς ἀποπάτους, ἀλλ' οὐ μιαίνεται, the sun shines into dung but is not tainted (attributed to Diogenes); TERTULLIAN *De Spectaculis* xx. *sane, sol et in cloacam radios suos defert nec inquinatur*, indeed the sun spreads his rays even into the sewer, and is not stained.] **1303** R. BRUNNE *Handlyng Synne* (EETS) l. 2299 The sunne, hys feyrnes neuer he tynes,[1] Thogh hyt on the muk hepe shynes. *c* **1390** CHAUCER *Parson's Tale* l. 911 Though that holy writ speke of horrible synne, certes holy writ may nat been defouled, na-moore

than the sonne that shyneth on the mixne.[2] **1578** LYLY *Euphues* I. 193 The Sun shineth vppon the dungehill and is not corrupted. **1732** T. FULLER *Gnomologia* no. 4776 The Sun is never the worse for shining on a Dunghill. **1943** E. M. ALMEDINGEN *Frossia* iv. Dreadful words did fly about then, but the sun loses nothing by shining into a puddle.

[1] loses. [2] midden, dunghill.

sun: *see also* happy is the BRIDE that the sun shines on; MAKE hay while the sun shines; there is NOTHING new under the sun.

sunny: *see* if CANDLEMAS day be sunny and bright, winter will have another flight.

supper: *see* AFTER dinner rest a while, after supper walk a mile; HOPE is a good breakfast but a bad supper.

He who SUPS with the Devil should have a long spoon

A proverb advocating caution when dealing with dangerous persons.

c **1390** CHAUCER *Squire's Tale* l. 602 Therfore bihoueth hire a ful long spoon That shal ete with a feend. **1545** R. TAVERNER tr. *Erasmus' Adages* (ed. 2) 9ᵛ He had nede to haue a longe spone that shuld eate with the deuyl. **1590** SHAKESPEARE *Comedy of Errors* IV. iii. 59 He must have a long spoon that must eat with the devil.—What tell'st thou me of supping? **1641** D. FERGUSSON *Scottish Proverbs* (STS) no. 350 He should have a long shafted spoon that sups kail[1] with the devil. **1721** J. KELLY *Scottish Proverbs* 147 He had need of a long Spoon that sups Kail with the Dee'l. He that has to do with wicked . . Men, had need to be cautious. **1840** R. H. BARHAM *Ingoldsby Legends* 1st Ser. 270 Who suppes with the Deville sholde have a longe spoone! **1920** 'D. YATES' *Berry & Co.* v. I'd rather watch from the stalk than assist him. . . Remember 'Who sups with the devil should hold a long spoon'. **1979** 'E. ANTHONY' *Grave of Truth* viii. Hindenburg and the army thought they could use [Hitler]. . . Who sups with the devil needs a long spoon.

[1] vegetable broth.

sure: see SLOW but sure.

suspicion: see CAESAR's wife must be above suspicion.

SUSSEX won't be druv
A local proverb which asserts that Sussex people have minds of their own, and cannot be forced against their will.

1910 in T. Wales *Sussex Garland* (1979) i. (*postcard*) Have got as fat as a Sussex [pig]—and 'wunt be druv'[1] from Brighton. **1924** H. DE SELINCOURT *Cricket Match* vi. 'Well, we'd better be going, I suppose,' Gauvinier announced . . well aware that 'Sussex won't be druv'. **1979** T. WALES *Sussex Garland* i. There ant no place like Sussex, Until ye goos above, For Sussex will be Sussex, And Sussex won't be druv!

[1] a dialectal variant of *drove* (standard English *driven*).

One SWALLOW does not make a summer
[Gr. μία χελιδὼν ἔαρ οὐ ποιεῖ, one swallow does not make a summer; ERASMUS *Adages* I. vii. *una hirundo non facit ver.*] **1539** R. TAVERNER tr. *Erasmus's Adages* 25 It is not one swalowe that bryngeth in somer. It is not one good qualitie that maketh a man good. **1546** J. HEYWOOD *Dialogue of Proverbs* II. v. H3 One swalow maketh not sommer. **1659** J. HOWELL *Proverbs* (English) 11 One Swallow doth not make a Summer. **1844** DICKENS *Martin Chuzzlewit* xliii. One foul wind no more makes a winter, than one swallow makes a summer. **1973** A. MACLEAN *Way to Dusty Death* vii. The first driver past the pits was Harlow. . . 'One swallow does not make a summer.'

It is idle to SWALLOW the cow and choke on the tail
It is senseless to give up when a great task is almost completed.

1659 J. HOWELL *Proverbs* (English) 13

To swallow an Ox, and be choaked with the tail. **1721** J. KELLY *Scottish Proverbs* 190 It is a Shame to eat the Cow, and worry[1] on the Tail. It is a Shame to perform a great Task all but a little, and then give over. **1915** J. BUCHAN *Salute to Adventurers* xviii. We had gone too far to turn back, and as our proverb says, 'It is idle[2] to swallow the cow and choke on the tail.'

[1] choke. [2] senseless, foolish.

swallow: see also (noun) the ROBIN and the wren are God's cock and hen.

swap: see don't CHANGE horses in midstream.

A SWARM in May is worth a load of hay; a swarm in June is worth a silver spoon; but a swarm in July is not worth a fly
1655 S. HARTLIB *Reformed Commonwealth of Bees* 26 It being a Proverb, that a Swarm of Bees in May is worth a Cow and a Bottle[1] of Hay, whereas a Swarm in July is not worth a Fly. **1710** Tusser *Redivivus* May 11 The Proverb says, 'A Swarm in May is worth a Load of Hay'. **1879** R. JEFFERIES *Wild Life in Southern County* vii. 'A swarm in May is worth a load of hay; a swarm in June is worth a silver spoon; but a swarm in July is not worth a fly'—for it is then too late . . to store up . . honey before the flowers begin to fade. **1945** F. THOMPSON *Lark Rise* v. As she reminded the children: A swarm in May's worth a rick of hay; And a swarm in June's worth a silver spoon; while A swarm in July isn't worth a fly.

[1] bundle (of hay): now somewhat local in use.

If every man would SWEEP his own doorstep the city would soon be clean
1624 T. ADAMS *Temple* 65 When we would haue the street cleansed, let euery man sweep his owne doore, and it is quickly done. **1666** G. TORRIANO *Italian Proverbs* 41 If every one will sweep his own house, the City will be clean. **1856** R. WHATELY *Bacon's Essays: with Annotations* 235 [We] ought to engage in the important work of *self-*

reformation. . . 'If each would sweep before his own door, we should have a clean street.' **1930** *Times* 25 Mar. 10 It appears to be hard to draw a clear distinction between deciding a question of right and wrong for one's self and deciding it for others. . . 'If every man would sweep his own doorstep the city would soon be clean.'

sweep: *see also* NEW brooms sweep clean.

sweet: *see* LITTLE fish are sweet; REVENGE is sweet; STOLEN fruit is sweet; STOLEN waters are sweet.

sweeter: *see* the NEARER the bone, the sweeter the meat.

From the SWEETEST wine, the tartest vinegar
    1578 LYLY *Euphues* I. 197 As the best wine doth make the sharpest vinaigar, so the deepest loue tourneth to the deadliest hate. **1637** J. HOWELL *Familiar Letters* 3 Feb. (1903) II. 140 He swears he had rather see a basilisk than her [his former love]. The sweetest wines may turn to the tartest vinegar. **1852** E. FITZ-

GERALD *Polonius* 9 'It is . . the sweetest wine that makes the sharpest vinegar,' says an old proverb. **1979** *Daedalus* Summer 121 The juxtaposition silently signals the cautionary maxim 'From the sweetest wine, the tartest vinegar'.

swift: *see* the RACE is not to the swift, nor the battle to the strong.

swim: *see* don't go near the WATER until you learn how to swim.

swine: *see* do not throw PEARLS to swine; on SAINT Thomas the Divine kill all turkeys, geese, and swine.

swing: *see* what you LOSE on the swings you gain on the roundabouts.

Swithun: *see* SAINT Swithun's day if thou be fair for forty days it will remain.

sword: *see* whosoever DRAWS his sword against the prince must throw the scabbard away; he who LIVES by the sword dies by the sword; the PEN is mightier than the sword.

T

table: *see* where MACGREGOR sits is the head of the table.

tail: *see* the HIGHER the monkey climbs the more he shows his tail; it is idle to SWALLOW the cow and choke on the tail.

tailor: *see* NINE tailors make a man.

TAKE the goods the gods provide
[PLAUTUS *Rudens* l. 1229 *habeas quod di dant boni*, you may keep what good the gods give.] **1697** DRYDEN *Alexander's Feast* 5 Lovely Thais sits beside thee, Take the good the gods provide thee. **1880** TROLLOPE *Duke's Children* III. xiv. 'It is only because I am the governor's son,' Silverbridge pleaded . . . 'What of that? Take the goods the gods provide you.' **1980** M. McMULLEN *Something of Night* viii. Take the goods the gods provide, and don't . . sulk when they are snatched away.

take: *see also* it takes ALL sorts to make a world; GIVE a thing, and take a thing, to wear the Devil's gold ring; GIVE and take is fair play; you can take a HORSE to the water, but you can't make him drink; you PAYS your money and you takes your choice; take care of the PENCE and the pounds will take care of themselves; it takes TWO to make a bargain; it takes TWO to make a quarrel; it takes TWO to tango.

A TALE never loses in the telling
Also used in the phrase *to lose* (or *grow*) *in the telling*: frequently implying exaggeration.
　1541 *Schoolhouse of Women* A4ᵛ What soeuer commeth to memorye Shall not be loste, for the tellinge. **1581** *Stationers' Register* (1875) II. 388 A good tale Cannot to[o] often be Tolde. **1609** S. HARWARD *MS* (Trinity College, Cambridge) 121 Tales lose nothing by the cariadge. **1721** J. KELLY *Scottish Proverbs* 44 A Tale never loses in the telling.

Fame or Report . . commonly receives an Addition as it goes from Hand to Hand. **1907** *Spectator* 16 Nov. 773 A story never loses in the telling in the mouth of an Egyptian. **1954** L. P. HARTLEY *White Wand* 15 No doubt Antonio was telling the story to his fellow-gondoliers and it would lose nothing in the telling. **1979** M. STEWART *Last Enchantment* 19 Like all strange tales, it will grow with the telling.

tale: *see* DEAD men tell no tales.

Never tell TALES out of school
The phrase *to tell tales out of school* is also used.
　1530 W. TYNDALE *Practice of Prelates* B1ᵛ So that what cometh once in may never out for feare of tellinge tales out of scole. **1616** J. WITHALS *Dict.* (rev. ed.) 573 You must not tel tales out of the Tauerne. **1721** J. KELLY *Scottish Proverbs* 303 Tell no School Tales. Do not blab abroad what is said in drink, or among Companions. **1876** I. BANKS *Manchester Man* I. xv. All attempts to make known school troubles and grievances were met with 'Never tell tales out of school'. **1963** A. CHRISTIE *Clocks* xxiv. 'Well—.' . . 'I understand. Mustn't tell tales out of school.'

TALK of the Devil, and he is bound to appear
Also abbreviated to *talk of the Devil!*, and used when a person just spoken of is seen.
　1666 G. TORRIANO *Italian Proverbs* 134 The English say, Talk of the Devil, and he's presently at your elbow. **1721** J. KELLY *Scottish Proverbs* 299 Speak of the Dee'l, and he'll appear. Spoken when they, of whom we are speaking, come in by Chance. **1773** R. GRAVES *Spiritual Quixote* II. VIII. v. 'How free he had made with the Devil's name.' . . 'Talk of the Devil, and he will appear.' **1830** MARRYAT *King's Own* II. v. The unex-

pected appearance of Mrs. Rainscourt made him involuntarily exclaim, 'Talk of the devil—' 'And she appears, Sir,' replied the lady. **1979** *Radio Times* 27 Oct.–2 Nov. 66 Talk of the Devil . . and he's bound to appear, they say.

talk: *see also* MONEY talks.

tango: *see* it takes TWO to tango.

tar: *see* do not spoil the SHIP for a ha'porth of tar.

Tartar: *see* SCRATCH a Russian and you find a Tartar.

tartest: *see* from the SWEETEST wine, the tartest vinegar.

taste: *see* there is no ACCOUNTING for tastes; EVERY man to his taste.

TASTES differ
Cf. *There is no* ACCOUNTING *for tastes.*
    1803 J. DAVIS *Travels in USA* ii. Tastes sometimes differ. **1868** W. COLLINS *Moonstone* I. xv. Tastes differ. . . I never saw a marine landscape that I admired less. **1924** H. DE SELINCOURT *Cricket Match* iii. It's no use arguing about that. . . Tastes differ. **1940** 'J. J. CON-NINGTON' *Four Defences* xii. Tastes differ. One has to admit it.

tattered: *see* there's many a GOOD cock come out of a tattered bag.

tax: *see* NOTHING is certain but death and taxes.

You cannot TEACH an old dog new tricks
    1530 J. FITZHERBERT *Husbandry* (ed. 2) G1ᵛ The dogge must lerne it when he is a whelpe, or els it wyl not be; for it is harde to make an olde dogge to stoupe. **1636** W. CAMDEN *Remains concerning Britain* (ed. 5) 300 It is hard to teach an old dog trickes. **1672** W. WALKER *English & Latin Proverbs* 46 An old dog will learn no new tricks. **1806** J. RANDOLPH *Letter* 15 Feb. (1834) 14 There is an old proverb, 'You cannot teach an old dog

new tricks.' **1979** F. OLBRICH *Sweet & Deadly* iv. It was not true that you could not teach an old dog new tricks; he had lost his teeth in his old master's service but was forced to perform at the new one's bidding.

Don't TEACH your grandmother to suck eggs
A caution against offering advice to the wise. The metaphorical phrase *to teach one's grandmother (to suck eggs)* is also found.
    1707 J. STEVENS tr. *Quevedo's Comical Works* IV. 403 You would have me teach my Grandame to suck Eggs. **1738** SWIFT *Polite Conversation* I. 57 'I'll mend it, Miss.'. . 'You mend it! go, Teach your Grannam to suck Eggs.' **1882** BLACK-MORE *Christowell* II. iii. A . . twinkle, which might have been interpreted—'instruct your grandfather in the suction of gallinaceous products'. **1967** RIDOUT & WITTING *English Proverbs Explained* 48 Don't teach your grandmother to suck eggs. **1979** 'S. WOODS' *This Fatal Writ* 110 'I'll remember that,' said Maitland gravely. Shacklock gave him a quick smile. 'I know, teaching my grand-mother,' he said.

teach: *see also* he who CAN does, he who cannot teaches.

teacher: *see* EXPERIENCE is the best teacher.

teeth: *see* the GODS send nuts to those who have no teeth.

TELL the truth and shame the Devil
    1548 W. PATTEN *Expedition into Scotland* A5 An Epigram . . the whiche I had, or rather (to saie truth and shame the deuel, for out it wool) I stale . . from a frende of myne. **1576** G. GASCOIGNE *Grief of Joy* II. 555, I will tell trewth, the devyll hymselfe to shame. **1597** SHAKESPEARE *Henry IV, Pt. 1* III. i. 58 And I can teach thee, coz,[1] to shame the devil By telling truth: tell truth, and shame the devil. **1639** J. CLARKE *Parœ-miologia Anglo-Latina* 316 Tell the truth, and shame the Devill. **1738** SWIFT *Polite*

Conversation I. 93 Well; but who was your Author? Come, tell Truth and shame the Devil. **1945** F. THOMPSON *Lark Rise* xiv. A few homely precepts, such as . . 'Tell the truth and shame the devil'. **1980** 'R. B. DOMINIC' *Attending Physician* iv. You'll get the opportunity to tell the truth and shame the devil.

¹ cousin: a fond or familiar term of address.

tell: *see also* BLOOD will tell; you can't tell a BOOK by its cover; CHILDREN and fools tell the truth; DEAD men tell no tales; every PICTURE tells a story; STRAWS tell which way the wind blows; never tell TALES out of school; TIME will tell.

telling: *see* a TALE never loses in the telling.

temper: *see* GOD tempers the wind to the shorn lamb.

ten: *see* one PICTURE is worth ten thousand words.

thaw: *see* ROBIN Hood could brave all weathers but a thaw wind.

themselves: *see* GOD helps them that help themselves; LISTENERS never hear any good of themselves.

thick: *see* YORKSHIRE born and Yorkshire bred, strong in the arm and weak in the head.

thicker: *see* BLOOD is thicker than water.

Set a THIEF to catch a thief

[Cf. CALLIMACHUS *Epigram* xliii. φωρὸς δ'ἴχνια φὼρ ἔμαθον, being a thief myself I recognized the tracks of a thief.] **1654** E. GAYTON *Pleasant Notes upon Don Quixote* IV. ii. As they say, set a fool to catch a fool; a Proverb not of that gravity (as the Spaniards are), but very usefull and proper. **1665** R. HOWARD *Four New Plays* 74 According to the old saying, Set a Thief to catch a Thief. **1812** M. EDGEWORTH *Tales of Fashionable Life* VI. 446 'You have all your life been evading the laws. . . Do you think this has qualified you peculiarly for being a

guardian of the laws?' Sir Terence replied, 'Yes, sure, set a thief to catch a thief is no bad maxim.' **1979** *Guardian* 5 July 9 'Set a thief to catch a thief.'. . She was implicitly condoning bent practices used by the police against . . The Underworld.

thief: *see also* HANG a thief when he's young, and he'll no' steal when he's old; there is HONOUR among thieves; LITTLE thieves are hanged, but great ones escape; OPPORTUNITY makes a thief; a POSTERN door makes a thief; PROCRASTINATION is the thief of time; if there were no RECEIVERS, there would be no thieves.

If a THING's worth doing, it's worth doing well

Job is sometimes used instead of *thing*. **1746** CHESTERFIELD *Letter* 9 Oct. (1932) III. 783 Care and application are necessary. . . In truth, whatever is worth doing at all is worth doing well. **1910** CHESTERTON *What's Wrong with World* IV. xiv. The elegant female, drooping her ringlets over her watercolours, . . was maintaining the prime truth of woman, the universal mother: that if a thing is worth doing, it is worth doing badly. **1915** H. G. WELLS *Bealby* v. 'If a thing's worth doing at all,' said the Professor . . 'it's worth doing well.' **1980** *Church Times* 22 Feb. 12 'Things worth doing are worth doing well,' was the motto at my first school.

thing: *see also* ALL good things must come to an end; ALL things are possible with God; ALL things come to those who wait; the BEST things come in small packages; the BEST things in life are free; FACTS are stubborn things; FIRST things first; there's no such thing as a FREE lunch; GIVE a thing, and take a thing, to wear the Devil's gold ring; KEEP a thing seven years and you'll always find a use for it; LITTLE things please little minds; there is MEASURE in all things; MODERATION in all things; things PAST cannot be recalled; to the PURE all things are pure; THREE things are not to be trusted;

you can have TOO much of a good thing; if you WANT a thing done well, do it yourself; the WORTH of a thing is what it will bring.

When THINGS are at the worst they begin to mend

1582 G. WHETSTONE *Heptameron of Civil Discourses* vi. Thinges when they are at the worst, begin again to amend. The Feauer giueth place to health, when he hath brought the pacyent to deathes door. **1600** *Sir John Oldcastle* H1ᵛ Patience good madame, things at worst will mend. **1623** WEBSTER *Duchess of Malfi* IV. i. Things being at the worst, begin to mend. **1748** RICHARDSON *Clarissa* III. liv. When things are at the worst they must mend. **1889** GISSING *Nether World* I. ii. When things are at the worst they begin to mend. . . It can't be much longer before he gets work.

THINK first and speak afterwards

1557 R. EDGEWORTH *Sermons* B6 Thinke well and thou shalt speak wel. **1623** W. PAINTER *Palace of Pleasure* B1ᵛ Thinke twise, then speak, the old Prouerbe doth say, Yet Fooles their bolts will quickely shoot away. **1639** J. CLARKE *Parœmiologia Anglo-Latina* 133 First thinke & then speak. **1640** R. BRATHWAIT *Art asleep Husband?* vii. You thinke twice before you speake, and may be demanded twice before you answer. **1855** H. G. BOHN *Hand-Book of Proverbs* 528 Think to-day and speak to-morrow. **1902** E. HUBBARD in *Philistine* May 192 Think twice before you speak and then talk to yourself. **1943** L. I. WILDER *Happy Golden Years* i. You must do your thinking first and speak afterward. If you will . . do that, you will not have any trouble. **1981** P. O'DONNELL *Xanadu Talisman* iv. Please think before you speak.

think: *see also* GREAT minds think alike; what MANCHESTER says today, the rest of England says tomorrow.

THIRD time lucky

c **1840** BROWNING *Letter* (1933) 5 'The luck of the third adventure' is proverbial. **1862** A. HISLOP *Proverbs of Scotland* 194 The third time's lucky. **1882** R. L. STEVENSON *New Arabian Nights* II. 59 'The next time we come to blows—' 'Will make the third,' I interrupted . . . 'Ay, true . . . Well, the third time's lucky.' **1942** N. MARSH *Death & Dancing Footman* vii. It was a glancing blow. . . It might have been my head. . . One of them's saying to himself: 'Third time lucky.' **1979** J. TATE tr. K. A. Blom's *Limits of Pain* ix. The expression third time lucky had something in it.

The THIRD time pays for all

1574 J. HIGGINS *Mirror for Magistrates* (1946) 93 The third payes home, this prouerbe is to true. **1599** *Warning for Fair Women* E3 The third time payes for all. **1855** GASKELL *North & South* I. xvii. 'This is th'third strike I've seen,' said she. . . 'Well, third time pays for all.' **1922** *Punch* 20 Dec. 594 Mrs. Ellison has already been twice married. The third time pays for all, so they say. **1978** S. KING *Stand* III. liv. If I could have brought myself to jump once . . I might not be here. Well, last time pays for all.

Thomas: *see* on SAINT Thomas the Divine kill all turkeys, geese, and swine.

thorn: *see* beware of an OAK it draws the stroke.

THOUGHT is free

c **1390** GOWER *Confessio Amantis* v. 4485, I have herd seid that thoght is fre. **1601** SHAKESPEARE *Twelfth Night* I. iii. 64 Fair lady, do you think you have fools in hand?—Now, sir, thought is free. **1874** G. MACDONALD *Malcolm* II. xvii. 'How do you come to think of such things?' 'Thought's free, my lord.'

thought: *see also* SECOND thoughts are best; the WISH is father to the thought.

thousand: *see* one PICTURE is worth ten thousand words.

THREATENED men live long

1534 LADY E. WHEATHELL in M. St. C. Byrne *Lisle Letters* II. ii. Ther es a nolde sayeng thretend men lyue long. *c* **1555** in H. L. Collman *Ballads of Elizabethan Period* (1912) 69 It is a true prouerbe: the threatned man lyues long. **1607** T. HEYWOOD *Fair Maid of Exchange* II. 68 Threatened men live long. **1655** T. FULLER *Church Hist. Britain* VIII. iii. Gardiner . . vowed . . to stop the sending of all supplies unto them. . . But threatned folke live long. **1865** G. W. THORNBURY *Haunted London* ii. Temple Bar was doomed to destruction by the City as early as 1790. . . 'Threatened men live long.'. . Temple Bar still stands. **1930** A. CHRISTIE *Murder at Vicarage* iv. Archer . . is vowing vengeance against me, I hear. Impudent scoundrel. Threatened men live long, as the saying goes. **1980** L. EGAN *Hunters & Hunted* i. The threat was an old one; and, the proverb ran, threatened men live long.

THREE may keep a secret, if two of them are dead

1546 J. HEYWOOD *Dialogue of Proverbs* II. v. G4^v We twayne are one to many (quoth I) for men saie, Three maie keepe a counsell, if two be awaie. *c* **1595** SHAKESPEARE *Romeo & Juliet* II. iv. 190 Is your man secret? Did you ne'er hear say Two may keep counsel, putting one away? **1735** B. FRANKLIN *Poor Richard's Almanack* (July) Three may keep a secret, if two of them are dead. **1979** D. CLARK *Heberden's Seat* ii. Two of everything . . two bodies, two causes of death. . . What was it? 'Three may keep a secret, if two of them are dead.'

THREE removals are as bad as a fire

1758 B. FRANKLIN *Poor Richard's Almanack* (Preface), I never saw an oft removed Tree, Nor yet an oft removed Family, That throve so well, as those that settled be. And again, Three Removes are as bad as a Fire. **1839** DICKENS *Letter* 14 Nov. (1965) I. 602 Did you ever 'move'? . . There is an old proverb that three removes are as bad as a fire.

1925 *Punch* 11 Nov. 505 'Three removals are as bad as a fire,' but you'd be surprised how difficult it is to make an insurance company see things in this light.

THREE things are not to be trusted: a cow's horn, a dog's tooth, and a horse's hoof

[Cf. 13th-cent. Fr. *dent de chael, pé de cheval, cul d'enfant ne sunt pas a crere,* a dog's tooth, a horse's hoof, and a baby's bottom are not to be trusted.] *c* **1383** JOHN OF FORDUN *Scotichronicon* (1759) II. XIV. xxxii. Till horsis fote thou never traist, Till hondis tooth, no womans faith. **1585** S. ROBSON *Choice of Change* K2 Trust not 3 things. Dogs teeth. Horses feete. Womens protestations. **1910** P. W. JOYCE *English as We speak it in Ireland* 110 Three things are not to be trusted—a cow's horn, a dog's tooth and a horse's hoof. **1948** T. H. WHITE *Elephant & Kangaroo* xiii. He was . . beginning to worry about being employed by a venomous Englishman. 'Four things not to trust,' said the Cashelmor proverb: 'a dog's tooth, a horse's hoof, a cow's horn, and an Englishman's laugh.'

three: see also two BOYS are half a boy, and three boys are no boy at all; from CLOGS to clogs is only three generations; one ENGLISHMAN can beat three Frenchmen; FISH and guests stink after three days; it takes three GENERATIONS to make a gentleman; from SHIRTSLEEVES to shirtsleeves in three generations; TWO is company, but three is none.

THRIFT is a great revenue

[CICERO *Paradox* 49 *non intellegunt homines quam magnum vectigal sit parsimonia,* men do not realize how great a revenue thrift is.] **1659** J. HOWELL *Proverbs* (French) 15 Parsimony is the best revenue. **1855** H. G. BOHN *Handbook of Proverbs* 530 Thrift is a good revenue. **1930** *Times* 10 Oct. 13 Thrift . . is not only a great virtue but also 'a great revenue'.

He that will THRIVE must first ask his wife

a **1500** in R. L. Greene *Early English Carols* (1935) 276 Hym that cast hym for to thryve, he must ask leve of his wyff. *c* **1549** J. HEYWOOD *Dialogue of Proverbs* I. xi. B8ᵛ He that will thryue, must aske leaue of his wyfe. *a* **1790** B. FRANKLIN *Autobiography* (1905) I. 324 He that would thrive, must ask his wife. It was lucky for me that I had one as much dispos'd to industry and frugality as myself. **1875** S. SMILES *Thrift* viii. There is an old English proverb which says, 'He that would thrive must first ask his wife'.

Don't THROW out your dirty water until you get in fresh

c **1475** in *Modern Philology* (1940) XXXVIII. 121 He ys a fole that castith a-way his olde water or he have new. **1623** W. PAINTER *Palace of Pleasure* C4ᵛ The wise prouerbe wish all men to saue Their foule water vntill they fayrer haue. **1710** S. PALMER *Proverbs* 89 Don't throw away Dirty Water till you have got Clean. . . The Man being possess'd with Avarice, throws away a Certain Benefit upon uncertain . . Expectations. **1842** S. LOVER *Handy Andy* xxix. 'I'll change my clothes.' . . 'You had better wait. . . You know the old saying, "Don't throw out your dirty wather until you get in fresh." ' **1911** G. B. SHAW *Fanny's First Play* III. 208 Dont you throw out dirty water til you get in fresh. Dont get too big for your boots.

Don't THROW the baby out with the bathwater

The proverb is often used allusively, especially in the metaphorical phrase *to throw* (or *empty) out the baby with the bath-water*.

[Cf. **1610** J. KEPLER *Tertius Interveniens* (sub-heading*) Das ist Warnung . . das sie . . nicht das Kindt mit dem Badt ausschüt-ten*, this is a caution . . lest you throw out the baby with the bath.] **1853** CAR-LYLE *Nigger Question* (ed. 2) 29 The Germans say, 'You must empty out the bathing-tub, but not the baby along with

it.'. . How to abolish the abuses of slav-ery, and save the precious thing in it: also, I do not pretend that this is easy. **1911** G. B. SHAW *Getting Married* Preface 186 We shall in a very literal sense empty the baby out with the bath by abo-lishing an institution [marriage] which needs nothing more than a little . . rationalizing to make it . . useful. **1937** M. WARD *Insurrection versus Resurrection* i. In their ardour to get rid of it [old-fashioned apologetic] they 'emptied out the baby with the bath-water'. **1979** J. P. YOUNG *Art of Learning to Manage* 91 Do be careful that you don't throw the baby out with the bath water, and find your-self with too many people who lack ex-perience.

throw: *see also* throw DIRT enough, and some will stick; those who live in GLASS houses shouldn't throw stones; do not throw PEARLS to swine.

Thursday: *see* Monday's CHILD is fair of face.

thyself: *see* KNOW thyself; PHYSICIAN, heal thyself.

thysen (yourself): *see* HEAR all, see all, say nowt.

tide: *see* TIME and tide wait for no man.

tiger: *see* he who RIDES a tiger is afraid to dismount.

There is a TIME and place for everything

Cf. *There is a* TIME *for everything.*

1509 A. BARCLAY *Ship of Fools* 94 Re-member: there is tyme and place for euery thynge. **1862** G. BORROW *Wild Wales* II. x. There is a time and place for everything, and sometimes the warm-est admirer of ale would prefer the lymph of the hill-side fountain to the choicest ale. **1979** *Guardian* 28 May 7 There is a time and a place for every-thing. Monetarism might work in ideal conditions.

TIME and tide wait for no man

c **1390** CHAUCER *Clerk's Tale* l. 118 For

thogh we slepe or wake, or rome, or ryde, Ay fleeth the tyme; it nil no[1] man abyde. *a* **1520** *Everyman* (1961) l. 143 The Tyde abydeth no man. **1592** GREENE *Disputation between He Cony-catcher & She Coney-catcher* X. 241 Tyde nor time tarrieth no man. **1639** J. CLARKE *Parœmiologia Anglo-Latina* 233 Time and tide tary on no man. **1767** 'A. BARTON' *Disappointment* II. i. Let's step into the state-room, and turn in: Time and tide waits for no one. **1822** SCOTT *Nigel* III. ii. Come, come, master, let us get afloat. . . Time and tide wait for no man. **1979** 'C. AIRD' *Some die Eloquent* x. Time and tide and newspapers wait for no man.

[1] *nil no*: will no (idiomatically expressed as a double negative).

TIME flies

[L. *tempus fugit*, time flies.] *c* **1390** CHAUCER *Clerk's Tale* l. 118 For though we slepe or wake, or rome, or ryde, Ay fleeth the tyme. **1639** J. CLARKE *Parœmiologia Anglo-Latina* 308 Time flyeth away without delay. **1776** J. W. FLETCHER *Letter* 21 Mar. in *Works* (1803) IX. 197 Time flies! Years of plenty . . disappear before the eternity to which we are all hastening. **1979** N. GOLLER *Tomorrow Silence* vi. It's *ten* years since I saw her, darling. Doesn't time fly?

There is a TIME for everything

Cf. *There is a* TIME *and place for everything.*

[ECCLESIASTES iii. 1 To euery thing there is a season.] *c* **1390** CHAUCER *Clerk's Prologue* l. 6 But Salomon seith 'every thyng hath tyme'. **1540** CRANMER *Bible* Prologue +3 Ther is tyme for euery thynge. **1590** SHAKESPEARE *Comedy of Errors* II. ii. 63 Well, sir, learn to jest in good time; there's a time for all things. **1818** AUSTEN *Northanger Abbey* xxx. Your head runs too much upon Bath; but there is a time for every thing—a time for balls . . and a time for work. **1980** 'M. INNES' *Going It Alone* I. x. There is a time for everything, and he hoped that, in the present exigency, Tim wasn't going to be . . frivolous.

TIME is a great healer

There are numerous expressions derived from the ancient concept that time heals, several of which are illustrated here. Predominantly used in the context of feelings and emotions rather than physical suffering.

[MENANDER *Fragments* dclxvii. (Kock) πάντων ἰατρός τῶν ἀναγκαίων κακῶν χρόνος ἐστιν, Time is the healer of all necessary evils.] *c* **1385** CHAUCER *Troilus & Criseyde* v. 350 As tyme hem[1] hurt, a tyme doth hem cure. **1591** J. HARINGTON tr. *Ariosto's Orlando Furioso* VI. ii. He hurt the wound which time perhaps had healed, weening[2] with greater sinne the lesse to mend. **1622** H. PEACHAM *Complete Gentleman* iv. Time, the Phisition of all. **1837** DISRAELI *Henrietta Temple* III. VI. ix. Time is the great physician. **1926** G. B. SHAW *Translations & Tomfooleries* 60 Time is the great healer. **1942** A. CHRISTIE *Body in Library* viii. He had a terrible shock and loss. . . But Time, as my dear mother used to say, is a great healer. **1974** W. GARNER *Big enough Wreath* xii. 'Your face, man. Can nothing be done?' 'Time,' said Smith tonelessly, 'heals all things.'

[1] i.e. them. [2] thinking, supposing.

TIME is money

[Gr. τὸ πολυτελέστατον. . ἀνάλωμα, τὸν χρόνον, the most costly outlay is time (attributed to Antiphon).] **1572** T. WILSON *Discourse upon Usury* 33 They saye tyme is precious. **1748** B. FRANKLIN *Papers* (1961) III. 306 Remember that Time is Money. He that can earn Ten Shillings a Day . . and . . sits idle one half of that Day . . has really . . thrown away Five Shillings. **1840** BULWER-LYTTON *Money* III. vi. 'You don't come often to the club, Stout?' . . 'No, time is money.' **1980** H. R. F. KEATING *Murder of Maharajah* xv. I can't wait here day after day. . . Time's money, you know.

No TIME like the present

1562 G. LEGH *Accidence of Armoury* 225[v] Mary[1] sir no time better then euen now. **1696** M. MANLEY *Lost Lover* IV. i. No time like the present. **1888** M.

OLIPHANT *Second Son* I. iv. 'If you were a-passing this way, sir, some time in the morning—.' 'There's no time like the present,' answered Roger. **1974** A. PRICE *Other Paths to Glory* II. ii. 'You wish to see the battlefield—Paul?' 'Why not?' No time like the present.
¹ to be sure.

TIME will tell

1539 R. TAVERNER tr. *Erasmus' Adages* 37 *Tempus omnia reuelat.* Tyme discloseth all thynges. **1616** T. DRAXE *Adages* 205 Time reuealeth all things. **1771** C. STUART *Letter* 15 Apr. in *Publications of Mississippi Hist. Society* (1925) V. 50 Time only will shew how far those Informations have been well founded. **1863** C. READE *Hard Cash* I. vi. I will answer . . that she will speak as distinctly to music as you do in conversation— Time will show, madam. **1913** E. H. PORTER *Pollyanna* xxiii. The doctor had looked very grave . . and had said that time alone could tell. **1929** J. J. CONNINGTON *Eye in Museum* xiv. 'I'm not . . bringing any charge.' . . 'Oh . . a bright idea, perhaps. Or perhaps not so bright.' 'Time will tell,' the Superintendent retorted. **1979** V. CANNING *Satan Sampler* vii. Somewhere in all this there is a very ancient and fish-like smell. . . I say no more. . . Time will or will not tell.

TIME works wonders

1588 A. MARTEN *Exhortation to defend Country* F2 You . . thinke that time will worke wonders, though you your selves follow your owne pleasures. **1815** BYRON *Letter* 7 Jan. (1975) IV. 252 Time does wonders. **1845** D. W. JERROLD (title) Time works wonders. **1872** G. J. WHYTE-MELVILLE *Satanella* II. xxiv. 'I want you to like me.'. . 'They say time works wonders . . and I feel I shall.'

time: *see also* NEVER is a long time; OTHER times, other manners; PARSLEY seed goes nine times to the Devil; PROCRASTINATION is the thief of time; one STEP at a time; a STITCH in time saves nine; THIRD time lucky; the THIRD time pays for all; WORK expands so as to fill the time available.

TIMES change and we with time

[Med. L. *omnia* (also *tempora*) *mutantur nos et mutamur in illis,* all things (*also* times) are changing and we with them.] **1578** LYLY *Euphues* I. 276 The tymes are chaunged as Ouid sayeth, and wee are chaunged in the times. **1666** G. TORRIANO *Italian Proverbs* 281 Times change, and we with them. . . The Latin says the same, *Tempora mutantur, et nos mutamur in illis.* **1943** C. MILBURN *Diary* 21 Feb. (1979) 168 In English cities the Red Flag has been flown. . . Times change indeed, and we with time. **1981** J. BINGHAM *Brock* 31 Times were changing and Melford with them.

tinker: *see* if IFS and ands were pots and pans, there'd be no work for tinkers' hands.

TODAY you; tomorrow me

[L. *hodie mihi, cras tibi,* today it is my turn, tomorrow yours.] *a* **1250** *Ancrene Wisse* (1962) 143 *Ille hodie, ego cras.* He to dei, & ich to marhen [he today, and I tomorrow]. **1620** T. SHELTON tr. *Cervantes' Don Quixote* II. lxv. To day for thee, and to-morrow for me. **1855** KINGSLEY *Westward Ho!* II. i. To-day to thee, to-morrow to me. **1906** A. CONAN DOYLE *Sir Nigel* xv. 'It is the custom of the Narrow Seas,' said they: 'To-day for them; to-morrow for us.' **1929** A. W. WHEEN tr. *E. M. Remarque's All Quiet on Western Front* ix. 'Comrade,' I say to the dead man, but I say it calmly, 'To-day you, to-morrow me.'

today: *see also* JAM tomorrow and jam yesterday, but never jam today; what MANCHESTER says today, the rest of England says tomorrow; never PUT off till tomorrow what you can do today.

Tom: *see* MORE people know Tom Fool than Tom Fool knows.

TOMORROW is another day

c **1527** J. RASTELL *Calisto & Melebea* C1ᵛ

Well mother to morrow is a new day.
1603 FLORIO tr. *Montaigne's Essays* II. iv.
A letter . . beeing delivered him . . at
supper, he deferred the opening of it,
pronouncing this by-word. To morrow
is a new day. **1824** SCOTT *St. Ronan's
Well* III. vii. We will say no more of it at
present. . . To-morrow is a new day.
1927 P. GREEN *Field God* I. 148 Go to it,
you Mag and Lonie! To-morrow's
another day, and you'll need all you can
hold. **1972** H. CARMICHAEL *Naked to
Grave* xii. There's an old adage that
tomorrow's another day. At this time of
night it's quite a comfort.

TOMORROW never comes

    1523 LD. BERNERS *Froissart* (1901) II.
309 It was sayde every day among
them, we shall fight tomorowe, the
whiche day came never. **1602** J. CHAM-
BERLAIN *Letter* 8 May (1939) I. 142
Tomorrow comes not yet. **1678** J. RAY
English Proverbs (ed. 2) 343 Tomorrow
come never. **1756** B. FRANKLIN *Poor
Richard's Almanack* (July) To-morrow,
every Fault is to be amended; but that
To-morrow never comes. **1889** GISSING
Nether World III. ix. 'It's probably as well
for you that *to-morrow* never comes.'
'Now just see how things turn out!'
went on the other. **1966** A. E. LINDOP
I start Counting xvii. 'It's late, honey.
Talk tomorrow.' 'Tomorrow never
comes.'

tomorrow: *see also* JAM tomorrow and jam
yesterday, but never jam today; what
MANCHESTER says today, the rest of England says tomorrow; never PUT off till
tomorrow what you can do today; TO-
DAY you, tomorrow me.

tongue: *see* a STILL tongue makes a wise
head.

TOO many cooks spoil the broth

    1575 ?J. HOOKER *Life of Carew* (1857) 33
There is the proverb, the more cooks the
worse potage. **1662** B. GERBIER *Principles of Building* 24 When . . an under-
taking hath been committed to many, it
caused but confusion, and therefore it is

a saying . . Too many Cooks spoils the
Broth. *c* **1805** AUSTEN *Watsons* (1972)
VI. 318 She professes to keep her own
counsel. . . 'Too many Cooks spoil the
Broth.' **1855** KINGSLEY *Westward Ho!* II.
vii. As Amyas sagely remarked, 'Too
many cooks spoil the broth, and half-a-
dozen gentlemen aboard one ship are as
bad as two kings of Brentford. **1979**
Guardian 7 Nov. 6 It was a great mistake
to think that administration was improved by taking on more admin-
istrators. . . 'Too many cooks spoil the
broth.'

You can have TOO much of a good thing

    1483 B. BURGH *Cato* in *Archiv* (1905)
CXV. 313 To much is nouht of any
maner thyng [too much of anything is
nothing]. **1546** J. HEYWOOD *Dialogue of
Proverbs* II. iv. G4ᵛ Well (quoth I) to
muche of one thyng is not good, Leaue
of this. **1611** R. COTGRAVE *Dict. French &
English* s.v. Manger, A man may take
too much of a good thing. **1738** SWIFT
Polite Conversation I. 77 Fie, Miss! you
said that once before; and, you know,
Too much of one Thing is good for
nothing. **1961** W. H. DUNN *J. A. Froude*
iii. Freedom is good, but you can have
too much even of a good thing.

tool: *see* a BAD workman blames his tools.

top: *see* there is always ROOM at the top.

touch: *see* if you gently touch a NETTLE it'll
sting you for your pains.

He that TOUCHES pitch shall be defiled

    Cf. *If you* PLAY *with fire you get burnt.*

    [APOCRYPHA *Ecclesiasticus* xiii. 1 He
that toucheth pitch, shal be defiled
therewith.] **1303** R. BRUNNE *Handlyng
Synne* (EETS) l. 6578 Who-so handlyth
pycche wellyng hote, He shal haue
fylthe therof sumdeyl.¹ **1578** LYLY
Euphues I. 250 He that toucheth
pitche shall be defiled. **1710** S. PALMER
Proverbs 249 Touch Pitch and you'll be
Defil'd. . . There is Danger every Way
in Ill Company. **1886** H. CONWAY *Living or Dead* II ix. The next two months of

my life . . made me take a lower and more debased view of the world. . . I was touching pitch, yet striving to keep myself from being defiled. **1979** *Listener* 13 Sept. 345 The makers of the series believe that those who meddle with pitch may be defiled.
[1] in some degree.

tough: *see* when the GOING gets tough, the tough get going.

town: *see* GOD made the country, and man made the town.

TRADE follows the flag
    1870 J. A. FROUDE in *Fraser's Mag.* Jan. 4 The removal of a million poor creatures to Canada and the establishment of them there . . would probably have turned out . . a profitable investment. Trade follows the flag. **1945** R. HARGREAVES *Enemy at Gate* 152 There is a glib saying . . that 'trade follows the flag'; an apophthegm that succeeds in putting the cart before the horse with greater aplomb than almost any other cant phrase in common use. **1979** in C. Allen *Tales from Dark Continent* i. There is a famous old quotation that 'Trade follows the Flag' but . . in West Africa . . the reverse was true.

trade: *see also* EVERY man to his trade; there are TRICKS in every trade; TWO of a trade never agree.

transplant: *see* you cannot SHIFT an old tree without it dying.

TRAVEL broadens the mind
    1933 A. POWELL *Venusberg* i. Seeing the world broadens the outlook. You can learn a lot abroad. **1949** N. STREATFEILD *Painted Garden* iii. Foreign travel broadens the mind . . and a broadened mind helps all art. **1969** 'E. LATHEN' *When in Greece* xvii. 'The Americans we have met compare very favorably with most other nationals.' Travel, after all, can broaden only so many minds. **1981** C. KING *Commonplace*

Book 22 It has so truly been said that travel—anywhere—broadens the mind.

travel: *see* (verb) BAD news travels fast; it is BETTER to travel hopefully than to arrive.

He TRAVELS fastest who travels alone
    [Cf. **1854** H. D. THOREAU *Walden* 78 The man who goes alone can start today; but he who travels with another must wait till that other is ready.] **1888** KIPLING *Story of Gadsby* (1889) 94 Down to Gehenna, or up to the Throne, He travels fastest who travels alone. **1921** E. WAUGH *Journal* 19 June in *Diaries* (1979) 129 Hale's gone already. I suppose he will have to. 'He travels fastest who travels alone' anyway. **1979** J. LEASOR *Love & Land Beyond* i. Perhaps . . she was basically a loner. He travels fastest who travels alone.

tread: *see* FOOLS rush in where angels fear to tread.

As a TREE falls, so shall it lie
One must not change long-established beliefs, etc., in the face of death.
    [ECCLESIASTES xi. 3 If the tree fall toward the South, or toward the North, in the place where the tree falleth, there it shall be.] **1549** LATIMER *Seven Sermons* IV. M3^v Wheresoeuer the tre falleth . . there it shall reste. **1578** LYLY *Euphues* I. 308 Where the tree falleth there it lyeth . . and every ones deathes daye is his domes day. **1678** J. RAY *English Proverbs* (ed. 2) 296 As a man lives so shall he die, As a tree falls so shall it lie. **1836** M. SCOTT *Cruise of Midge* II. ii. It is of no use. . . As the tree falls, so must it lie – it is a part of my creed. **1921** W. H. HUDSON *Traveller in Little Things* iii. She sent a message . . to the old father to come and see her before she died. . . His answer was, 'As a tree falls so shall it lie.'

The TREE is known by its fruit
    [MATTHEW xii. 33 The tree is knowen by his fruit.] **1528** W. TYNDALE *Obedience of Christian Man* 88^v Judge the tre by his frute, and not by his leves. **1597-8**

SHAKESPEARE *Henry IV, Pt. 1* II. iv. 414
If then the tree may be known by the
fruit . . there is virtue in that Falstaff.
1670 J. RAY *English Proverbs* 11 A tree is
known by the fruit, and not by the
leaves. **1896** J. A. FROUDE *Council of
Trent* iv. Lutherans said the tree
is known by its fruit. Teach a pure
faith, and abuses will disappear. **1928**
D. H. LAWRENCE *Lady Chatterley's Lover*
iv. The mental life . . [is] rooted in spite
and envy. . . Ye shall know the tree by
its fruit.

tree: *see also* the APPLE never falls far from
the tree; he that would EAT the fruit
must climb the tree; you cannot SHIFT an
old tree without it dying; as the TWIG is
bent, so is the tree inclined; WALNUTS
and pears you plant for your heirs; a
WOMAN, a dog, and a walnut tree, the
more you beat them the better they be.

trick: *see* you cannot TEACH an old dog new
tricks.

There are TRICKS in every trade

1632 M. PARKER (*title*) Knavery in all
Trades. **1654** *Mercurius Fumigosus* 12–
19 July 49 If there be not Knavery in All
Trades, I shrewdly am mistaken. **1692**
R. L'ESTRANGE *Fables of Aesop* clxxxiii.
Jupiter appointed Mercury to make him
a Composition of Fraud and Hypocrisie,
and to give Every Artificer his Dose
on't. . . Mercury . . gave the Taylors the
Whole Quantity that was Left; and from
hence comes the Old Saying, There's
Knavery in All Trades, but Most in
Taylors. **1857** E. BENNETT *Border Rover*
vi. 'I would be willing to swear you had
bewitched this rifle.' . . 'Thar's tricks to
all trades 'cept ourn.' **1948** F. P. KEYES
Dinner at Antoine's xx. There's tricks to
all trades. Running down murderers is
one way of making a living, so that
makes it a trade. **1978** L. BLOCK *Burglar
in Closet* xvii. You age them [bills, paper
money] . . by cooking them with a little
coffee—well, there are tricks in every
trade—and I don't . . know some of the
ones the counterfeiters have come up
with.

A TROUBLE shared is a trouble halved

1931 D. L. SAYERS *Five Red Herrings* ix.
'Unbosom yourself,' said Wimsey.
'Trouble shared is trouble halved.'
1966 A. CARTER *Shadow Dance* viii. He
found he wanted to share the experience of the previous night with
Edna (a trouble shared is a trouble
halved). **1980** F. CRADOCK *Thunder over
Castle Rising* i. They do say that a
trouble shared is a trouble halved, but
still I'm very much afraid this is a matter
for his Lordship now.

Never TROUBLE trouble till trouble troubles you

Cf. *Do not* MEET *troubles half-way.*

1884 *Folk-Lore Journal* II. 280 Never
trouble trouble, till trouble troubles
you. **1972** J. GATHORNE-HARDY *Rise &
Fall of British Nanny* 332 Never trouble
trouble till trouble troubles you.

trouble: *see also* (noun) do not MEET
troubles half-way.

Many a TRUE word is spoken in jest

c **1390** CHAUCER *Monk's Prologue* l.
1964 Be nat wrooth, my lord, though
that I pleye. Ful ofte in game a sooth[1] I
have herd seye! *a* **1628** J. CARMICHAELL
Proverbs in Scots no. 1099 Manie suith
word said in bourding.[2] *c* **1665** in *Roxburghe Ballads* (1890) VII. 366 Many a
true word hath been spoke in jest. **1738**
SWIFT *Polite Conversation* I. iii. 'I did a
very foolish thing yesterday.' . . 'They
say, many a true Word's spoken in
Jest.' **1898** G. B. SHAW *Widower's Houses*
I. in *Plays Pleasant & Unpleasant* I. 10
There actually are Johannis churches
here . . as well as Apollinaris ones. . .
There is many a true word spoken in
jest. **1979** D. LESSING *Shikasta* 356 By
the time we have finished I expect
we shall have a dozen or more
[children]. . . Many a true word is
spoken in jest.

[1] truth. [2] jesting.

true: *see also* the COURSE of true love never
did run smooth; what EVERYBODY says

must be true; MORNING dreams come true; what is NEW cannot be true.

Put your TRUST in God, and keep your powder dry

Attributed to Oliver Cromwell.

1834 COLONEL BLACKER *Oliver's Advice* in E. Hayes *Ballads of Ireland* (1856) I. 192 Put your trust in God, my boys, and keep your powder[1] dry. **1856** E. HAYES *Ballads of Ireland* (ed. 2) I. 191 Cromwell . . when his troops were about crossing a river . . concluded an address . . with these words—'put your trust in God; but mind to keep your powder dry.' **1908** *Times Literary Supplement* 6 Nov. 383 In thus keeping his powder dry the bishop acted most wisely, though he himself ascribes the happy result entirely to the observance of the other half of Cromwell's maxim. **1979** V. CANNING *Satan Sampler* iv. God . . created us for a better end. . . We must put our trust in Him and keep our powder dry.

[1] i.e. gunpowder.

There is TRUTH in wine

[Gr. ἐν οἴνῳ ἀλήθεια, there is truth in wine (attributed to Alcaeus); L. *in vino veritas*.] **1545** R. TAVERNER tr. *Erasmus' Adages* (ed. 2) H5ᵛ In wyne is trouthe. **1659** T. PECKE *Parnassi Puerperium* 5 Grant but the Adage true, that Truth's in wine. **1869** TROLLOPE *He knew He was Right* II. li. There is no saying truer than that . . there is truth in wine. Wine . . has the merit of forcing a man to show his true colours. **1934** R. GRAVES *Claudius the God* ix. The man who made the proverb 'There's truth in wine' must have been pretty well soaked when he made it. [Cf. **1973** G. BLACK *Bitter Tea* i. Though, when sober next day, he denied this, I believe wine tells the truth.]

TRUTH is stranger than fiction

Cf. FACT *is stranger than fiction*.

1823 BYRON *Don Juan* XIV. ci. Truth is always strange, Stranger than Fiction. **1863** C. READE *Hard Cash* II. xv. Sampson was greatly struck with the revelation: he . . said truth was stranger than fiction. **1979** *Daily Telegraph* 29 Dec. 18,

I was interested to read . . that clementines are seedless. I was eating one at that moment and I felt a pip in my mouth. Would it be too unkind to say 'Truth is stranger than fiction'?

TRUTH lies at the bottom of a well

[Gr. ἐτεῇ δὲ οὐδὲν ἴσμεν, ἐν βυθῷ γὰρ ἡ ἀλήθεια, we know nothing certainly, for truth lies in the deep (attributed to Democritus); LACTANTIUS *Institutiones Divinae* III. xxviii. *in puteo* . . *veritatem iacere demersam*, truth lies sunk in a well.] **1562** J. WIGAND *De Neutralibus* G6ᵛ The truth lyeth yet still drowned in the depe. **1578** H. WOTTON tr. *J. Iver's Courtly Controversy* 90, I shall conduct you . . vnto the Mansion where the truth so long hidden dothe inhabite, the which sage Democritus searched in the bottom of a well. *a* **1721** M. PRIOR *Dialogues of Dead* (1907) 225 You know the Antient Philosophers said Truth lay at the bottom of a Well. **1887** J. R. LOWELL *Democracy* 30 Truth . . is said to lie at the bottom of a well. **1933** V. BRITTAIN *Testament of Youth* III. xii. The proverbial well is too shallow a hiding-place in which to look for the imprisoned Truth.

TRUTH will out

Cf. MURDER *will out*.

1439 LYDGATE *Life of St. Alban* (1974) 203 Trouthe wil out. . . Ryghtwysnesse may nat ben hid. **1596** SHAKESPEARE *Merchant of Venice* II. ii. 73 Truth will come to light; murder cannot be hid long; a man's son may, but in the end truth will out. **1822** M. EDGEWORTH *Letter* 17 Jan. (1971) 324 Whether about a novel or a murder the truth will out. **1979** S. ALLAN *Mortal Affair* x. The time would come when she . . would want to cash in on the Trust fund and truth would out.

truth: *see also* CHILDREN and fools tell the truth; the GREATER the truth, the greater the libel; HALF the truth is often a whole lie; TELL the truth and shame the Devil.

try: *see* you never KNOW what you can do

till you try; if at first you don't SUCCEED, try, try, try, again.

Every TUB must stand on its own bottom
A proverb advocating independence and initiative.

1564 W. BULLEIN *Dialogue against Fever* 48[v] Let euery Fatte[1] stande vpon his owne bottome. 1639 J. CLARKE *Parœmiologia Anglo-Latina* 66 Every tub must stand on his owne bottome. 1721 C. CIBBER *Refusal* v. 72, I have nothing to do with that. . . Let every Tub stand on its own Bottom. 1866 C. READE *Griffith Gaunt* I. vi. There is an old saying, 'Let every tub stand on its own bottom.' 1948 F. THOMPSON *Still glides Stream* iv. 'Every tub must stand on its own bottom,' was one of his homely ways of expressing the individual independence desirable in children.

[1] i.e. vat.

Tuesday: *see* Monday's CHILD is fair of face.

tug: *see* when GREEK meets Greek, then comes the tug of war.

tune: *see* why should the DEVIL have all the best tunes?; a DRIPPING June sets all in tune; when the FURZE is in bloom, my love's in tune; there's many a GOOD tune played on an old fiddle; he that LIVES in hope dances to an ill tune; he who PAYS the piper calls the tune.

TURKEY, heresy, hops, and beer came into England all in one year

1599 H. BUTTES *Diet's Dry Dinner* G4 I know not how it happened (as he merrily saith) that herisie & beere came hopping into England both in a yeere. 1643 R. BAKER *Chronicle* Henry VIII 66 About [1524] . . it happened that divers things were newly brought into England, whereupon this Rime was made: 'Turke[y]s, Carps, Hoppes, Piccarell,[1] and Beere, Came into England all in one yeere.' 1906 KIPLING *Puck of Pook's Hill* 235 We say – 'Turkey, Heresy, Hops, and Beer Came into England all in one

year.' 1979 *Observer* 16 Dec. 56 'Turkeys, heresies, hops and beer All came to England in the one year' says the rhyme, but the Romans gave us hops.

[1] young pike.

turkey: *see also* on SAINT Thomas the Divine kill all turkeys, geese, and swine.

TURN about is fair play

1755 *Life of Captain Dudley Bradstreet* 338 Hitherto honest Men were kept from shuffling the Cards, because they would cast knaves out from the Company of Kings, but we would make them know, Turn about was fair Play. 1854 R. S. SURTEES *Handley Cross* xviii. 'Turn about is fair play,' as the devil said to the smoke-jack.[1] 1979 B. PETERSON *Peripheral Spy* iv. Since turn about is fair play, let me tell you about the additions to *my* knowledge.

[1] an apparatus for turning a roasting-spit.

turn: *see also* (noun) one GOOD turn deserves another; (verb) a BAD penny always turns up; CLERGYMEN'S sons always turn out badly; even a WORM will turn.

turneth: *see* a SOFT answer turneth away wrath.

turning: *see* it is a LONG lane that has no turning.

twelve: *see* it is not SPRING until you can plant your foot upon twelve daisies.

twice: *see* he GIVES twice who gives quickly; LIGHTNING never strikes the same place twice; ONCE bitten, twice shy; OPPORTUNITY never knocks twice at any man's door.

As the TWIG is bent, so is the tree inclined

[1530 J. PALSGRAVE *L'éclaircissement de la Langue Française* 161 A man may bende a wande[1] while it is grene[2] & make it strayght though it be neuer so croked.] 1732 POPE *Epistles to Several Persons* I. 102 'Tis Education forms the common mind, Just as the Twig is bent, the Tree's inclined. **1818** T. G. FESSENDEN *Ladies Monitor* 75 'Tis education forms the tender mind, Just as the twig is bent the tree's inclined.' This hacknied adage, not more trite than true. **1940** P. McGINLEY *Primary Education* in *Pocketful of Wry* 16 As bends the twig, thus grows the el-em . . So, twice a month, we're bound to sell 'em The doctrine of Impartial Minds. **1979** 'C. AIRD' *Some die Eloquent* viii. 'Nature, not nurture?' murmured the biologist. . . 'As the twig is bent,' Sloan came back.

¹ shoot or stick.   ² young, pliant.

TWO blacks don't make a white

Cf. TWO *wrongs don't make a right.*

1721 J. KELLY *Scottish Proverbs* 321 Two Blacks make no White. An Answer to them who, being blam'd, say others have done as ill or worse. **1822** SCOTT *Letter* 14 Mar. (1934) VII. 96 To try whether I cannot contradict the old proverb of 'Two blackies [Lockhart *Life*: blacks] not making a white'. **1882** A. AINGER *Charles Lamb* vii. As two blacks do not make a white, it was beside the mark to make laborious fun over Southey's youthful ballads. **1932** G. B. SHAW *Adventures of Black Girl* 28 Never forget that two blacks do not make a white. **1966** A. E. LINDOP *I start Counting* viii. 'What's the modern murderer got to fear? . . They'll only go to prison.' . . 'Two blacks don't make a white.'

TWO heads are better than one

Cf. FOUR *eyes see more than two.*

c **1390** GOWER *Confessio Amantis* I. 1021 Tuo han more wit then on. **1530** J. PALSGRAVE *L'éclaircissement de la Langue Française* 269 Two wyttes be farre better than one. **1546** J. HEYWOOD *Dialogue of Proverbs* I. ix. C2ᵛ But of these two thynges he wolde determyne none Without ayde. For two hedds are better than one. **1778** S. FOOTE *Nabob* I. 5 Here comes brother Thomas; two heads are better than one; let us take his opinion. **1817** SCOTT *Rob Roy* I. viii. 'Francis . .

was likely to be as effectually . . supported by my presence than by yours.' 'Two heads are better than one, you know.' **1979** J. RATHBONE *Euro-killers* xviii. Two heads are better than one. . . I'd value your advice.

TWO is company, but three is none

The proverb is often used with the alternative ending *three's a crowd.*

1706 J. STEVENS *Spanish & English Dict.* s.v. Compañia, A Company consisting of three is worth nothing. It is the Spanish Opinion, who say that to keep a Secret three are too many, and to be Merry they are too few. **1860** T. C. HALIBURTON *Season Ticket* viii. Three is a very inconvenient limitation, constituting, according to an old adage, 'no company'. **1869** W. C. HAZLITT *English Proverbs* 442 Two is company, but three is none. **1944** *Modern Language Notes* LIX. 517 Two's company, three's a crowd. **1979** J. LEASOR *Love & Land Beyond* viii. Two's company and three's none, so one of the three has been taken out of the game.

TWO of a trade never agree

1630 DEKKER *Second Part of Honest Whore* II. 154 It is a common rule, and 'tis most true, Two of one trade never loue. **1673** E. RAVENSCROFT *Careless Lovers* A2ᵛ Two of a Trade can seldome agree. **1727** GAY *Fables* I. xxi. In every age and clime we see, Two of a trade can ne'er agree. **1887** MEREDITH *Poems* (1978) I. 148 Two of a trade, lass, never agree! Parson and Doctor!—don't they love rarely, Fighting the devil in other men's fields! **1914** 'SAKI' *Beasts & Super-Beasts* 96 The snorts and snarls . . went far to support the truth of the old saying that two of a trade never agree. **1981** E. LONGFORD *Queen Mother* vii. There is an old adage, 'Two of a kind never agree.'

If TWO ride on a horse, one must ride behind

1598–9 SHAKESPEARE *Much Ado about Nothing* III. v. 34 An two men ride of a horse, one must ride behind. *c* **1628** J. SMYTH *Berkeley MSS* (1885) III. 32 If two ride upon an horse one must sit behinde; meaninge, That in each contention one must take the foile.[1] **1874** G. J. WHYTE-MELVILLE *Uncle John* I. x. There is an old adage . . 'When two people ride on a horse, one must ride behind.' **1942** V. RATH *Posted for Murder* VI. iii. There comes a point when you are very exasperating. . . 'When two ride on one horse, one must ride behind.' But I'm getting off for a while.

[1] repulse or defeat.

There are TWO sides to every question

[PROTAGORAS *Aphorism* (in Diogenes Laertius *Protagoras* IX. li.) καὶ πρῶτος ἔφη [Protagoras] δύο λόγους εἶναι περὶ παντὸς πράγματος ἀντικειμένους ἀλλήλοις, Protagoras was the first to say that there are two sides to every question, one opposed to the other.] **1802** J. ADAMS *Autobiography* (1966) III. 269 There were two Sides to a question. **1817** T. JEFFERSON *Letter* 5 May in L. J. Cappon *Adams-Jefferson Letters* (1959) II. 513 Men of energy of character must have enemies: because there are two sides to every question, and . . those who take the other will of course be hostile. **1863** KINGSLEY *Water Babies* vi. Let them recollect this, that there are two sides to every question, and a downhill as well as an uphill road. **1940** O. NASH *Face is Familiar* 42 And if your devoted mother suggests that you will some day be rich and famous, why perish the suggestion . . if you are afflicted with the suspicion that there are two sides to every question. **1957** E. SITWELL *Letter* in V. Glendinning *Edith Sitwell* (1981) xxviii. It's more than platitudinous to say that there are two sides to every question but there is something to be said on both sides.

It takes TWO to make a bargain

1598 *Mucedorus* B2 Nay, soft, sir, tow words to a bargaine. *a* **1637** MIDDLETON et al. *Widow* v. i. There's two words to a bargain ever . . and if love be one, I'm sure money's the other. **1766** GOLD-SMITH *Vicar of Wakefield* II. xii. 'Hold, hold, Sir,' cried Jenkinson, 'there are

two words to that bargain.' **1943** M. FLAVIN *Journey in Dark* iv. Takes two to make a bargain, and you both done mighty wrong.

It takes TWO to make a quarrel

1706 J. STEVENS *Spanish & English Dict.* s.v. Barajar, When one will not, two do not Quarrel. **1732** T. FULLER *Gnomologia* no. 4942 There must be two at least to a Quarrel. **1859** H. KINGSLEY *Geoffrey Hamlyn* II. xiii. It takes two to make a quarrel, Cecil, and I will not be one. **1979** *Times* 3 Dec. 13 If it were not for the truism that it takes at least two to make a quarrel, the French and the Germans . . could fairly claim that the fault lay wholly with the United Kingdom.

It takes TWO to tango

1952 HOFFMAN & MANNING *Takes Two to Tango* (song title) 2 There are lots of things you can do alone! But, takes two to tango. **1965** *Listener* 24 June 923 As for negotiation . . the President has a firm, and melancholy, conviction: it takes two to tango. **1974** G. JENKINS *Bridge of Magpies* vii. 'We're not getting anywhere.' 'It takes two to tango,' I said. 'I'll listen.' **1979** *Guardian* 4 Apr. 12 It takes two to tango. . . Mrs Thatcher has turned Mr Callaghan down.

TWO wrongs don't make a right

Cf. TWO *blacks don't make a white.*

1783 B. RUSH *Letter* 2 Aug. (1951) I. 308 Three wrongs will not make one right. **1814** J. KERR *Several Trials of David Barclay* 249 Two wrongs don't make one right. **1905** S. WEYMAN *Starvecrow Farm* xxiv. He ought to see this! . . After all, two wrongs don't make a right. **1979** *Daily Telegraph* 24 May 14 If two wrongs do not make a right, three rights can make a terrible wrong.

two: *see also* BETTER one house spoiled than two; BETWEEN two stools one falls to the ground; a BIRD in the hand is worth two in the bush; two BOYS are half a boy, and three boys are no boy at all; of two EVILS choose the less; FOUR eyes see more than two; one HOUR's sleep before midnight is worth two after; NO man can serve two masters; ONE for sorrow, two for mirth; if you can't RIDE two horses at once, you shouldn't be in the circus; if you RUN after two hares you will catch neither; THREE may keep a secret, if two of them are dead; one VOLUNTEER is worth two pressed men.

U

undone: *see* what's DONE cannot be undone.

The UNEXPECTED always happens
Cf. NOTHING *is certain but the unforeseen.*

[PLAUTUS *Mostellaria* I. iii. *insperata accidunt magis saepe quam quae speres*, unexpected things happen more often than those you hope for.] **1885** E. J. HARDY *How to be Happy though Married* xxv. A woman may have much theoretical knowledge, but this will not prevent unlooked-for obstacles from arising. . . It is the unexpected that constantly happens. **1909** *Times Weekly* 12 Nov. 732 No place in the world is more familiar than the House of Commons with 'the unforeseen that always happens'. **1938** E. WAUGH *Scoop* I. iii. Have nothing which in a case of emergency you cannot carry in your own hands. But remember that the unexpected always happens. **1977** L. J. PETER *Peter's Quotations* 296 Peter's Law—The unexpected always happens.

unforeseen: *see* NOTHING is certain but the unforeseen.

UNION is strength
Unity is a popular alternative for *union*, especially when used as a trade-union slogan.

[HOMER *Iliad* XIII. 237 συμφερτὴ δ'ἀρετὴ πέλει ἀνδρῶν καὶ μάλα λυγρῶν, even weak men have strength in unity; L. *vis unita fortior*, force united is stronger. Cf. *c* **1527** T. BERTHELET tr. *Erasmus' Sayings of Wise Men* A4ᵛ Concorde maketh those thynges that are weake, mighty and stronge.] **1654** R. WILLIAMS *Complete Writings* (1963) VI. 280 Union strengthens. **1837** in D. Porter *Early Negro Writing* (1971) 228 In Union is strength. **1848** S. ROBINSON *Letter* 29 Dec. in *Indiana Hist. Collections* (1936) XXII. 178 'Union is strength,' and that is the only kind that can control the floods of such a 'great father of rivers [the

Mississippi]'. **1877** E. WALFORD *Tales of Great Families* I. 264 The prosperity of the House of Rothschild [is due to] the unity which has attended the co-partnership of its members, . . a fresh example of the saying that 'union is strength'. **1933** H. ADAMS *Strange Murder of Hatton* xxix. Union is strength. We, by pooling our resources, . . are able . . to secure a steady income. **1981** E. AGRY *Assault Force* ix. This unfortunate misunderstanding; we must clear it up. . . After all, unity is strength.

UNITED we stand, divided we fall
1768 J. DICKINSON *Liberty Song* in *Boston Gazette* 18 July, Then join Hand in Hand brave Americans all, By uniting we stand, by dividing we fall. **1849** G. P. MORRIS *Flag of our Union* in *Poems* (1853) 41 'United we stand—divided we fall!'—It made and preserves a nation! **1894** J. JACOBS *Fables of Aesop* 122 Then Lion attacked them one by one and soon made an end of all four [oxen]. United we stand, divided we fall. **1974** D. G. COMPTON *Continuous Katherine Mortenhoe* iv. Establishing a higher rate for the job. United we stand, divided we fall.

unity: *see* UNION is strength.

unlucky: *see* LUCKY at cards, unlucky in love.

What goes UP must come down
Commonly associated with wartime bombing and anti-aircraft shrapnel.

1939 L. I. WILDER *By Shores of Silver Lake* xxviii. 'But you can't unload the stove by yourself,' Ma objected. 'I'll manage,' Pa said. 'What goes up must come down.' **1949** N. MAILER *Naked & Dead* III. vi. Gravity would occupy the place of mortality (what goes up must come down). **1967** F. J. SINGER *Epigrams at Large* 57 'What goes up, must come down' is really a time-worn statement which wore out after the Venus and

Mars probes. **1980** R. L. DUNCAN *Brimstone* ix. They maintain systems to measure radioactivity when above-ground testing ended years ago. . . Bring me some black coffee. . . All that goes up must come down.

upright: *see* EMPTY sacks will never stand upright.

use: *see* KEEP a thing seven years and you'll always find a use for it.

V

vacuum: *see* NATURE abhors a vacuum.

vain: *see* in vain the NET is spread in the sight of the bird.

valet: *see* NO man is a hero to his valet.

valour: *see* DISCRETION is the better part of valour.

VARIETY is the spice of life

1785 W. COWPER *Task* II. 76 Variety's the very spice of life, That gives it all its flavour. 1854 'M. LANGDON' *Ida May* vi. Take all de wives you can get,—bariety am de spite of life. 1954 'M. COST' *Invitation from Minerva* 174 'Your signal is different from ours?' . . 'Variety is the spice of life,' he retorted. 1974 J. I. M. STEWART *Gaudy* v. 'It sounds . . as if the owner switches his interests during vacations.' 'As he well may do, if there is anything in the saying that variety is the spice of life.'

varlet: *see* an APE's an ape, a varlet's a varlet, though they be clad in silk or scarlet.

vengeance: *see* REVENGE is a dish that can be eaten cold.

venture: *see* NOTHING venture, nothing gain; NOTHING venture, nothing have.

vessel: *see* EMPTY vessels make the most sound.

view: *see* DISTANCE lends enchantment to the view.

vinegar: *see* HONEY catches more flies than vinegar; from the SWEETEST wine, the tartest vinegar.

VIRTUE is its own reward

[OVID *Tristia* v. xiv. *pretium sibi virtus*, virtue is its own reward.] 1509 A. BARCLAY *Ship of Fools* 10ᵛ Vertue hath no

rewarde. 1596 SPENSER *Faerie Queene* III. xii. Your vertue selfe her owne reward shall breed, Euen immortall praise, and glory wyde. 1642 BROWNE *Religio Medici* I. 87 That vertue is her owne reward, is but a cold principle. 1673 DRYDEN *Assignation* III. i. Virtue . . is its own reward: I expect none from you. 1844 DICKENS *Martin Chuzzlewit* xv. It *is* creditable to keep up one's spirits here. Virtue's its own reward. 1979 'J. GASH' *Grail Tree* vii. Don't you try telling me that virtue is or has its own reward because it's not and it hasn't.

virtue: *see also* PATIENCE is a virtue.

The VOICE of the people is the voice of God

[*a* 804 ALCUIN *Letter* clxiv. in *Works* (1863) I. 438 *solent dicere: vox populi, vox Dei*, they often say: the voice of the people is the voice of God.] *c* 1412 HOCCLEVE *Regimen of Princes* (EETS) 104 Peples vois is goddes voys, men seyne. 1450 in T. Wright *Political Poems* (1861) II. 227 The voice of the people is the voice of God. 1646 BROWNE *Pseudodoxia Epidemica* I. iii. Though sometimes they are flattered with that Aphorisme, [they] will hardly beleeve the voyce of the people to bee the voyce of God. 1822 C. C. COLTON *Lacon* II. 266 The voice of the People is the voice of God; this axiom has manifold exceptions. 1914 G. B. SHAW *Misalliance* p. lxxii. An experienced demagogue comes along and says, 'Sir: *you* are the dictator: the voice of the people is the voice of God.'

One VOLUNTEER is worth two pressed men

1705 T. HEARNE *Journal* 31 Oct. in *Remarks & Collections* (1885) I. 62 'Tis sᵈ my Lᵈ Seymour presently after Mʳ. Smith was pronounc'd Speaker, rose up, and told them, Gentlemen; you have got a Low Church man; but pray remember that

100 Voluntiers are better than 200 press'd[1] men. **1834** MARRYAT *Jacob Faithful* I. xiii. 'Shall I give you a song?' 'That's right, Tom; a volunteer's worth two pressed men.' **1837** F. CHAMIER *Arethusa* I. iii. Don't fancy you will be detained against your will; one volunteer is worth two pressed men. **1897** KIPLING *Captains Courageous* x. He presumed Harvey might need a body-servant some day or other, and . . was sure that one volunteer was worth five hirelings. **1979** M. M. KAYE *Shadow of Moon* (rev. ed.) iv. The Earl could not be persuaded to send her away. . . In any case, said the Earl, Winter would need a personal maid, and in his opinion one volunteer was worth three pressed men.

[1] forced into military service.

vomit: *see* the DOG returns to his vomit.

W

wag: *see* it is MERRY in hall when beards wag all.

wait: *see* ALL things come to those who wait; TIME and tide wait for no man.

waiting: *see* it's ILL waiting for dead men's shoes.

We must learn to WALK before we can run

The metaphorical phrase *to run before one can walk* is also common.

c **1350** *Douce MS* 52 no. 116 Fyrst the chylde crepyth and after gooth.[1] c **1450** *Towneley Play of First Shepherds* (EETS) l. 100 Ffyrst must vs crepe and sythen[2] go. **1670** J. RAY *English Proverbs* 75 You must learn to creep before you go. **1794** G. WASHINGTON *Letter* 20 July in *Writings* (1940) XXXIII. 438 We must walk as other countries have done before we can run. **1851** G. BORROW *Lavengro* II. ii. Ambition is a very pretty thing; but sir, we must walk before we run. **1876** J. PLATT *Business* 124 More fail from doing too much than too little. We must learn and be strong enough to walk before we can run. **1947** M. PENN *Manchester Fourteen Miles* xv. Mrs. Winstanley reproved her for being impatient. She pointed out . . that everybody must learn to walk before they could run. **1980** K. AMIS *Russian Hide & Seek* iv. At the moment we can't leave it to the English to do anything. We must learn to walk before we can run.

[1] walks.   [2] afterwards, then.

walk: *see also* AFTER dinner rest a while, after supper walk a mile.

wall: *see* the WEAKEST go to the wall.

WALLS have ears

Cf. FIELDS *have eyes, and woods have ears.*

1575 G. GASCOIGNE *Supposes* I. i. The tables . ., the portals, yes and the cupbords them selves have eares. **1592** G. DELAMOTHE *French Alphabet* II. 29 The walles may have some eares. . . *Les murailles ont des aureilles.* **1620** T. SHELTON tr. *Cervantes' Don Quixote* II. xlviii. They say Walls haue eares. **1766** D. GARRICK *Neck or Nothing* II. i. Not so fast and so loud, good master of mine— walls have ears. **1822** SCOTT *Nigel* I. vi. It is not good to speak of such things. . . Stone walls have ears. **1958** L. DURRELL *Mountolive* XII. 232 She lay in the silence of a room which had housed (if walls have ears) their most secret deliberations.

walnut: *see* a WOMAN, a dog, and a walnut tree, the more you beat them the better they be.

WALNUTS and pears you plant for your heirs

[Cf. CICERO *Cato Major* vii. 24 'serit arbores, quae alteri saeclo prosint,' ut ait Statius noster in Synephebis, 'he plants trees, which will be of use to another age,' as Statius says in his *Synephebi*.] **1640** G. HERBERT *Outlandish Proverbs* no. 198 The tree that growes slowly, keepes it selfe for another. **1732** T. FULLER *Gnomologia* no. 2401 He who plants a Walnut-Tree, expects not to eat of the Fruit. **1863** A. SMITH *Dreamthorp* xi. My oaks are but saplings; but what undreamed-of English kings will they not outlive? . . A man does not plant a tree for himself; he plants it for posterity. **1941** C. MACKENZIE *Red Tapeworm* xv. 'Better to plant them promptly,' said Miss Quekett. 'It's only walnuts and pears you plant for your heirs.'

If you WANT a thing done well, do it yourself

Cf. *If you would be well* SERVED, *serve yourself.*

1541 M. COVERDALE tr. *H. Bullinger's Christian State of Matrimony* xix. If thou wilt prospere, then loke to euery thynge thyne owne self. **1616** T. DRAXE *Adages*

163 If a man will haue his business well done, he must doe it himselfe. **1858** LONGFELLOW *Poems* (1960) 160 That's what I always say; if you want a thing to be well done, You must do it yourself. **1927** *Times* 14 Nov. 15 Lastly there is the illustration of the great principle: if you want a thing done, do it yourself. **1975** 'E. LATHEN' *By Hook or by Crook* xxi. Do you know how I got it done in the end? I went down to Annapolis myself. I always say, if you want a thing done well, do it yourself!

For WANT of a nail the shoe was lost; for want of a shoe the horse was lost; and for want of a horse the man was lost
The proverb is found in a number of forms.

[c **1390** GOWER *Confessio Amantis* v. 4785 For sparinge of a litel cost Fulofte time a man hath lost The large cote for the hod.[1]] **1629** T. ADAMS *Works* 714 The French-men haue a military prouerbe, The losse of a nayle, the losse of an army. The want[2] of a nayle looseth the shooe,[3] the losse of a shooe troubles the horse, the horse indangereth the rider, the rider breaking his ranke molests the company, so farre as to hazard the whole Army. **1640** G. HERBERT *Outlandish Proverbs* no. 499 For want of a naile the shoe is lost, for want of a shoe the horse is lost, for want of a horse the rider is lost. **1880** S. SMILES *Duty* x. 'Don't care' was the man who was to blame for the well-known catastrophe:—'For want of a nail the shoe was lost, for want of a shoe the horse was lost, and for want of a horse the man was lost.' **1925** S. O'CASEY *Juno & Paycock* I. 16 You bring your long-tailed shovel, an' I'll bring me navvy.[4] We mighten' want them, an', then agen, we might: for want of a nail the shoe was lost, for want of a shoe the horse was lost, an' for want of a horse the man was lost—aw, that's a darlin' proverb, a daarlin'. **1979** M. McCARTHY *Missionaries & Cannibals* viii. No detail . . was too small to be passed over. . . 'For want of a nail,' as the proverb said.

[1] i.e. hood   [2] lack.   [3] i.e. horseshoe.
[4] a device for excavating earth.

want: *see also* (verb) the MORE you get, the more you want; if you want PEACE, you must prepare for war; WASTE not, want not.

WANTON kittens make sober cats
1732 T. FULLER *Gnomologia* no. 5415 Wanton[1] Kitlins may make sober old Cats. **1832** A. HENDERSON *Scottish Proverbs* 97 Wanton kittens mak douce[2] cats. **1855** H. G. BOHN *Hand-Book of Proverbs* 551 Wanton kittens may make sober cats. **1975** J. O'FAOLAIN *Women in Wall* i. I was fleshy . . in my youth. Carnal. But wanton kittens make sober cats.

[1] frisky, frolicsome; (of persons) lascivious.
[2] quiet, sedate.

war: *see* COUNCILS of war never fight; all's FAIR in love and war; when GREEK meets Greek, then comes the tug of war; if you want PEACE, you must prepare for war.

warling (one who is despised or disliked): *see* BETTER be an old man's darling, than a young man's slave.

warm: *see* COLD hands, warm heart.

One does not WASH one's dirty linen in public
It is unwise to publicize private disputes or scandals. The saying is also used in the metaphorical phrase *to wash one's dirty linen in public*.

[Cf. Fr. *c'est en famille, ce n'est pas en publique, qu'on lave son linge sale*, one washes one's dirty linen amongst the family, not in public (attributed to Napoleon).] **1809** T. G. FESSENDEN *Pills* 45 The man has always had a great itch for scribbling, and has mostly been so fortunate as to procure somebody who pitied his ignorance, to 'wash his dirty linen'. **1867** TROLLOPE *Last Chronicle of Barset* II. xliv. I do not like to trouble you with my private affairs;—there is nothing . . so bad as washing one's dirty linen in public. **1886** E. J. HARDY *How to be Happy though Married* i. Married people . . should remember the proverb about the home-washing of soiled linen. **1942** 'P. WENTWORTH' *Danger*

Point xlviii. The case . . will be dropped. . . There's nothing to be gained by washing a lot of dirty linen in public. **1980** T. HOLME *Neapolitan Steak* 199 Her look raked him from head to toe. 'One does not wash one's Dirty Linen in Public, commissario.'

wash: *see also* one HAND washes the other.

WASTE not, want not
Want is variously used in the senses 'lack' and 'desire'.

1772 J. WESLEY *Letter* 10 Aug. (1931) V. 334 He will waste nothing; but he must want nothing. **1800** M. EDGEWORTH *Parent's Assistant* (ed. 3) V. 136 The following words . . were written . . over the chimney-piece, in his uncle's spacious kitchen—'Waste not, want not'. **1872** HARDY *Under Greenwood Tree* I. I. viii. Helping her to vegetable she didn't want, and when it had nearly alighted on her plate, taking it across for his own use, on the plea of waste not, want not. **1941** C. MACKENZIE *Red Tapeworm* xxii. 'The lorry's full of children as well as rubbish.'. . 'And what is printed on the banner?' . . 'Waste Not Want Not.' **1979** M. KAUFMAN *Container* xi. Don't forget to keep a check on what you spend. Waste not, want not.

waste: *see also* (noun) HASTE makes waste.

A WATCHED pot never boils
1848 E. GASKELL *Mary Barton* II. xiv. What's the use of watching? A watched pot never boils. **1880** M. E. BRADDON *Cloven Foot* III. viii. Don't you know that vulgar old proverb that says that 'a watched pot never boils'? **1940** C. BOOTHE *Europe in Spring* x. 'He [Mussolini] is waiting to see how the next battle turns out,' they said. . . 'A watched pot never boils,' they said— only this one finally did. **1979** A. MORICE *Murder in Outline* ii. I . . remained seated . . fixing my eyes hungrily on the rear mirror for signs of other guests arriving. It was a case of the watched pot, however.

Don't go near the WATER until you learn how to swim
1855 H. G. BOHN *Hand-Book of Proverbs* 459 Never venture out of your depth till you can swim. **1975** D. BAGLEY *Snow Tiger* xv. 'There I was. . . Over-protected and regarded as a teacher's pet into the bargain.' ' "Don't go near the water until you learn how to swim," ' quoted McGill.

water: *see also* BLOOD is thicker than water; DIRTY water will quench fire; you can take a HORSE to the water, but you can't make him drink; the MILL cannot grind with the water that is past; you never MISS the water till the well runs dry; STILL waters run deep; STOLEN waters are sweet; don't THROW out your dirty water until you get in fresh.

The WAY to a man's heart is through his stomach
1814 J. ADAMS *Letter* 15 Apr. in *Works* (1851) VI. 505 The shortest road to men's hearts is down their throats. **1845** R. FORD *Hand-Book for Travellers in Spain* I. i. The way to many an honest heart lies through the belly. **1857** D. M. MULOCK *John Halifax, Gentleman* xxx. 'Christmas dinners will be much in request.' 'There's a saying that the way to an Englishman's heart is through his stomach.' **1975** A. PRICE *Our Man in Camelot* v. The way to a man's heart wasn't through his stomach, it was through an appreciation of what interested him.

way: *see also* the LONGEST way round is the shortest way home; LOVE will find a way; OTHER times, other manners; STRAWS tell which way the wind blows; a WILFUL man must have his way; where there's a WILL there's a way.

There are more WAYS of killing a cat than choking it with cream
See also the following two entries.

1839 S. SMITH *John Smith's Letters* 91 There's more ways to kill a cat than one. **1855** KINGSLEY *Westward Ho!* II. xii. Hold on yet awhile. More ways of kill-

ing a cat than choking her with cream.
1941 'R. WEST' *Black Lamb* I. 506 Now I
see the truth of the old saying that there
are more ways of killing a cat than
choking it with cream. In Bosnia the
Slavs did choke the Turk with cream,
they glutted him with their wholesale
conversions. . . But here cream just did
not come into the question. **1974** T.
SHARPE *Porterhouse Blue* ii. I have yet to
meet a liberal who can withstand the
attrition of prolonged discussion of the
inessentials. . . There are more ways of
killing a cat than stuffing it with. . .

**There are more WAYS of killing a dog
than choking it with butter**
See also the adjacent entries.
   1845 W. T. THOMPSON *Chronicles of
Pineville* 35 There's more ways to kill a
dog besides choking him with butter.
1945 F. THOMPSON *Lark Rise* xvi. A
proverb always had to be capped. No
one could say, 'There's more ways of
killing a dog than hanging it' without
being reminded, 'nor of choking it with
a pound of fresh butter.' **1955** W. C.
MACDONALD *Destination, Danger* x. It
[liquor] was a lifesaver and I'm much
obliged. But you can kill a dog without
choking him with butter.

**There are more WAYS of killing a dog
than hanging it**
See also the two preceding entries.
   1678 J. RAY *English Proverbs* (ed. 2) 127
There are more ways to kill a dog then
hanging. **1721** J. KELLY *Scottish Proverbs*
253 Many ways to kill a Dog, and not to
hang him. There be many ways to bring
about one and the same Thing, or
Business. **1725** SWIFT *Drapier's Letters* X.
165, I know that very homely Proverb,
more ways of killing a Dog than hanging
him. **1945** F. THOMPSON *Lark Rise* xvi. A
proverb always had to be capped. No
one could say, 'There's more ways of
killing a dog than hanging it' without
being reminded, 'nor of choking it with
a pound of fresh butter.'

weak: *see* YORKSHIRE born and Yorkshire

bred, strong in the arm and weak in the
head.

The WEAKEST go to the wall
Usually said to derive from the installa-
tion of seating (around the walls) in the
churches of the late Middle Ages. *To go to
the wall* means figuratively 'to succumb
in a conflict or struggle'.
   a **1500** *Coventry Plays* (EETS) 47 The
weykist gothe eyuer to the walle.
c **1595** SHAKESPEARE *Romeo & Juliet* I. i.
14 That shows thee a weak slave; for the
weakest goes to the wall. **1714** DEFOE
(*title*) The weakest go to the wall, or the
Dissenters sacrific'd by all parties. **1834**
MARRYAT *Peter Simple* I. v. You will be
thrashed all day long. . . The weakest
always goes to the wall there. **1888**
C. M. DOUGHTY *Travels in Arabia Deserta* I.
x. There perished many among them;
. . it is the weak which go to the wall.
1916 'J. OXENHAM' *My Lady of Moor* i. He
saw to it that I had a good education, . .
knowing the necessity and value of it in
these strenuous days of the 'weak to the
wall'.

weakest: *see also* a CHAIN is no stronger
than its weakest link.

wealthy: *see* EARLY to bed and early to rise,
makes a man healthy, wealthy, and
wise.

wear: *see* BETTER to wear out than to rust
out; if the CAP fits, wear it; CONSTANT
dropping wears away a stone; GIVE a
thing, and take a thing, to wear the Devil's gold ring.

weather: *see* ROBIN Hood could brave all
weathers but a thaw wind.

wed: *see* BETTER wed over the mixen than
over the moor.

One WEDDING brings another
Cf. *One* FUNERAL *makes many.*
   1634 M. PARKER in *Roxburghe Ballads*
(1880) III. 54 'Tis said that one wedding
produceth another. **1713** GAY *Wife of
Bath* I. i. One Wedding, the Proverb

says, begets another. **1885** C. H. SPURGEON *Salt-Cellars* I. 88 Bridesmaids may soon be made brides. One wedding . . brings on another. **1929** S. T. WARNER *True Heart* I. 54 Cheer up, Suke! I dare say you'll get a boy in time—they do say one wedding brings another.

Wednesday: *see* Monday's CHILD is fair of face.

weed: *see* ILL weeds grow apace.

weeding: *see* ONE year's seeding means seven years' weeding.

week: *see* if you would be HAPPY for a week take a wife.

weep: *see* LAUGH and the world laughs with you, weep and you weep alone.

weeper: *see* FINDERS keepers (losers weepers).

welcome: *see* when all FRUIT falls, welcome haws.

WELL begun is half done

[HORACE *Epistles* I. ii. *dimidium facti qui coepit habet*, he who has made a beginning, has half done.] *c* **1415** *Middle English Sermons* (EETS) 148 The wise man seth that halfe he hath don that wel begynneth is werke. **1542** N. UDALL *Erasmus' Apophthegms* I. 16 Laertius ascrybeth to hym [Socrates] this saiyng also: to haue well begoone is a thyng halfe dooen. **1616** J. WITHALS *Dict.* (rev. ed.) 555 Well begun, is halfe done. **1703** P. A. MOTTEUX *Don Quixote* IV. xli. Let me . . get . . ready for our Journey. . . 'Twill be soon done, and A Business once begun, you know, is half ended. **1883** C. S. BURNE *Shropshire Folklore* 273 They also account it very unlucky to give trust[1] for the first article sold. 'Well begun is half done,' is evidently their principle. **1907** A. MACLAREN *Acts* I. 176 Satan spoils many a well–begun work. . . Well begun is half—but only half—ended. **1981** P. O'DONNELL *Xanadu Talisman* iv. The

nannie-like proverbs . .Well begun is half done, The early bird catches the worm.

[1] credit.

All's WELL that ends well

[*c* **1250** *Proverbs of Hending* in *Anglia* (1881) IV. 182 Wel is him that wel ende mai.] **1381** in J. R. Lumby *Chronicon Henrici Knighton* (1895) II. 139 If the ende be wele, than is alle wele. *c* **1530** R. HILL *Commonplace Book* (EETS) 110 'All ys well that endyth well,' said the gud wyff. **1602** SHAKESPEARE *All's Well that ends Well* IV. iv. 35 All's Well That Ends Well. Still the fine's[1] the crown. **1836** MARRYAT *Midshipman Easy* I. vi. I had got rid of the farmer, . . bull, and the bees—all's well that ends well. **1979** G. HAMMOND *Dead Game* xviii. My rank's been confirmed. So all's well that ends well.

[1] end's.

well: *see also* (noun) you never MISS the water till the well runs dry; the PITCHER will go to the well once too often; TRUTH lies at the bottom of a well.

well: *see also* (noun) LET well alone; (adjective) the DEVIL was sick, the Devil a saint would be, the Devil was well, the devil a saint was he!; (adverb) he LIVES long who lives well; if you would be well SERVED, serve yourself; SPARE well and have to spend; everyone SPEAKS well of the bridge which carries him over; if a THING's worth doing, it's worth doing well; if you WANT a thing done well, do it yourself.

west: *see* EAST, west, home's best.

wet: *see* the CAT would eat fish, but would not wet her feet.

what: *see* what goes UP must come down; what MUST be must be.

wheel: *see* the SQUEAKING wheel gets the grease.

while: *see* while there's LIFE there's hope.

whirlwind: *see* they that sow the wind shall reap the whirlwind.

whistle: *see* don't HALLOO till you are out of the wood; a sow may whistle, though it has an ill mouth for it.

A WHISTLING woman and a crowing hen are neither fit for God nor men

1721 J. KELLY *Scottish Proverbs* 33 A crooning cow, a crowing Hen and a whistling Maid boded never luck to a House. The two first are reckoned ominous, but the Reflection is on the third. 1850 *Notes & Queries* 1st Ser. II. 164 A whistling woman and a crowing hen, Is neither fit for God nor men. 1891 J. L. KIPLING *Beast & Man* ii. 'A whistling woman and a crowing hen are neither fit for God nor men,' is a mild English saying. 1917 J. C. BRIDGE *Cheshire Proverbs* 28 A whistling woman and a crowing hen will fear the old lad[1] out of his den. 1933 L. I. WILDER *Farmer Boy* xi. Royal teased her, Whistling girls and crowing hens Always come to some bad ends. 1981 R. ENGEN *Kate Greenaway* i. Mary was . . always ready with a relevant proverb . . 'a whistling woman and a crowing hen . . are neither good for God nor men.'

[1] the Devil.

white: *see* TWO blacks don't make a white.

whole: *see* the HALF is better than the whole; HALF the truth is often a whole lie.

whore: *see* ONCE a whore, always a whore.

wife: *see* a BLIND man's wife needs no paint; CAESAR's wife must be above suspicion; a DEAF husband and a blind wife are always a happy couple; if you would be HAPPY for a week take a wife; the HUSBAND is always the last to know; my SON is my son till he gets him a wife, but my daughter's my daughter all the days of her life; he that will THRIVE must first ask his wife.

A WILFUL man must have his way

1816 SCOTT *Antiquary* I. vi. A wilful man must have his way. 1907 W. DE MORGAN *Alice-for-Short* xxxvii. 'A wilful man will have his way,' says Peggy, laughing. . . Alice replies: 'Never mind!' 1931 J. BUCHAN *Blanket of Dark* xii. 'Take one of my men with you.'. . He shook his head. . . 'A wilful man must have his way,' she said.

He that WILL not when he may, when he will he shall have nay

a 1000 in *Anglia* (1889) XI. 388 Nu sceal ælc man efsten, thæt he to gode gecerre tha hwile the he muge, thelæste, gyf he nu nelle tha hwile the he muge, eft thone he wyle, he ne mæig [now shall each man hasten to turn to God while he may, lest if he will not now while he may, later when he will, he may not.] 1303 R. BRUNNE *Handlyng Synne* (EETS) l. 4795 He that wyl nat when he may, He shal nat, when he wyl. *c* 1450 in Brown & Robbins *Index of Middle English Verse* (1943) 186 He that will not when he may, When he will he shall have nay.[1] 1621 BURTON *Anatomy of Melancholy* (ed. 2) III. ii. They omit oportunities. . . He that will not when he may, When he will he shall haue nay. 1893 R. L. STEVENSON *Catriona* xix. That young lady, with whom I so much desired to be alone again, sang . . 'He that will not when he may, When he will he shall have nay.' 1935 N. MITCHISON *We have been Warned* III. 297 'She that will not when she may, When she will she shall have nay.' Aren't you feeling a bit like that?

[1] denial or refusal.

Where there's a WILL, there's a way

1640 G. HERBERT *Outlandish Proverbs* no. 730 To him that will, waìs are not wanting. 1822 HAZLITT in *New Monthly Magazine* Feb. 102 Where there's a will, there's a way.—I said so to myself, as I walked down Chancery-lane . . to inquire . . where the fight the next day was to be. 1849 BULWER-LYTTON *Caxtons* III. xviii. v. I must consider of all this, and talk it over with Bolt. . . Meanwhile, I fall back on my favourite proverb,—'Where there's a will there's a way'. **1979** E. KOCH *Good Night Little Spy* xi. I've no idea how it can be done. But where there's a will, there's a way.

will: *see also* (noun) he that COMPLIES against his will is of his own opinion still; (verb) if ANYTHING can go wrong, it will; there's none so BLIND as those who will not see; there's none so DEAF as those who will not hear; what MUST be must be.

He who WILLS the end, wills the means

Cf. *Where there's a* WILL, *there's a way*.

1692 R. SOUTH *Twelve Sermons* 497 That most true aphorism, that he who wills the end, wills also the means. **1910** *Spectator* 29 Oct. 677 We won at Trafalgar . . because we not only meant to win, but knew how to win—because we understood . . the maxim, 'He who wills the end wills the means.' **1980** *Listener* 13 Mar. 332, I could offer a text . . from Aneurin Bevan: 'It's no good willing the end unless you're also ready to will the means.'

You WIN a few, you lose a few

An expression of consolation or resignation of American origin, also found in the form *you win some, you lose some*. See also the next entry.

[**1897** KIPLING *Captains Courageous* x. 'Thirty million dollars' worth of mistake, wasn't it? I'd risk it for that.' 'I lost some; and I gained some.'] **1966** P. O'DONNELL *Sabre-Tooth* xiv. You win a few, you lose a few, and it's no good getting sore. **1969** D. E. WESTLAKE *Up your Banners* xxvi. I go home. . . You win a few, you lose a few. **1971** C. WILLIAMS *And Deep Blue Sea* xiv. She stopped, arrested by something in the attitude of the two figures leaning on the rail, and shrugged. You won a few, you lost a few. **1976** *Times* 23 Nov. 14 You look like being saddled with the uninspiring Willy. . . On the other hand, you seem to have got your way over Mrs. Thatcher's nominee. . . You win some, you lose some.

You can't WIN them all

See also the preceding entry.

1953 R. CHANDLER *Long Good-bye* xxiv. Wade took him by the shoulder and spun him round. 'Take it easy, Doc. You can't win them all.' **1962** J. D. MACDONALD *Girl, Gold Watch & Everything* vii. 'Well, hell,' she said wistfully. 'You can't win 'em all.' **1974** D. FRANCIS *Knock Down* ii. 'We treat people with a problem. Yes.' 'Successfully?' . . 'Some.' 'You can't win them all.'

win: *see also* let them LAUGH that win; *also* WON.

When the WIND is in the east, 'tis neither good for man nor beast

1600 R. CAWDREY *Treasury of Similes* 750 The East wind is accounted neither good for man or beast. **1609** S. HARWARD *MS* (Trinity College, Cambridge) 86 The wind East is neither good for man nor beast. **1659** J. HOWELL *Proverbs* (English) 19 When the wind is in the east it is good for neither man nor beast. **1670** J. RAY *English Proverbs* 41 When the wind's in the East, It's neither good for man nor beast. . . The East-wind with us is commonly very sharp, because it comes off the Continent. **1825** W. HONE *Every-Day Book* I. 670 When the wind's in the east, It's neither good for man nor beast. **1927** *Times* 21 Nov. 15 Science is beginning a new incursion into . . such wisdom as is contained in the lines When the wind is in the East 'Tis neither good for man nor beast.

wind: *see also* APRIL showers bring forth May flowers; GOD tempers the wind to the shorn lamb; it's an ILL wind that blows nobody any good; ONE for the mouse, one for the crow; a REED before the wind lives on, while mighty oaks do fall; ROBIN Hood could brave all weathers but a thaw wind; they that sow the wind shall reap the whirlwind; STRAWS tell which way the wind blows.

window: *see* the EYES are the window of the soul; when POVERTY comes in at the door, love flies out of the window.

When the WINE is in, the wit is out

c **1390** GOWER *Confessio Amantis* VI. 555 For wher that wyn doth wit aweie,[1] Wisdom hath lost the rihte weie.[2] **1529** MORE *Dialogue of Images* III. xvi. Whan the wyne were in and the wyt[3] out, wolde they take vppon them . . to handle holy scrypture. **1560** T. BECON *Works* I. 536[v] When the wine is in, the wit is out. **1710** S. PALMER *Proverbs* 18 When the Wine's In, the Wit's Out. **1854** J. W. WARTER *Last of Old Squires* vi. None is a Fool always, every one sometimes. When the Drink goes in, then the Wit goes out. **1937** V. WILKINS *And so—Victoria* iii. Remember what I told you last night—that with wine in, wits go out.

[1] i.e. away. [2] i.e. way. [3] wisdom, intelligence.

wine: *see also* GOOD wine needs no bush; you can't put NEW wine in old bottles; from the SWEETEST wine, the tartest vinegar; there is TRUTH in wine.

wing: *see* a BIRD never flew on one wing; the MOTHER of mischief is no bigger than a midge's wing.

wink: *see* a NOD's as good as a wink to a blind horse.

winter: *see* if CANDLEMAS day be sunny and bright, winter will have another flight; the RICH man has his ice in the summer and the poor man gets his in the winter.

wisdom: *see* EXPERIENCE is the father of wisdom.

It is easy to be WISE after the event

c **1590** G. HARVEY *Marginalia* (1913) 99 It is good, to be wise before the Mischiff. **1616** JONSON *Epicœne* II. iv. Away, thou strange iustifier of thy selfe, to bee wiser then thou wert, by the euent. **1717** R. WODROW *Letter* 28 Sept. (1843) II. 319 Had we not verified the proverb of being wise behind the time, we might for ever [have] been rid of them. **1900** A. CONAN DOYLE *Great Boer War* xix. It is easy to be wise after the event, but it does certainly appear that . . the action at Paardeberg was as unnecessary as it was expensive. **1943** 'P. WENTWORTH' *Chinese Shawl* xliv. I committed an error of judgment, but it is easy to be wise after the event.

It is a WISE child that knows its own father

1584 J. WITHALS *Dict.* (rev. ed.) L4 Wise sonnes they be in very deede, That knowe their Parentes who did them breede. **1589** GREENE *Menaphon* VI. 92 Wise are the Children in these dayes that know their owne fathers, especially if they be begotten in Dogge daies,[1] when their mothers are frantick with love. **1596** SHAKESPEARE *Merchant of Venice* II. ii. 69 It is a wise father that knows his own child. **1613** G. WITHER *Abuses* I. ii. Is't not hence this common Prouerbe growes, 'Tis a wise child that his owne father knowes? **1762** GOLDSMITH *Mystery Revealed* 21 She called her father John instead of Thomas . . but perhaps she was willing to verify the old proverb, that It is a wise child that knows its own father. **1823** SCOTT *Peveril* III. x. I only laughed because you said you were Sir Geoffrey's son. But no matter—'tis a wise child that knows his own father. **1967** D. MORRIS *Naked Ape* iii. It may be a wise child that knows its own father, but it is a laughing child that knows its own mother.

[1] *Dogge daies*: those hot days about the time of the visible rising of the Dog-star before sunrise (according to most systems, in July and August) which were thought to have a pernicious influence.

wise: *see also* EARLY to bed and early to rise, makes a man healthy, wealthy, and wise; a FOOL may give a wise man counsel; FOOLS ask questions that wise men cannot answer; FOOLS build houses and wise men live in them; where IGNORANCE is bliss, 'tis folly to be wise; one cannot LOVE and be wise; PENNY wise and pound foolish; a STILL tongue makes a wise head; a WORD to the wise is enough.

The WISH is father to the thought

1598 SHAKESPEARE *Henry IV, Pt.* 2 IV. v. 93, I never thought to hear you speak again.—Thy wish was father, Harry, to that thought. **1783** P. VAN SCHAACK *Letter* 5 Jan. in H. C. Van Schaack *Life* (1842) 321 My 'wish is father to the thought'. **1860** TROLLOPE *Framley Parsonage* III. xiv. The wish might be father to the thought . . but the thought was truly there. **1940** E. F. BENSON *Final Edition* iii. She spied a smallish man . . walking away from us. The wish was father to the thought. 'Ah, there is Lord Ripon,' she said. . . He turned round. It wasn't Lord Ripon at all.

If WISHES were horses, beggars would ride

a **1628** J. CARMICHAELL *Proverbs in Scots* no. 140 And wishes were horses pure[1] men wald ryde. **1721** J. KELLY *Scottish Proverbs* 178 If Wishes were Horses, Beggars would ride. **1844** J. O. HALLIWELL *Nursery Rhymes of England* (ed. 4) 501 If wishes were horses, Beggars would ride; If turnips were watches, I would wear one by my side. **1912** *British Weekly* 18 Jan. 480 If wishes were horses Unionists would ride rapidly into office. **1959** M. SCOTT *White Elephant* xii. I agree that she'd be an ideal wife . . but if wishes were horses, then beggars would ride. **1979** S. RIFKIN *McQuaid in August* vi. The kind of thinking which always brought out the same response from my mother: if wishes were horses, beggars would ride. I had better dismount.

[1] i.e. poor.

wit: *see* BREVITY is the soul of wit; when the WINE is in, the wit is out.

wolf: *see* HUNGER drives the wolf out of the wood.

A WOMAN, a dog, and a walnut tree, the more you beat them the better they be

'The old custom of beating a walnut-tree was carried out firstly to fetch down the fruit and secondly to break the long shoots and so encourage the production of short fruiting spurs': M. Hadfield *British Trees* (1957).

[L. *nux, asinus, mulier verbere opus habent,* a nut tree, an ass, and a woman need a beating.] **1581** G. PETTIE tr. *S. Guazzo's Civil Conversation* III. 20, I have redde, I know not where, these verses. A woman, an asse, and a walnut tree, Bring the more fruit the more beaten they bee. **1670** J. RAY *English Proverbs* 50 A spaniel, a woman and a walnut tree, The more they're beaten the better still they be. **1836** T. C. HALIBURTON *Clockmaker* 1st Ser. xxv. There was an old sayin there [Kent], which . . is not far off the mark: A woman, a dog, and a walnut tree, The more you lick[1] 'em, the better they be. **1929** E. LINKLATER *Poet's Pub* xii. A woman, a dog, and a walnut tree, The more you beat 'em, the better they be. **1945** F. THOMPSON *Lark Rise* v. A handsome pie was placed before him . . such as seemed to . . illustrate the old saying, 'A woman, a dog and a walnut tree, the more you beat 'em the better they be'. **1981** *Daily Telegraph* 5 Feb. 17 It's not right to batter your wife. It is like the old saying, 'A wife, a dog and a walnut tree, the more you beat them the better they be.' It's just not true.

[1] beat, thrash.

A WOMAN and a ship ever want mending

[Cf. PLAUTUS *Poenulus* ll. 210–15 *negoti sibi qui volet vim parare, navem et mulierem, haec duo comparato. . . Neque umquam satis hae duae res ornantur,* whoever wants to acquire a lot of trouble should get himself a ship and a woman. For neither of them is ever sufficiently equipped, and there is never enough means of equipping them.] **1578** J. FLORIO *First Fruits* 30 Who wil trouble hym selfe all dayes of his life, Let hym mary a woman, or buy hym a shyp. **1598** *Mirror of Policy* (1599) X2 Is it not an old Prouerbe. That Women and Shippes are neuer so perfect, but still there is somewhat to bee amended. **1640** G. HERBERT *Outlandish Proverbs* no. 780 A shippe and a woman are ever repairing. **1840** R. H. DANA *Two Years before Mast* iii. As has often been said, a ship is like a lady's watch, always out of repair. **1928** A. T. SHEPPARD *Here comes Old Sailor* II. vi. There are special proverbs for us shipmen: . . 'A woman and a ship ever want mending.'

A WOMAN's place is in the home

1844 'J. SLICK' *High Life* II. 121 A woman's place is her own house, a taking care of the children. **1897** 'S. GRAND' *Beth Book* (1898) xix. If we had . . done as we were told, the woman's-sphere-is-home would have been as ugly and comfortless a place for us to-day as it used to be. **1936** R. A. J. WALLING *Corpse with Dirty Face* iv. Mrs. Franks, being a dutiful wife, was always on the premises. 'Ah, yes—woman's place is in the home,' said Pierce. **1943** A. CHRISTIE *Moving Finger* vi. I go up in arms against the silly old-fashioned prejudice that women's place is always the home. **1979** G. WAGNER *Barnardo* v. Barnardo . . firmly believed that a woman's place was in the home.

A WOMAN's work is never done

1570 T. TUSSER *Husbandry* (rev. ed) 26 Some respite to husbands the weather doth send, but huswiues affaires haue never none ende. **1629** in *Roxburghe Ballads* (1880) III. 302 *(title)* A woman's work is never done. **1722** B. FRANKLIN *Papers* (1960) I. 19 If you go among the Women, you will learn . . that a Woman's Work is never done. **1920** *Times Weekly* 12 Mar. 209 'Women's work is never done.'. . We shall never hear the whole of woman's work during the war. **1981** 'G. GAUNT' *Incomer* xiv. My grannie used to say, A woman's work is never done when it never gets started!

woman: *see also* HELL hath no fury like a woman scorned; a MAN is as old as he feels, and a woman as old as she looks; SILENCE is a woman's best garment; SIX hours' sleep for a man, seven for a woman, and eight for a fool; a WHISTLING woman and a crowing hen are neither fit for God nor men.

women: *see* never CHOOSE your women or your linen by candlelight.

won: *see* FAINT heart never won fair lady.

wonder: *see* TIME works wonders.

WONDERS will never cease

1776 H. BATES in T. Boaden *Private Correspondence of D. Garrick* (1823) II. 174 You have heard, no doubt, of his giving me the reversion of a good living in Worcestershire. . . Wonders will never cease. **1843** C. J. LEVER *Jack Hinton* I. xx. The by-standers . . looked from one to the other, with expressions of mingled surprise and dread. . . 'Blessed hour. . . Wonders will never cease.' **1974** A. PRICE *Other Paths to Glory* I. vii. Wonders will never cease. . . Early Tudor—practically untouched.

wood: *see* FIELDS have eyes, and woods have ears; don't HALLOO till you are out of the wood; HUNGER drives the wolf out of the wood.

Many go out for WOOL and come home shorn

Many seek to better themselves or make themselves rich, but end by losing what they already have.

1599 J. MINSHEU *Dialogues in Spanish* 61 You will goe for wooll, and returne home shorne. **1612** T. SHELTON tr. *Cervantes' Don Quixote* I. vii. To wander through the world . . without once considering how many there goe to seeke for wooll, that returne againe shorne themselues. **1678** J. RAY *English Proverbs* (ed. 2) 220 Many go out for wool and come home shorn. **1858** S. A. HAMMETT *Piney Woods Tavern* xxiii. There's a proverb about going out after wool, and coming home shorn. **1910** G. W. E. RUSSELL *Sketches & Snapshots* 315 Some go [to Ascot] intent on repairing the ravages of Epsom or Newmarket; and in this speculative section not a few . . who go for wool come away shorn. **1981** N. FREELING *One Damn Thing after Another* iii. One always comes back tired from holidays. 'Go for wool and

come back—?' 'Shaved—no, cropped.'
'Sheared. Yes.'

wool: *see also* MUCH cry and little wool.

A WORD to the wise is enough
Now often abbreviated to *a word to the
wise.*
[L. *verbum sat sapienti,* a word is sufficient to a wise man; also *verb. sap.*]
a **1513** DUNBAR *Poems* (1979) 206 Few
wordis may serve the wyis. **1546** J.
HEYWOOD *Dialogue of Proverbs* II. vii. I4ᵛ
Fewe woords to the wise suffice to be
spoken. *a* **1605** W. HAUGHTON *Englishmen for My Money* (1616) D3 They say, a
word to the Wise is enough: so by this
little French that he speakes, I see he is
the very man I seeke for. **1768** STERNE
Sentimental Journey III. 164 A word,
Mons. Yorick, to the wise . . is enough.
1841 DICKENS *Old Curiosity Shop* ii.
'Fred!' cried Mr. Swiveller, tapping his
nose, 'a word to the wise is sufficient for
them—we may be good and happy
without riches, Fred.' *a* **1947** F. THOMP-
SON *Still Glides Stream* (1948) vi. I advise
you to keep an eye on that eldest daughter of yours. . . You know what they
say, a word to the wise. **1968** M. WOOD-
HOUSE *Rock Baby* xi. Some undesirable
elements around. Word to the wise,
eh?

word: *see also* ACTIONS speak louder than
words; an ENGLISHMAN'S word is his
bond; FINE words butter no parsnips;
HARD words break no bones; one PIC-
TURE is worth ten thousand words;
STICKS and stones may break my bones,
but words will never hurt me; many a
TRUE word is spoken in jest.

All WORK and no play makes Jack a dull boy
1659 J. HOWELL *Proverbs* (English) 12
All work and no play, makes Jack a dull
boy. **1825** M. EDGEWORTH *Harry & Lucy
Concluded* II. 155 All work and no play
makes Jack a dull boy. All play and no
work makes Jack a mere toy. **1859** S.
SMILES *Self-Help* xi. 'All work and no
play makes Jack a dull boy'; but all play

and no work makes him something
greatly worse. **1979** R. MUTCH *Gemstone*
xi. 'All work and no play makes Jack a
dull boy,' he observed, pouring the
champagne into a glass.

WORK expands so as to fill the time available
Commonly known as 'Parkinson's
Law', after Professor C. Northcote Parkinson, who first propounded it.
1955 C. N. PARKINSON in *Economist* 19
Nov. 635 It is a commonplace observation that work expands so as to fill the
time available for its completion. **1972**
M. ARGYLE *Social Psychology of Work* viii.
'Parkinson's Law' is that 'work expands
so as to fill the time available'. **1976**
Scotsman 25 Nov. 14 Though there are
fewer Bills than usual, MPs, being well
known as exemplars of Parkinson's law,
can be relied on to stretch their work
to fill all the time available and
more.

It is not WORK that kills, but worry
1879 D. M. MULOCK *Young Mrs. Jardine*
III. ix. Working . . all day, writing . . at
night . . Roderick had yet . . never spent
a happier three months . . for it is not
work that kills, but 'worry'. **1909** *British
Weekly* 8 July 333 It is worry that
kills, they say, and not work. . . The
canker of care seems to eat the life
away.

If you won't WORK you shan't eat
[2 THESSALONIANS iii. 10 If any would
not work, neither should he eat.]
c **1535** D. LINDSAY *Satire of Three Estates*
(EETS) 475 *Qui non laborat, non man-
ducet.* . . Quha labouris nocht he sall not
eit. **1624** CAPT. J. SMITH *General Hist.
Virginia* III. x. He that will not worke
shall not eate. **1891** KIPLING *Life's
Handicap* 362 If you won't work you
shan't eat. . . You're a wild elephant, and
no educated animal at all. Go back to your
jungle. **1938** N. STREATFEILD *Circus is
Coming* v. Proper termagent she was,
bless her. 'Them as can't work can't eat,'
she always said. **1981** J. STUBBS *Iron-*

master xx. I say them as don't work shan't eat.

work: *see also* (noun) the DEVIL finds work for idle hands to do; the END crowns the work; the EYE of a master does more work than both his hands; FOOLS and bairns should never see half-done work; if IFS and ands were pots and pans, there'd be no work for tinkers' hands; MANY hands make light work; a WOMAN's work is never done; (verb) TIME works wonders.

workman: *see* a BAD workman blames his tools.

workshop: *see* an IDLE brain is the Devil's workshop.

world: *see* it takes ALL sorts to make a world; all's for the BEST in the best of all possible worlds; BETTER be out of the world than out of the fashion; GOD's in his heaven, all's right with the world; one HALF of the world does not know how the other half lives; the HAND that rocks the cradle rules the world; LAUGH and the world laughs with you, weep and you weep alone; LOVE makes the world go round.

Even a WORM will turn
Even the humblest will strike back if harassed or borne upon too far.
    1546 J. HEYWOOD *Dialogue of Proverbs* II. iv. G4ᵛ Treade a worme on the tayle, & it must turne agayne. **1592** GREENE *Groatsworth of Wit* XII. 143 Stop shallow water still running, it will rage. Tread on a worme and it will turne. **1854** 'M. LANGDON' *Ida May* xi. Even the worm turns when he is trodden upon. **1889** W. JAMES in *Mind* XIV. 107 Since even the worm will 'turn', the space-theorist can hardly be expected to remain motionless when his Editor stirs him up. **1962** A. CHRISTIE *Mirror Crack'd* xii. He's a very meek type. Still, the worm will turn, or so they say. **1967** RIDOUT & WITTING *English Proverbs Explained* 52 'This morning that cantankerous wife of

his tried to pick a quarrel with the conductor, and Percy told her . . to shut up.' 'Good for Percy! Even a worm will turn.'

worm: *see also* the EARLY bird catches the worm.

worry: *see* it is not WORK that kills, but worry.

worse: *see* NOTHING so bad but it might have been worse.

worst: *see* HOPE for the best and prepare for the worst; when THINGS are at the worst they begin to mend.

The WORTH of a thing is what it will bring
    [Cf. 15th-cent. Fr. *tant vault la chose comme elle peut estre vendue,* a thing is worth just so much as it can be sold for.] **1569** J. SANFORDE tr. *H. C. Agrippa's Vanity of Arts & Sciences* xci. The thinge is so muche worthy as it maye be solde for. **1664** BUTLER *Hudibras* II. i. For what is Worth in any thing, But so much Money as 'twill bring? **1813** SOUTHEY *Life of Nelson* I. ii. Vouchers, he found in that country were no check whatever; the principle was, that 'a thing was always worth what it would bring'. **1847** J. O. HALLIWELL *Dict.* II. 864 The worth of a thing is what it will bring. **1908** *Spectator* 4 Apr. 535 'The real worth of anything Is just as much as it will bring.' You cannot get beyond that piece of ancient wisdom as to the determination of value.

worth: *see also* a BIRD in the hand is worth two in the bush; an OUNCE of practice is worth a pound of precept; a PECK of March dust is worth a king's ransom; one PICTURE is worth ten thousand words; a SWARM in May is worth a load of hay; if a THING's worth doing, it's worth doing well; one VOLUNTEER is worth two pressed men.

worthy: *see* the LABOURER is worthy of his hire.

wrath: *see* a SOFT answer turneth away wrath.

wren: *see* the ROBIN and the wren are God's cock and hen.

wrong: *see* (noun) the KING can do no wrong; TWO wrongs don't make a right; (adverb) if ANYTHING can go wrong, it will.

Y

year: *see* there are no BIRDS in last year's nest; a CHERRY year a merry year; KEEP a thing seven years and you'll always find a use for it; you should KNOW a man seven years before you stir his fire; ONE year's seeding means seven years' weeding; TURKEY, heresy, hops, and beer came into England all in one year.

yesterday: *see* JAM tomorrow and jam yesterday, but never jam today.

YORKSHIRE born and Yorkshire bred, strong in the arm and weak in the head
The names of other (chiefly, northern) English counties and towns are also used instead of *Yorkshire*.

1852 *Notes & Queries* 1st Ser. V. 573 Derbyshire born and Derbyshire bred, Strong i' th' arm, and weak i' th' head. **1869** W. C. HAZLITT *English Proverbs* 273 Manchester bred: Long in the arms, and short in the head. **1920** C. H. DOUGLAS *Credit-Power & Democracy* vi. Organised labour at this time shows considerable susceptibility to the Border gibe of being 'strong i' th' arm and weak i' th' head'. **1966** J. BINGHAM *Double Agent* ii. He thought, Yorkshire born and Yorkshire bred, strong in th' arm and weak in't head; but it wasn't true, most of them were as quick as weasels and sharp as Sheffield steel.

YOUNG folks think old folks to be fools, but old folks know young folks to be fools
1577 J. GRANGE *Golden Aphroditis* O2ᵛ Yong men thinks old men fooles, but old men knoweth well, Yong men are fooles. **1605** W. CAMDEN *Remains concerning Britain* 221 Wise was that saying of Doctor Medcalfe: You Yong men do thinke vs olde men to be fooles, but we olde men do know that you yong men are fooles. **1790** R. TYLER *Contrast* v. ii. Young folks think old folks to be fools; but old folks know young folks to be fools. **1850** F. E. SMEDLEY *Frank Fairlegh*

xxx. 'Young folks always think old ones fools, they say.' 'Finish the adage, Sir, that old folks know young ones to be so, and then agree with me that it is a saying founded on prejudice.'

YOUNG men may die, but old men must die
1534 MORE *Dialogue of Comfort* (1553) II. ii. As the younge man maye happe some time to die soone, so the olde man can never liue long. **1623** W. CAMDEN *Remains concerning Britain* (ed. 3) 276 Yong men may die, but old men must die. **1758** LADY M. W. MONTAGU *Letter* 5 Sept. (1967) III. 174 According to the good English Proverb, young people may die, but old must. **1863** B. I. WILEY *Life of Billy Yank* (1952) xii. That is the Way of the World. The old must die and the young may die.

young: *see also* BETTER be an old man's darling than a young man's slave; whom the GODS love die young; the GOOD die young; HANG a thief when he's young, and he'll no' steal when he's old; you cannot put an OLD head on young shoulders.

YOUTH must be served
1829 P. EGAN *Boxiana* 2nd Ser. II. 60 Tom Cannon made his appearance in the Prize Ring rather too late in life, under the idea that 'Youth must be served'. **1900** A. CONAN DOYLE *Green Flag* 125 There were . . points in his favour. . . There was age—twenty-three against forty. There was an old ring proverb that 'Youth will be served'. **1941** G. HEYER *Envious Casca* iv. You're just an old curmudgeon, and you're upset because you didn't like young Roydon's play. . . But, my dear chap, youth must be served!

Yule: *see* a GREEN Yule makes a fat churchyard.

BIBLIOGRAPHY

OF MAJOR PROVERB COLLECTIONS AND
WORKS CITED FROM MODERN EDITIONS

QUOTATIONS are taken from the first edition of the work in question unless otherwise stated. Standard modern editions of several major authors, particularly from the medieval and Renaissance periods, have been used for ease of reference.

MAJOR PROVERB COLLECTIONS

Apperson, G. L., *English Proverbs and Proverbial Phrases* (London, 1929).

Bohn, H. G., *Hand-Book of Proverbs* (London, 1855).

Carmichaell, J., *James Carmichaell Collection of Proverbs in Scots*, ed. M. L. Anderson (Edinburgh, 1957).

[Clarke, J.,] *Parœmiologia Anglo-Latina . . or Proverbs English, and Latine, methodically disposed according to the Common-place heads, in Erasmus his Adages* (London, 1639).

Denham, M. A., *Collection of Proverbs and Popular Sayings relating to the seasons, the weather, and agricultural pursuits* (London, 1846).

Draxe, T., *Bibliotheca Scholastica Instructissima, or, a Treasurie of ancient Adagies, and sententious Prouerbes, selected out of the English, Greeke, Latine, French, Italian, and Spanish* (London, 1616).

Dykes, O., *English Proverbs, with Moral Reflexions* (London, 1709).

Fergusson, D., *Fergusson's Scottish Proverbs from the Original Print of 1641 together with a larger Manuscript Collection of about the same period hitherto unpublished*, ed. E. Beveridge (Edinburgh, 1924).

Franklin, B., *Poor Richard's Almanack: Sayings of Poor Richard*, ed. P. L. Ford (Brooklyn, 1890).

Fuller, T., *Gnomologia: Adagies and Proverbs; Wise Sentences and Witty Sayings, Ancient and Modern, Foreign and British* (London, 1732).

Hazlitt, W. C., *English Proverbs and Proverbial Phrases* (London, 1869).

Henderson, A., *Scottish Proverbs* (Edinburgh, 1832).

H[erbert]., G., *Outlandish Proverbs* (1640), and *Jacula Prudentum* (1651), in *Works*, ed. F. E. Hutchinson (Oxford, 1941).

Heywood, J., *Dialogue conteinyng the number in effect of all the prouerbes in the englishe tongue* (London, 1546). Later editions are also cited.

Hislop, A., *Proverbs of Scotland* (Glasgow, 1862).

H[owell]., J., *Paroimiographia. Proverbs, or Old sayed sawes & adages in English, . . Italian, French, and Spanish. whereunto the British . . are added* (London, 1659).

Kelly, J., *Complete Collection of Scotish Proverbs Explained and made Intelligible to the English Reader* (London, 1721).

Lean, V. S., *Collectanea* (5 vols., Bristol, 1902–4).

[Mapletoft, J.,] *Select Proverbs, Italian, Spanish, French, English, Scotish, British, &c. Chiefly Moral* (London, 1707).

Oxford Dictionary of English Proverbs, ed. F. P. Wilson (ed. 3, Oxford, 1970).

Palmer, S., *Moral Essays on some of the most Curious and Significant English, Scotch, and Foreign Proverbs* (London, 1710).

R[ay]., J., *Collection of English Proverbs* (Cambridge, 1670). Later editions are also cited.

Stevenson, B., *Home Book of Proverbs, Maxims, and Familiar Phrases* (rev. ed., New York, 1961).

Taverner, R., *Proverbes or adagies with newe addicions gathered out of the Chiliades of Erasmus* (London, 1539).

Taylor, A., and B. J. Whiting, *Dictionary of American Proverbs and Proverbial Phrases 1820–1880* (Cambridge, Massachusetts, 1958).

Tilley, M. P., *Dictionary of the Proverbs in England in the Sixteenth and Seventeenth Centuries* (Ann Arbor, 1950).

Torriano, G., *Dictionary English & Italian, with severall Proverbs*, first published with J. Florio's *Vocabolario Italiano & Inglese, a Dictionary Italian & English* (rev. ed., London, 1659).

—— *Piazza universale di proverbi Italiani; or, a common place of Italian proverbs and proverbial phrases* (London, 1666).

—— *Second alphabet consisting of proverbial phrases* (London, 1662).

—— *Select Italian Proverbs* (Cambridge, 1642).

Tusser, T., *Husbandry: Hundreth good pointes of husbandrie* (London, 1557), *Five hundreth points of good husbandry united to as many of good huswiferie* (1573, various revised and augmented editions).

Whiting, B. J., *Early American Proverbs and Proverbial Phrases* (Cambridge, Massachusetts, 1977).

—— *Proverbs, Sentences, and Proverbial Phrases from English Writings mainly before 1500* (Cambridge, Massachusetts, 1968).

WORKS CITED FROM MODERN EDITIONS

Beaumont, F., and J. Fletcher, *Works*, ed. A. Glover and A. R. Waller (10 vols., Cambridge, 1905–12).

Bible: Authorised Version of the English Bible 1611, ed. W. A. Wright (5 vols., Cambridge, 1909).

Chaucer, G., *Works*, ed. F. N. Robinson (ed. 2, London, 1966).

Dekker, T., *Dramatic Works*, ed. F. T. Bowers (4 vols., Cambridge, 1953–61). Non-dramatic works are cited from the first edition.

Douce MS 52: Förster, M., 'Die Mittelenglische Sprichwörtersammlung in Douce 52', *Festschrift zum XII. Allgemeinen Deutschen Neuphilologentage in München, Pfingsten 1906* (Erlangen, 1906), 40–60.

Ford. J., *Dramatic Works*, ed. W. Bang and H. de Vocht (2 vols., Louvain, 1908, 1927).

Gascoigne, G., *Complete Works*, ed. J. W. Cunliffe (2 vols., Cambridge, 1907–10).

Gower, J., *English Works*, ed. G. C. Macaulay (2 vols., London, 1900–1).

Greene, R., *Life and Complete Works in prose and verse*, ed. A. B. Grosart (15 vols., London, 1881–6).

Heywood, T., *Dramatic Works*, ed. R. H. Shepherd (6 vols., London, 1874). *Captives* is cited from the first publication of 1885.

Jonson, B., *Works*, ed. C. H. Herford, P. and E. M. Simpson (11 vols., Oxford, 1925–52). *Cynthia's Revels* is cited from the first edition.

Langland, W., *Vision of William concerning Piers Plowman*, ed. W. W. Skeat (5 vols., London, 1867–85).

Lyly, J., *Complete Works*, ed. R. W. Bond (3 vols., Oxford, 1902). Principal works cited are *Euphues: the anatomy of wit* (1578) and *Euphues and his England* (1580).

Marlowe, C., *Works and Life*, ed. R. H. Case et al. (6 vols., London, 1930–3).

Marston, J., *Works*, ed. A. H. Bullen (3 vols., London, 1887).

Massinger, P., *Plays and Poems*, ed. P. Edwards and C. Gibson (Oxford, 1976).

Middleton, T., *Works*, ed. A. H. Bullen (8 vols., London, 1885–6).

Milton, J., *Complete Prose Works*, ed. D. M. Wolfe et al. (8 vols., New Haven, 1953–).

—— *Works*, ed. F. A. Patterson (18 vols., New York, 1931–8). Cited for poetical works.

Nashe, T., *Works*, ed. R. B. McKerrow (5 vols., London, 1904–10; corrected reprint 1958, 1966).

Pope, A., *Twickenham edition of the poems*, ed. J. Butt et al. (10 vols., London, 1939–67). The *Dunciad* and the translation of the *Odyssey* are cited from the first edition; *Letters* are cited as marked in the text.

Porter, H., *Pleasant Historie of the two angrie women of Abington*, ed. W. W. Greg (Oxford, for the Malone Society, 1912).

Romaunt of the Rose: in Chaucer, G., *Works*, supra.

Shakespeare, W., *Complete Works*, ed. P. Alexander (London, 1951).

Skelton, J., *Poems*, ed. R. S. Kinsman (Oxford, 1969). Items not found here are cited from *Poetical Works*, ed. A. Dyce (2 vols., London, 1843).

Spenser, E., *Works: a variorum edition*, ed. E. Greenlaw et al. (10 vols., Baltimore, 1932–57).

Swift, J., *Poems*, ed. H. Williams (3 vols., ed. 2, Oxford, 1958).

—— *Prose Works*, ed. H. Davis (14 vols., Oxford, 1939–68). The *Polite Conversation* ('S. W[agstaff].', *Complete collection of genteel and ingenious conversation, according to the most polite mode and method now used at court, and in the best companies of England* (London, 1738)) is cited from the first edition.

Thompson, F., *Lark Rise to Candleford*: cited from the first collected edition of 1945.

Webster, J., *Complete Works*, ed. F. L. Lucas (4 vols., London, 1927).